NITROGEN NUTRITION AND PLANT GROWTH

Editors

**H.S. SRIVASTAVA
RANA P. SINGH**

Science Publishers, Inc.
U.S.A.

SCIENCE PUBLISHERS, INC.
Post Office Box 699
Enfield, New Hampshire 03748
United States of America

Internet site: http://www.scipub.net

Sales@scipub.net (marketing department)
editor@scipub.net (editorial department)
info@scipub.net (for all other enquiries)

ISBN 1-57808-032-0

© 1999, COPYRIGHT RESERVED

All rights reserved. No part of this publication may be reproduced, stored in a retrieval system, or transmitted in any form or by any means, electronic, mechanical, photocopying or otherwise, without the prior permission of the copyright owner. Applications for such permission, with a statement of the purpose and extent of the reproduction, should be addressed to the publisher.

Published by Science Publishers, Inc., Enfield, NH USA.

Printed in India

PREFACE

Nitrogen is the most important mineral nutrient for the existence and maintenance of a plant's life. The juvenile life of higher plants, however, is self-supporting as far as the nitrogen requirement is concerned. During seed germination, the nitrogen stored in the cotyledons or endosperm in the form of storage proteins is released by the catalytic action of proteases, in the form of smaller peptides and amino acids. The proteolytic enzymes are usually synthesized during seed germination, although a few long-lived proteases may lie dormant in the dry seed and be activated during germination. It is also believed that there is deamidation of storage proteins prior to their hydrolytic degradation. This increases their solubility and susceptibility to the protease action. The peptides and amino acids released from storage proteins are transported to the embryonic axis (root and shoot) and support its growth until the seedling is fully established in the soil. Some of the amino acids from the storage amino acid proteins in the storage organs are also utilized in the synthesis of new enzymes and perhaps they also help in the transport of stored food and nutrients from the storage organs to the embryonic axis. Chapter 1 on nitrogen nutrition during seedling formation deals with these aspects of nitrogen nutrition in higher plants.

Once the seedlings are established, the roots start acquiring nitrogen and other minerals from the soil; nitrate being the predominant form of soil nitrate. The uptake of nitrate occurs by means of carriers in the root plasmalemma. There are at least two carrier systems, one of which is an inducible high-affinity carrier and the other a constitutive low-affinity carrier. Both work by a nitrate/H^+ symport process. Ammonium is taken up by a separate high-affinity and a low-affinity system, the second of which has both inducible and constitutive properties. Inside the root, nitrate is either stored, loaded into the xylem or reduced to ammonium by the enzymes nitrate and nitrite reductases. All ammonium in the roots, either arising directly from the external medium or from the reduction of nitrate is assimilated in the roots primarily by the action of enzymes glutamine synthetase and glutamate synthase. Most of the enzymes of nitrogen assimilation in roots are inducible by nitrogen supply as are the carriers, and the carbohydrate level controls both uptake and root assimilation of nitrate and ammonium. Chapter 2 in this book, 'Root Absorption and Assimilation of Inorganic Nitrogen' describes these aspects of nitrogen nutrition.

Symbiotic nitrogen fixation is an important mode of nitrogen nutrition in legumes and also in some non-leguminous plants. Three important bacterial-plant symbiotic systems, i.e., *Rhizobium*-leguminous plants, *Frankia*-actinorhizal plants and *Anabaena-Azolla* symbiosis have been studied in great detail, which contribute a major share in nitrogen fixation. The *Rhizobium*-legume symbiosis is the most significant in terms of global nitrogen fixation with representatives in tropical and temperate zones, in pastures, in aerable lands and in forests. The chapter on symbiotic nitrogen fixation covers some of the physiological and regulatory aspects of *Rhizobium*-legume symbiosis.

The chapter on *Casuarina-Frankia* symbiosis deals with the physiology of nitrogen fixing *Casuarina-Frankia* symbiosis. The events during early formation of the symbiosis are very important. Different methods to measure nitrogenase activity are also discussed in this chapter, as is the process of nitrogen fixation and hydrogen metabolism as well as the factors affecting symbiosis and nitrogen fixation. Since the members of the Casuarinaceae family could be important in soil reclamation and nitrogen cycle, plant selection and breeding methods for increased symbiosis and the current importance of Casuarinaceae are also described.

The role of nitrogen oxides in the atmosphere as an air pollutant is well known. However, their role as nitrogen nutrient is evaluated in the chapter 'Foliar Absorption and Use of Airborne Oxidized Nitrogen by Terrestrial Plants'. Plants are able to absorb and assimilate nitrogen oxides to a limited extent only. They cannot be grown with nitrogen oxides as the sole nitrogen source. This is apparently because the oxides exert phytotoxic effects at physiological and biochemical levels. The mechanism of phytotoxicity is not fully understood, however.

The importance of nitrogen forms in *in-vitro* plant propagation and morphogenesis is described in the sixth chapter of this book. An adequate supply of usable nitrogen is essential for culturing cells and tissues. Supplementation of the culture media with organic nitrogen, especially certain amino acids, e.g. glutamine, alanine, proline, serine, glutamate, etc., plays an important role in the improvement of somatic embryogenesis, shoot formation and subsequent development of the regenerants. Some of the amino acids have regulatory roles also in the *in-vitro* plant morphogenesis. Polyamines have also been shown to be involved in morphogenetic response. However, more in-depth studies are required to trace the signal transduction mechanism of the regulation by these nitrogenous metabolites during *in-vitro* cell and tissue culture.

The role of nitrogen in carbon metabolism is an important aspect of the nitrogen nutrition in higher plants. Leaf photosynthetic capacity usually increases with leaf nitrogen level, although the C:N relationship depends on the growth potential of the species. The importance of various nitrogen sources in carbon and nitrogen metabolism and their interactive balances particularly in woody plants are emphasized in the chapter 'Role of Nitrogen Source in Carbon Balance'. The influence of nitrogen sources in carbon partitioning between roots and shoots and on comparative energetic costs of root and shoot nitrogen assimilation are also discussed in this chapter.

Chapter 8 relates to nitrogen, stress and plant growth regulation. It describes plant growth rate, biomass production and yield and capacity of the plant to transport nitrate from the root to the shoot. Nitrate transport to the shoot is a function of the activity of the K^+ shuttle, which allows co-ordination of nitrate uptake and nitrate consumption while promoting rapid growth rates. Some but not all trees have a limited K^+ shuttle activity inhibiting the transport of nitrate and cytokinin to the shoot, while enhancing reduction of nitrate in the root and assimilation of the resultant ammonium. Salinity therefore causes a change of the main location of nitrate assimilation from the shoot to the root, followed by a corresponding reduction in growth rate. Ammonium assimilation under saline conditions is accompanied by enhanced CO_2 fixation by the root by enhanced phosphoenolpyruvic carboxylase to generate oxaloacetate. Xylem loading of cytokinins and abscisic acid are involved in the regulation of massive nitrate fluxes through xylem parenchyma loading and transport to the shoot of amides and nitrate. The balance of cytokinin/abscisic acid in the xylem sap correlates with changes in the expression of Mo containing enzymes aldehyde oxidase and nitrate reductase. Ammonium and stress increase the level of root aldehyde oxidase while nitrate depresses it. The amide/nitrate ratio correlates with the determination of assimilate allocation to vegetative or reproductive sinks determining the harvest index of the crops.

The last chapter in this book deals with a novel approach to manage the supply of added nitrogen fertilizers in the rhizosphere. Excessive input of nitrogen fertilizers in the green revolution package increased plant productivity and subsequently food security for the world's ever-increasing population was achieved. However, it simultaneously increased the energy input in agriculture practices and due to leaching of the excess nitrogen to the groundwater and surface water bodies, many environmental problems were aggravated, e.g. nitrate and nitrite pollution, eutrophication in water channels, volatilization of NH_4^+, emissions of NO_x etc. Chapter 9 'Slow-Release Nitrogen Fertilizer and Plant Nutrition' describes various kinds of slow-release fertilizers in use these days. These fertilizers can provide a solution to environmental problems by decreasing the nitrogen load in agriculture and subsequently effecting environmental upgradation, while maintaining high plant productivity. To reduce the cost of this technology, low-cost degradable non-toxic waste materials can be a better option for the support material to prepare slow-release fertilizers. They provide a sustained nitrogen input to the plants and minimize the loss of N in the soil.

Bareilly, U.P., India   **H.S. Srivastava**
31 March 1997   **R.P. Singh**

CONTENTS

Preface   iii

List of Contributors   ix

1. Nitrogen Nutrition During Seed Germination and Seedling Formation   1
   H.S. Srivastava

2. Root Absorption and Assimilation of Inorganic Nitrogen   23
   D.J. Pilbeam and A.U. Jan

3. Rhizobium-Legume Association   45
   S.N. Mishra, P.K. Jaiwal, Rana P. Singh and H.S. Srivastava

4. Foliar Absorption and Use of Airborne Oxidized Nitrogen by Terrestrial Plants   103
   S. Nussbaum, M. Ammann and J. Fuhrer

5. Physiology of Nitrogen-Fixing *Casuarina-Frankia* Symbiotic Association   173
   Anita Sellstedt

6. Role of Nitrogen in Plant Morphogenesis *in vitro*   205
   Rana P. Singh, Susan J. Murch and Praveen K. Saxena

7. Role of Nitrogen Source in Carbon Balance   231
   Maria Amélia Martins-Loução and Cristina Cruz

8. Nitrogen, Stress and Plant Growth Regulation   283
   S. Herman Lips

9. Slow-Release Nitrogen Fertilizers and Plant Nutrition   305
   Masahiko Saigusa

Index   337

THE CONTRIBUTORS

Amman, M., Air Pollution and Plant Ecology Section
Swiss Federal Research Station for Agricultural Chemistry
and Environmental Hygiene, CH-3097 Liebefeld-Bern, Switzerland.

Cruz, C., Departmento De Biologia Vegetal
Faculdade de Ciencias, Universidade de Lisboa
Bloco C 2, Piso 4, Campo Grande, 1700, Lisboa, Portugal.

Fuhrer, J., Air Pollution and Plant Ecology Section
Swiss Federal Research Station for Agricultural Chemistry and
Environmental Hygiene, CH-3097 Liebefeld-Bern, Switzerland.

Jaiwal, P.K., Department of Biosciences
M.D. University, Rohtak (124001), India.

Jan, A.U., Department of Pure and Applied Biology
University of Leeds, Leeds, LS2 9JT, U.K.

Lips, S.H., Biostress Research Laboratory
J. Blaustien Institute for Desert Research
Ben Gurion University of Negev, Sede Boqer 84993, Israel.

Martins-Loucao, M.A., Departmento De Biologia Vegetal
Faculdade De Ciencias Universídade de Lisboa
Bloco C 2, Piso 4, Campo Grande, 1700 Lisboa, Portugal.

Mishra, S.N., Department of Biosciences
M.D. University, Rohtak (124001), India.

Murch, S.J., Department of Horticultural Sciences
University of Guelph, Guelph, Ontario N1G 2W1, Canada.

Nussbaum, S., Institute of Environmental Protection and Agriculture
Schwarzenburgstrasse 155, CH-3097 Liebefeld-Bern, Switzerland.

Pilbeam, D.J., Department of Pure and Applied Biology
University of Leeds, Leeds, LS2 9JT, U.K.

Saigusa, M., Experimental Farm of Tohoku University
Kawatabi, Naruko, Tmatsukuri, Miyagi Pref., 989-67 Japan.

Saxena, P.K., Department of Horticultural Sciences
University of Guelph, Guelph, Ontario N1G 2W1, Canada.

Sellstedt, Anita, Department of Plant Physiology
University of Umea S-901 87 Umea, Sweden.

Singh, R.P., Department of Biosciences
M.D. University, Rohtak (124001), India.

Srivastava, H.S., Department of Plant Science
Rohilkhand University, Bareilly (243006), India.

Chapter 1

NITROGEN NUTRITION DURING SEED GERMINATION AND SEEDLING FORMATION

H.S. Srivastava

I. Introduction

II. Seed Storage Proteins
   A. Protein Bodies
   B. Protein Types

III. Proteolytic Activity
   A. Types of Proteases
   B. Protein Deamidation
   C. Tentative Model of Protein Hydrolysis

IV. Nitrogen Transfer to Embryonic Axis

V. Factors Affecting Proteolysis and Nitrogen Transfer
   A. Light
   B. Atmospheric Environments
   C. Nitrogenous Salts
   D. Plant Growth Regulators
   E. Miscellaneous

VI. Conclusions and Future Prospects

Literature Cited

I. INTRODUCTION

Seed germination is the initial event in the life of a higher plant. The dormant seeds during this phase imbibe water, which causes a rapid burst in respiratory and related metabolic activities, resulting in the emergence of the root in the form of a radicle. The plumule also emerges simultaneously to form a shoot. Until the roots and shoots are fully established and become functional, the germinating seed or the seedling depends entirely on stored food products in the endosperm (monocots) or cotyledons (dicots) for its nourishment (see Bewely and Black, 1994). Human beings also rely extensively on reserve food in seeds for their own nutrition. The seed proteins of legumes and cereals are the principal source of dietary proteins. The protein content of cereals is usually 8–17% of the total dry weight of the seed; wheat may have a higher protein content. Legume seeds have an even higher (up to 50%) protein content. It is for this reason that legume seeds have been used since prehistoric times in many different cultures (see Miege, 1982). From the human nutrition point of view, cereal proteins have usually low lysine and isoleucine contents and legumes have low methionine and cysteine, which are essential amino acids. Essential in the sense that they have to be supplied in the diet, as they cannot be synthesized from their precursors in the human body. Traditionally, breeding techniques have been applied to improve the amino acid composition and nutritive value of seed proteins. However, only limited success has been achieved using this approach. Recently, the techniques of protein engineering and gene transfer have been tried (for example, Ohtani et al., 1991; Dryer et al., 1993; Zheng et al., 1995), which appear to be quite successful.

II. SEED STORAGE PROTEINS

Few proteins from plants have been studied as extensively as the seed storage proteins. Seed proteins are synthesized and accumulate during the formation and development of seed on its parental plant, and are specially degraded during seed germination and seedling formation to provide nitrogen and carbon to the growing root and shoot. Storage proteins and their genome may play roles other than the nutrition of root and shoot. For example, in high lysine opaque-2 mutant of maize (*Zea mays* L.), the opaque-2 locus controls at least two distinct classes of genes in maize endosperm and it has been suggested that the opaque-2 protein plays a more general role in endosperm development (Neto et al., 1995).

A. Protein Bodies

Seed proteins are usually stored in the protein bodies of storage parenchymal cells, although some proteins are also stored in the endosperm outside the protein bodies. Protein bodies are the vacuoles transformed to store proteins. Protein storage vacuoles are generated by the fragmentation of parenchyma vacuoles in the late stage of seed maturation (Hara-Nishimura et al., 1987). When seeds germinate and the proteins are hydrolyzed, the protein storage vacuoles fuse to form a single large vacuole. For this reason, protein bodies are also known as protein storage vacuoles (PSV).

Most plant seeds have only one type of protein bodies, which are usually spherical. However, rice (*Oryza sativa*) seeds which contain 80% glutenin and 5% prolamine and other proteins, have two types of protein bodies (Yamagata et al., 1982). Type I protein bodies are spherical with a concentric ring structure and contain prolamine whereas type II protein bodies do not have this ring structure and are rich in glutenin and globulin (Tanaka et al., 1980; Krishnan et al., 1986).

Besides storing protein, protein bodies also function as a store for macronutrients such as P, Ca, Mg and K. These nutrients are stored in the form of a complex molecule, the phytin. The phytin is usually a Ca or Mg salt of inositol hexaphosphate, or sometimes as a K or Mg salt. Besides acting as a storage organ, the protein bodies are also involved in some active metabolic roles, such as in post-translational modifications of polypeptides. They may also contain proteolytic enzymes. For example, in barley both aspartate and cysteine proteases have been demonstrated to be located inside the aleurone layers of protein bodies (Benthke et al., 1996). In addition, the protein bodies of legumes accumulate certain toxic proteins, such as inhibitors of proteases and amylases, ribosomal inactivating proteins and lectins or lectin-related proteins (Richardson, 1991). Like other proteins of protein bodies, the toxic proteins are also synthesized on the endoplasmic reticulum and are transported to their final destination through the Golgi bodies (Chrispeels, 1983; Santino et al., 1992). These toxic proteins may play a role of defence against predators in the leguminous seeds.

B. Protein Types

Seed proteins are usually classified on the basis of their solubility, which is in fact determined by the amino acid composition of the proteins. Thus, there are four types of proteins: albumins, globulins, prolamines and glutenins (Miege, 1982). *Albumins* are soluble in water and usually have low molecular weight. They are rich in ionizable amino acids such as arginine, lysine, glutamic acid, tryptophan, etc. and hence are able to form polar bonds with water molecules. *Globulins* are soluble in dilute salt solutions. Although they have ionizable amino acids, they are large molecular weight molecules and hence require salts to spread the water dipoles around their molecules. *Prolamines* are soluble in polar organic solvents such as ethanol. They are rich in polar amino acids with hydroxyl or amide groups such as proline or glutamine. They are also rich in aliphatic hydrophobic side-chain amino acids such as leucine and alanine. *Glutenins* are soluble in acid, alkali, ionic detergents or urea-containing solutions. They have high molecular weight and have disulphide bonds with electrostatic linkages which break during solubilization with acid or alkali.

The major storage proteins of legumes and other dicotyledonous plants are globulins and those of monocotyledons are prolamines and glutenins (Higgins, 1984; Shotwell and Larkins, 1989). In castor bean (*Ricinus communis*) seeds, 40% of the total protein is albumin (Youle and Huang, 1978), which is usually considered to be an enzymatic and metabolic protein rather than a storage protein. These proteins, however, undergo rapid hydrolysis during seed germination. Smaller quantities of albumins are present in legumes and cereals also. A small albumin polypeptide (Mr 1–18 kD) is present in sunflower seeds, which is rich in methionine (Kortt and Caldwell, 1990b). Rice is an important cereal, containing both prolamines and glutenins, these two together constituting more than 80% of the total grain protein. The amino acid sequence of the rice glutenin is similar to that of 11 S globulin found in legumes. As described earlier, the prolamines and the glutenins are stored in different types of protein bodies. In wheat seeds, glutenins and gliadins are principal storage proteins located in the starchy endosperm (Kent and Evers, 1994). Glutenins are monomeric units with Mr 30–60 kD and are soluble in alcohol-ether solution. Glutenins exist as large polymers linked by non-covalent and disulphide bond interactions and are largely insoluble in alcohol-water solutions. The Mr of the subunits are 30 to 90 kD.

Some important proteins of legumes and cereals are described below.

1. *Zein*: This is an ethanol soluble protein (a prolamine) of maize and makes up about 50 to 70% of the total endosperm protein. Four types of zein have been recognized (Esen, 1986): Alpha (α) zein, Beta (β) zein, Gama (γ) zein and Delta (δ) zein. Alpha zein is readily soluble in 70% ethanol and makes up to 80% of the total zein. It has an Mr of 20–24 kD. Beta zein makes about 10 to 15% of total zein and is a 17 kD protein. Gama zein is 5 to 10% of the total protein and is a 22–27 kD protein. Delta zein has been detected in a few cases, which is a small protein

with Mr 10 kD and is soluble in ethanol only in the presence of reducing agents.

2. *12 S/11 S Globulins*: Classification of proteins on the basis of their sedimentation coefficients has also been attempted. 11 S globulins (sometimes detected as 12 S globulins) are quite common in legumes and are often termed as legumins (Kosanke et al., 1990). They are also named according to the source: the one from soybean (*Glycine max*) is called glycinin, from field bean (*Vicia faba*) vicinin and from sunflower (*Helianthus annuus*) helianthinin, etc. 11 S globulins are usually hexameric heteropolymers containing an acidic polymer (Mr 35–45 kD) and basic polypeptide (Mr 17–35 kD); the total Mr of the protein is more than 300 kD. The helianthinin from sunflower has Mr in the range of 300 to 350 kD. It is an oligomeric protein with 6 subunits, each subunit having a large acidic and a small basic chain (Kortt and Caldwell, 1990a). Glycinin, which is the major storage protein in soybean, constituting as much as 20% of the total seed weight in some cultivars, is also a hexamer and like helianthinin each subunit has an acidic and a basic chain (Staswick et al., 1981).

3. *Beta-Conglycinin*: Conglycinin is a glycoprotein (conglutenin) stored abundantly in the protein bodies of soybean seeds (Higgins, 1984). It is composed of three major subunits, α (76 kD), β (72 kD) and γ (53 kD). These subunits assemble into a trimeric (7 S) to a hexameric (9 S) form. Beta conglycinin along with another major storage protein, glycinin (11 S globulin), accumulates in developing soybean embryos during early to late maturation stages of seed development. Several other types of conglutenins are also found in leguminous seeds.

4. *7 S Globulins*: The most common 7 S globulin is phaseolin, which is the most abundant (about 50% of the total seed protein), protein in *Phaseolus vulgaris*. It is a trimeric protein with polypeptides of 47 to 50 kD (Bollini and Chrispeels, 1978). The vicilin from *Vicia faba* is also a 7 S globulin with Mr 18.6 kD.

5. *2 S Globulins*: 2 S globulins are quite common in cereals. For example, napin from *Brassica napus* is a 2 S globulin. It is synthesized as a proprotein from which N-terminal amino acid and an internal peptide is removed (Muren et al., 1995). 2 S globulins have been isolated from gymnosperm *Pinus pinaster* also (Allona et al., 1994). These are dimeric with a large subunit and a smaller one linked by disulphide bridges. They are rich in arginine and glycine content. A comparison of their characteristics with those of angiosperm 2 S proteins suggests that there is a homology between them.

III. PROTEOLYTIC ACTIVITY

During seed germination and seedling formation, the initial growth of the root (radicle) and the shoot (plumule) is self-supported by the nitrogen present in the embryonic axis. Following imbibition and respiratory upsurge, the proteins in the axis are hydrolyzed. The pool of the amino acids formed supplies precursors for the formation of new proteins, which are needed for the growth of roots and shoots (Srivastava et al., 1976; Vigil and Fang, 1995). The contribution of nitrogen from the nitrogen stored in the cotyledons or endosperm is negligible during this period. This period of nitrogen self-sufficiency of the embryonic axis depends on the species and the germination conditions. In maize seeds (kernels) germinated at 26°C, this period is about 48 h; during this period there is no increase in the total nitrogen content of the embryonic axis (Srivastava et al., 1976). Similarly, in cotton seeds the initial 24 h of growth of radicles or hypocotyls is maintained by the embryo nitrogen (Vigil and Fang, 1995).

As the seedling develops, the proteins in the storage organs are hydrolyzed and the amino acids are transported to roots and shoots. A sizeable amount of amino acids is utilized in protein synthesis in storage organs also, as many enzymes are synthesized *de novo* in the storage organs during seedling growth. The hydrolysis of storage proteins is not uniform; different proteins are hydrolyzed at different rates. In lupines for example, α, β and γ conglutenins are subjected to differential proteolysis during seed germination and seedling growth (Ferreira et al., 1995). In wheat, gliadins are especially degraded during germination, ω-gliadin fraction being the first to be degraded (Bigiarni et al., 1995). There is also a co-ordination between hydrolysis of proteins in different cells and tissues in the storage organs. In cereals, there is a substantial amount of protein in the starchy endosperm besides that in the protein bodies. In these seeds there is a co-ordinated hydrolysis of proteins at two sites (Enari and Saponen, 1986). First the storage proteins in the protein bodies are hydrolyzed to produce amino acids. These amino acids are then used for the synthesis of proteases and other hydrolytic enzymes. The proteases are then secreted to the starchy endosperm where they hydrolyze proteins to produce amino acids which are then transported to the embryonic axis via the scutellum.

A. Types of Proteases

Protein hydrolyzing enzymes are known as proteases (proteinases) or peptidases depending on the size of the protein they hydrolyze. Peptidases are then classified as exopeptidases or endopeptidases depending on whether they hydrolyze the terminal peptide bond or an intermediary bond. The important proteases acting on large protein molecules may be classified as: 1) Aspartate protease, 2). Cysteine protease, 3) Serine protease and 4) Metalloprotease.

1. *Aspartate Protease*: Aspartate proteases (E.C. 3.4.23) have usually two aspartate residues at their active centre. They are usually active in conditions of strong acidity and are present in dry seeds. Thus, they are involved in early degradation of storage proteins during seed germination. An aspartate protease purified from wheat bran has an optimum pH of 3.3 and K_m of 0.375 mM (Galleschi and Felicioli, 1994). The apparent Mr is 66.5 kD. It hydrolyzes endogenous globulin releasing fragments with Mr over 20 kD. The enzyme is inhibited by pepstatin (aspartate inhibitor), but other inhibitors have no effect (Belozersky et al., 1989; Galleschi and Felicioli, 1994). Aspartate protease of dry or imbibed barley seeds is similar to cathepsin D, a mammalian lysosomal enzyme (Sarkkinen et al., 1992).

Aspartate proteases are present in developing seeds and also in flowers, stems, leaves and roots of barley (Tomakangas et al., 1994). This suggests that they might have other functions also appropriate to the needs of different tissues.

2. *Cysteine Protease*: Cysteine proteases (E.C. 3.4.22) have a cysteine residue at their active centre and act in an acidic range. They are present in dormant seeds and also appear *de novo* during seed germination (Poulle and Jones, 1988; Benthke et al., 1996). Cysteine inhibiting reagents such as iodoacetate, N-ethylmaleimide (EMI) and trans-epoxysuccinyl-L-leucylamido (4-guanidino) butane (E64) inhibit the activity of cysteine proteases (Holwedra and Rogers, 1992; Wrobel and Jones, 1992). A cysteine protease from *Vigna radiata*, which actually acts as an endopeptidase has a pH optimum of 5.5 and Mr of 33 kD (Yamaoka et al., 1990). A barley aleurone layer cysteine protease (endopeptidase) has an Mr 37 kD and optimum pH 5.0 (Koehler and Ho, 1988). Holwedra and Rogers (1992) have demonstrated that the barley aleurone protease, which acts as an aminopeptidase (and termed as aleurain) has many similarities with mammalian lysosomal enzyme cathepsin H, which include heterogeneity of charge forms, position of the NH_2 terminus of the mature protein and pH activity profile (6.5–7.0 pH). It indicates that the genes of both the enzymes aleurain and cathepsin H have common ancestory. In fact, it has been demonstrated that there is approximately 65% similarity between the c-DNA of both the enzymes (Rogers et al., 1985).

In maize there are at least 4 types of cysteine proteases, A_1, A_2, A_3 and A_4 which hydrolyze α, β, γ and δ zeins respectively (De Barros and

Larkins, 1990). The optimum pH for their activity is highly acidic, i.e., 3.8. A cysteine endopeptidase from castor endosperm shows absolute specificity towards an asparagine bond in various polypeptide substrates (Cornel and Plaxton, 1994). The asparaginyl endopeptidase obtained from the seeds of jack-bean and many other species appears to be a cysteine proteinase (Abe et al., 1993). The enzyme is inhibited by thiol reagents N-ethyl maleimide and parachloro-mercuric benzoate and is strongly reactivated by dithiothritol suggesting that a free sulphhydryl group is involved in enzyme activity (Bottari et al., 1996a). The proteinase from wheat endosperm has an optimum pH of 4.25 and Mr 30 kD (Bottari et al., 1996 b). It is activated by the sulphhydryl compounds.

3. *Serine Protease*: Serine proteases contain serine at the active site and hence are inhibited by serine inhibitors. These are largely responsible for proteolytic activity in legumes (Baumgartner and Chrispeels, 1977; Csoma and Polgar, 1984). Protease C_1 of soybean is also similar to serine proteases. It has an Mr of 70 kD and hydrolyzes α and α^1 subunits of *b*-conglycinin of soybean (Shuttuck-Eidens and Beachy, 1985; Tan-Wilson et al., 1996).

4. *Metalloprotease*: Metalloproteases require metallic ions for their catalytic activity. A metalloprotease from the dry seeds of buckwheat (*Fagopyrum esculentum*) causes limited proteolysis of 13 S globulin (Dunaevsky and Belozersky, 1989). It has an optimum pH of 5.0 and is completely inhibited by EDTA and 1, 10-phenanthroline, suggesting that the enzyme is a metalloprotease.

B. Protein Deamidation

In several leguminous and cereal species, there is extensive deamidation of the storage proteins prior to their proteolytic breakdown during seed germination and seedling growth. In the process ammonia is liberated from the amides present in the protein (Daussant et al., 1969). The process of deamidation has been particularly demonstrated in wheat (Vaintraub et al., 1981), wherein the enzyme deamidase (one removing amide group) has also been characterized (Vaintraub et al., 1992). This enzyme is able to deamidate storage proteins from leguminous species and also from several other species, although only glutamyl residue in the protein seems to be deamidated. The enzyme differs from transglutaminase and is a true deamidase. The hydroxamates, which are characteristic products of transglutaminase are not formed by the action of deamidases. Recently, Vaintraub et al. (1996) purified and characterized deamidases from kidney bean (*Phaseolus vulgaris*) and squash (*Cucurbita pepo*). The characteristics of the deamidases obtained from these species seem to be similar to those from wheat. They have an Mr of about 18 kD and an optimum pH of 7.0. Calcium causes partial inhibition whereas sulphhydryl reagents have no significant effects. In this respect they differ from transglutaminase, which requires Ca^{2+} for its activity and is inhibited by sulphhydryl reagents (Mycek et al., 1961).

The presence of protein deamidases in several phylogenetically unrelated species and their striking similarities suggest that they have a role in seed germination and seedling growth. Vaintraub et al. (1996) suggested that deamidation increases the negative charge of the proteins which effects further proteolytic breakdown of modified proteins both by increasing their solubility and by changing the physical organization of the molecule. In earlier studies on wheat, Vaintraub et al. (1981) found only a slight decrease in storage protein during the first three days of germination, although the amide content of the proteins during the same period decreased by about 16%. Later studies revealed that while deamidase activity peaked on the second day, proteinase was most active on the fourth day (Vaintraub et al., 1992). Thus, it may be suggested that deamidation usually precedes protein hydrolysis during seed germination.

C. Tentative Model of Protein Hydrolysis

Based on studies with legumes and other dicotyledonous species over the past 25 years, a model for protein hydrolysis during seed germination may be proposed. It is proposed that the storage proteins are first attacked by deamidases, which remove the amide nitrogen and increase the solubility and susceptibility of storage proteins to further action by proteases. In the next step, the endopeptidases are activated which break the larger proteins into smaller fragments.

The endopeptidases are presumed to act in two steps, A and B, and perhaps the enzymes for the two steps are different. Proteases B are active only after proteases A have acted upon the protein. For example, an asparaginyl endopeptidase from durum wheat is unable to hydrolyze the native protein of wheat. However, once the native protein has been hydrolyzed by a cysteine protease, which acts as protease A, the enzyme degrades the protein (Shutov and Vaintraub, 1987). The smaller fragments are then hydrolyzed to produce amino acids by the action of exopeptidases, which include carboxypeptidases and amino peptidases (Enari and Saponen, 1986). Sometimes dipeptides are produced as intermediaries, which are hydrolyzed by the dipeptidases (Fig. 1.1).

The protein hydrolyzing enzymes may be present in dry and dormant seeds. However, many proteases are synthesized *de novo* during seed germination and seedling formation. The existence of proteases in

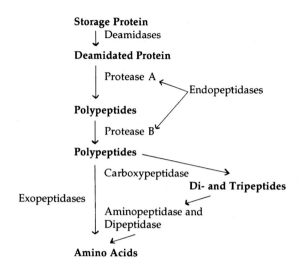

Fig. 1.1: Tentative model of protein hydrolysis during seed germination.

quiescent seeds inside protein bodies has been reported in many species, including pumpkin cotyledons (Hara and Matsubara, 1980), lupines (Duranti et al., 1987; Scarafoni et al., 1992), buckwheat (Elpidina et al., 1991), kidney bean (Mikkonen et al., 1986), hempseed (St. Angelo et al., 1969), barley (Benthke et al., 1996), etc. The activities of these proteases usually decline during seed germination (Wrobel and Jones, 1992). It is presumed that some natural inhibitors of the proteases also exist which prevent them from degrading the proteins in the dormant seeds. Several types of proteases may exist in dormant seeds. An acidic and a neutral endopeptidase system was identified at the quiescent stage of seeds of white lupine (*Lupinus albus*) (Duarte et al., 1996). The acidic endopeptidase was much more active than the neutral one. Evidence for the *de novo* synthesis of several types of proteases during seed germination has also been obtained. This activity was detected in maize after just a few hours (24 to 48 h) of imbibition, which peaked after 6–8 days (Harvey and Oaks 1974; Bose et al., 1984). The protein synthesis inhibitor cycloheximide inhibited development of protease activity in studies done by Bose et al. (1982), indicating *de novo* synthesis of the enzyme during seed germination. In the cotyledons of *Vigna radiata* seeds also, total proteolytic activity increased markedly after the first day of imbibition and peaked on the 3rd day, declining thereafter (Yamaoka et al., 1990).

Proteases may exist either inside the protein bodies or in discrete enzyme particles called *proteosomes* (Schliephacke et al., 1991). The seed proteosome from pea is a multimeric protein of 600 kD and possesses three characteristic partial activities—trypsin-like, chymotrypsin-like and peptidyl glutamyl peptidase (Skoda and Malek, 1992). These activities perhaps degrade unwanted proteins produced in the seeds during stress metabolism.

IV. NITROGEN TRANSFER TO EMBRYONIC AXIS

Solubilization of nitrogen in the storage tissues (endosperm or cotyledons) facilitates its transport to the embryonic axis (root + shoot), where it is utilized in the formation of proteins, nucleic acids and other complex nitrogenous molecules. As a consequence of this transport the nitrogen content of the storage tissue declines and that of the embryonic axis increases during seed germination and seedling formation (Fig. 1.2).

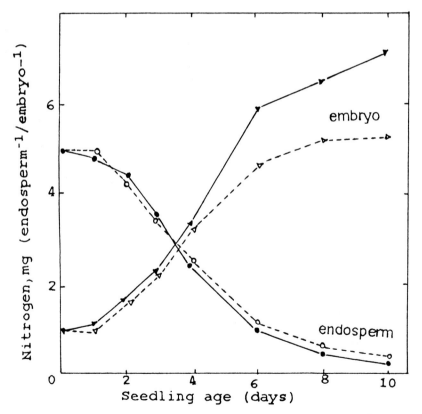

Fig. 1.2: Nitrogen mobilization from endosperm to embryonic axis during seedling formation in the presence (—) or absence (- - -) of nutrient nitrogen (Srivastava, 1976).

Nitrogen is transported to the embryonic axis in the form of amino acids and small peptides. Thus there is usually an increase in proteolytic activity and a parallel decrease in protein content during seed germination (Hay et al., 1991). This also causes a temporary increase in peptide and amino acid content of the storage tissue. Thus, in barley endosperm, the peptide pool increases during the first three days of germination but subsequently decreases (Higgins and Payne, 1981). The amino acid composition of these peptides is similar to that of storage proteins. Furthermore, the amino acid composition of the storage proteins within the same species seems to have little effect on the rate of nitrogen transfer to the embryonic axis. In maize, for example, the loss of total endosperm nitrogen and the concurrent gain in embryonic axis nitrogen during early seedling growth are similar in the normal and the high lysine opaque-2 mutant (Srivastava, 1976). Contrarily, the pattern of nitrogen mobilization observed in maize differed from that observed in pea (Nikolova et al., 1993).

Nitrogen content and nitrogen transfer from storage tissues to the embryonic axis is the main determinant of seedling growth in several species. Barley cultivars containing high nitrogen germinate more rapidly and produce larger seedlings than the cultivars of low nitrogen (Metivier and Dale, 1977). Further, in the same cultivar two groups of seedlings differing in seed protein revealed different rates of seedling growth (Ching and Rynd, 1978).

It has also been realized that about 5 to 20% of the total nitrogen lost from the endosperm or cotyledons is not gained in the embryonic axis; it is leached in the medium instead. In a natural environment it might play an important role in the establishment of symbiotic and non-symbiotic microflora in the root environment of the seedling. Many chemicals and environmental factors influence the leaching of nitrogen from seeds (as shown in section V below).

V. FACTORS AFFECTING PROTEOLYSIS AND NITROGEN TRANSFER

Several plant and environmental factors influence seed germination and seedling growth through their effects on proteolysis and nitrogen transfer from storage tissues to the embryonic axis. A brief account of some important factors is given below.

A. Light

The effect of light on seed germination and seedling formation is well documented—the effect varying from promotory to inhibitory, to no effect at all. These effects may be manifested at the levels of protein hydrolysis and nitrogen transfer, as has been observed in several species. In bean, hydrolysis of storage proteins is more rapid in light than in darkness, which is apparently due to increased proteolytic activity in light (Forest and Wightman, 1971). Increase in soluble or non-proteinaceous nitrogen content of storage organs in light, over that in darkness, has been observed in squash seeds (Willey and Ashton, 1967), rice grain (Palmiano and Juliano, 1972), maize kernels (Bose and Srivastava, 1980), etc.

Light seems to promote nitrogen transfer from the storage organ to the embryonic axis. In bean, after 8 days of seedling growth, about 60% of the cotyledon nitrogen had been transported to the embryonic axis in dark-grown seedlings, whereas almost 90% had been similarly moved during the same period in light-grown seedlings (Forest and Wightman, 1971). Similarly, in maize the loss of total nitrogen from the endosperm was almost double in light than in darkness during a 5-day period, although the gain in total nitrogen in the embryo was only about 10% higher in light than in darkness (Bose and Srivastava, 1980). Obviously, a major part of nitrogen coming out of the endosperm in light is leached out in the medium. It is also pertinent to note that light is slightly inhibitory to the germination of maize seeds. Germination is faster in darkness than in light. These observations indicate that changes in endosperm nitrogen are not linked with the speed of seed germination, strengthening the earlier stated view that the storage organs contribute very little nitrogen during seed germination and seedling formation. Nevertheless, the stimulatory effect of light on endosperm proteolysis is carried over to seedling growth. In maize, seedlings obtained from seeds imbibed in light have better growth, nitrate reductase activity and organic nitrogen content than those from dark imbibed (Bose and Srivastava, 1980).

B. Atmospheric Environments

The composition of atmospheric environment also influences proteolysis and nitrogen transfer during seed germination and seedling formation. In maize, solubilization of endosperm protein and transfer of endosperm nitrogen to the embryo was significantly lower in nitrogen or CO_2 atmosphere than in normal atmosphere (Bose and Srivastava, 1982). Seed germination was also significantly lower in nitrogen or CO_2 atmosphere. Similar effects were observed for protease activity also. Such results are rather expected because germination and associated metabolic activities depend on the energy derived from the aerobic oxidation of the respiratory substrates.

C. Nitrogenous Salts

Among nitrogenous salts, nitrates are known to promote germination of both dormant and non-dormant seeds. This promotion is specific for nitrate ion and is not linked with the cationic components of the nitrate salts (see Srivastava et al., 1976; Saini et al., 1987). Recent studies with *Avena fatua* seeds have demonstrated that nitrate is taken up and accumulated in the germinating seeds and treatment with nitrate also increases the amino nitrogen component of the embryonic axis (McIntyre et al., 1996). This nutritional effect of nitrate supply, however, has not been reported in many other species, in which nitrate supply does not increase organic nitrogen components of germinating seeds (Srivastava et al., 1976). In those cases in which nitrate increases germination without being assimilated, it appears to be effective through its osmotic role (Blom-Zandstra and Lampe, 1985). Hilhorst and Karssen (1989) demonstrated that endogenous nitrate in the seeds of hedge mustard (*Sisymbrium officinale*) could induce germination without being reduced. It was shown that no reduction of nitrate occurred during an 8-h period of germination induction by red light, and also that metabolic inhibitors of nitrate reduction failed to prevent germination. In experiments by McIntyre et al. (1996) in which an increase in amino acids with nitrate supply was demonstrated, the amino acids also may have played an inductive role in germination through their osmotic effects, as suggested by Takeba (1980). The amino acid produced may also ensure continued growth of the radicle and plumule, independent of nitrogen transfer from the storage organ. It has also been suggested that the ability of nitrate to induce germination is due to promotion of respiration, as it may act as an alternate (to O_2) electron acceptor (Adkins et al., 1984; Hilton and Thomas, 1986). This transfer of electron to nitrate is apparently catalyzed by the enzyme nitrate reductase (NR), which exists in at least two isoforms, NADH:NR. (EC 1.6.6.1) and NAD(P)H:NR (EC 1.6.6.2). In germinating maize seeds, in which both the isoforms of nitrate reductase are present in roots and scutella, their activity is induced by nitrate supply and at the same time the nitrate supply causes a decline in uptake of exogenous O_2 (Shankar and Srivastava, 1997). It is also possible that enhancement in respiration, usually associated with nitrate supply, is due to the osmotic effect of the nitrate and amino acids through their effects on water uptake. It is an established fact that the respiration rate is linked with tissue hydration during imbibition of seeds. The regulatory role of nitrate at the molecular level has also been suggested (Redinbaugh and Campbell, 1991; Champigny and Foyer, 1992). The ion *per se* besides inducing genes for nitrate uptake and assimilation also induces some regulatory proteins involved in complex phenomena such as root proliferation, enhancement of respiration and other physiological processes.

D. Plant Growth Regulators

Many plant growth regulators induce seed germination and seedling growth apparently through general metabolic enhancement. However, the effect of growth regulators on proteolysis and nitrogen transfer are often noticed which may not be linked with other metabolic effects of the growth regulators.

Gibberellins are often considered to be critical hormones in cereals, regulating the expression of hydrolytic enzymes in the endosperm during seed germination. The hormone is secreted from the embryo and stimulates the aleurone cells of the endosperm to produce and secrete numerous hydrolytic enzymes, including proteases (Fincher, 1989). Stimulation of protease activity by exogenously applied gibberellins has been demonstrated in cereals (Hammerton and Ho, 1986; Koehler and Ho, 1988). Even the isolated protoplasts from barley aleurone layers responded positively to GA_3 treatment by increased cysteine protease activity (Benthke et al., 1996). However, the aspartate protease activity in the same system was not affected by GA treatment.

Cytokinins also play a vital role in seed germination and seedling formation, although their role in nitrogen mobilization is not clearly understood. The induction of seed germination has been reported in many species, which is apparently due to the promotory effects on nucleic acid and protein synthesis and on cell division (Van Staden, 1983; Zarain et al., 1987). However, from the studies on the influence of embryonic axis on reserve mobilization, it has been generally agreed that the embryonic axis creates a kind of 'sink effect' on proteolysis and nitrogen transfer from storage tissue, especially in dicotyledonous seeds (Chapman and Davies, 1983). During seed germination the cytokinins are synthesized in the embryonic axis and from there transported to the cotyledons, where proteolysis and nitrogen mobilization are affected (Palni et al., 1990). The effect of embryonic axis may be mimicked by the exogenous application of cytokinins in some cases (Nandi et al., 1995). The induced proteolytic activity and nitrogen mobilization with the exogenous application of cytokinins has been demonstrated in species such as bean (Gopstein and Ilan, 1980), chick-pea (Munoz et al., 1990), lupines (Nandi et al., 1995). As indicated earlier, gibberellins might play a prominent role in protein hydrolysis and nitrogen mobilization in cereals.

Abscisic acid (ABA) is usually a germination inhibiting hormone; it prevents precocious seed germination. In a study with barley seeds, it was found that dormant barley grains have a higher endogenous ABA level and also that they are more sensitive to ABA than non-dormant ones (Wang et al., 1995). Exogenously applied ABA has little effect, however, on aspartate or cysteine protease activities in barley aleurone layers (Benthke et al., 1996). But it inhibits protease activity in many other systems, including bean (Yomo and Srinivasan, 1973) and apple (Ranjan and Lewak, 1995), and antagonizes the promotory effect of GA_3 on protease production in barley seeds (Fincher, 1989).

Jasmonic acid and methyl jasmonate have been proposed as naturally occurring growth regulators because of their wide distribution and also because of their varied physiological effects on plants (Sembdner and Parthier, 1993). Methyl jasmonate was found to inhibit protease activity and mobilization of reserves from the endosperm to the embryo in rice (Tsai and Kao, 1996). However, jasmonic acid increased protease activity in rice (Ranjan and Lewak, 1995).

Salicylic acid is also known to influence nitrogen transfer during seedling formation. In maize it accelerated transport of nitrogen in the form of amino acids from the endosperm to the embryonic axis (Jain and Srivastava, 1981). However, most of the nitrogen is eventually leached out in the medium and is not gained in the embryonic axis. Thus, the phenolic inhibits growth of the embryonic axis even though it accelerates nitrogen mobilization from the endosperm.

E. Miscellaneous

Several other environmental and chemical factors are also known to influence proteolysis and nitrogen transfer during seed germination and seedling formation. Inhibition of nitrogen transfer by water stress (Khademi et al., 1991; De and Kar, 1994) and by organophosphate insecticide 'rogor' (dimethoalate) (Mathur et al., 1988) has been demonstrated. However, the imposition of salinity stress increases protease activity in several species, such as bajra (Reddy and Vora, 1985), mung bean (Sheoran and Garg, 1978), soybean (Durgaprasad et al., 1996), etc. Some amino acids, such as proline, may have some specific role also in saline conditions.

The pH of the external medium also influences seed germination and proteolytic activity. In fact, several proteases and alpha-amylase have acidic pH, and Ca^{2+} liberation and metabolite uptake by the scutellar epithelium are enhanced at low pH (Hamabata et al., 1988). It is also likely that at low acidic pH nitrogen transfer to the embryo will be accelerated. In fact, the opposing effects of GA and ABA on seed germination and seedling growth may be explained in terms of their opposing effects on the pH of the medium. In barley protoplasts, while ABA increased pH, GA decreased it (Heimovaara-Dijkstra et al., 1994).

VI. CONCLUSIONS AND FUTURE PROSPECTS

The physiology and biochemistry of protein hydrolysis and nitrogen transfer from storage tissue to the embryonic axis during seed germination and seedling formation are fairly well understood. However, the involvement of various types of proteases, the presence of some of these in dormant seeds, and *de novo* synthesis of others, make the process rather complicated. The role of some of the proteases in the processes other than protein hydrolysis and nitrogen mobilization is to be explored. A study of the molecular biology and some of the regulatory aspects of these enzymes might help in exploring these possibilities. The possible involvement of some molecules or the biophysical factors in the transport of nitrogen from storage tissues to the embryonic axis needs also to be investigated.

LITERATURE CITED

Abe, Y., Shirane, K., Yokosawa, H., Matsushita, H., Mitta, M., Kato, I. and Ashii, S. 1993. Asparaginyl endopeptidase of jack bean seed. Purification, characterisation and high utility in protein sequence analysis. *J. Biol. Chem.* 268: 3525–3529.

Adkins, S.W., Simpson, G.W. and Taylor, J.M. 1984. The physiological basis of seed dormancy in *Avena fatua*. IV. Alternative respiration and nitrogenous compounds. *Physiol. Plant.* 60: 234–238.

Allona, J., Collada, C., Casado, R. and Angincillo, C. 1994. 2 S arginine rich proteins from *Pinus pinaster* seeds. *Tree Physiol.* 14: 211–218.

Baumgartner, B. and Chrispeels, M.J. 1977. Purification and characterization of vicilin peptidohydrolase, the major endopeptidase in the cotyledons of mungbean seedlings. *Eur. J. Biochem.* 77: 223–233.

Belozersky, M.A., Sarabakanova, S.T. and Dunavaesky, Y.E. 1989. Aspartic proteinase from wheat seeds. Isolation, properties and action on gliadin. *Planta* 177: 321–326.

Benthke, P.C., Hillmer, S. and Jones, R.L. 1996. Isolation of intact protein storage vacuoles from barley aleurone. Identification of aspartic and cysteine proteases. *Plant Physiol.* 110: 521–529.

Bewely, J.D. and Black, M. 1994. *Seeds. Physiology of Development and Germination.* Plenum Press, New York.

Bigiarini, L., Pieri, N., Grilli, I., Galeschi, L., Capochi, A. and Fontanini, D. 1995. Hydrolysis of gliadin during germination of wheat seeds. *J. Plant Physiol.* 147: 161–167.

Blom-Zandstra, M. and Lampe, J.E.M. 1985. The role of nitrate in the osmoregulation of lettuce (*Lactuca sativa* L.) grown at different light intensities. *J. Exp. Bot.* 36: 1043–1052.

Bollini, R. and Chrispeels, M.J. 1978. Characterization and sub-cellular localization of vicilin and phytohaemagglutin, the two major reserve proteins of *Phaseolus vulgaris* L. *Planta* 142: 291–298.

Bose, B. and Srivastava, H.S. 1980. Proteolytic activity and nitrogen transfer in maize seeds during imbibition. *Biol. Plant.* 22: 414–419.

Bose, B. and Srivastava, H.S. 1982. Nitrogen metabolism, protease activity and germination of maize seeds under different atmospheric environments. *Indian J. Plant Physiol.* 25: 313–316.

Bose, B., Srivastava, H.S. and Mathur, S.N. 1982. Effect of antibiotics on the germination and protease activity of maize seeds. *Indian J. Plant Physiol.* 25: 271–275.

Bose, B., Srivastava, H.S. and Mathur, S.N. 1984. Partial purification and properties of protease from maize endosperm. *Beitr. Biol. Pflanzen.* 58: 383–391.

Bottari, A., Capocchi, A., Galeschi, L., Japova, A. and Saviozzi, F. 1996a. Asparaginyl endopeptidase during maturation and germination of durum wheat. *Physiol. Plant.* 97: 475–480.

Bottari, A., Capocchi, A., Fontanini, D. and Galeschi, L. 1996b. Major proteinase hydrolysing gliadin during wheat germination. *Phytochemistry* 43: 39–44.

Champigny, M.L. and Foyer, C. 1992. Nitrate activation of cytosolic protein kinases diverts photosynthetic carbon from sucrose to amino acid biosynthesis. *Plant Physiol.* 100: 7–12.

Chapman, J.M. and Davies, H.V. 1983. Control of breakdown of food reserves in germinating dicotyledonous seeds, a reassessment. *Annal. Bot.* 52: 593–595.

Ching, T.M. and Rynd, L. 1978. Developmental differences in embryos of high and low protein wheat seeds during germination. *Plant Physiol.* 62: 866–870.

Chrispeels, M.J. 1983. The Golgi apparatus mediates the transport of phytohaemagglutinin in the protein bodies in bean cotyledons. *Planta* 158: 140–151.

Cornel, F.A. and Plaxton, W.C. 1994. Characterization of peptidase activity in developing and germinating castor oil seeds. *Physiol. Plant* 91: 599–604.

Csoma, C. and Polgar, L. 1984. Proteinase from germinating bean cotyledons. Evidence for involvement of a thiol group in catalysis. *Biochem. J.* 222: 769–776.

Daussant, J.M., Neucere, N.J. and Conkerton, E.J. 1969. Immunological studies on *Arachis hypogaea* protein. Protein modification during germination. *Plant Physiol.* 44: 480–484.

De, R. and Kar, R.K. 1994. Effect of water stress on protein amino acids and protein contents in germinating mung bean seeds. *Indian J. Plant Physiol.* 37: 116–118.

De Barros, E.G. and Larkins, B.A. 1990. Purification and characterization of zein degrading proteases from endosperm of germinating maize seeds. *Plant Physiol.* 94: 297–303.

Dryer, J.M., Nelson, J.W. and Mural, N. 1993. Strategies for selecting mutation sites for methionine enhancement in the bean seed storage protein phaseolin. *J. Protein Chem.* 12: 545–560.

Duarte, I., Ricardo, C.P.P. and Duque-Magalhaes, M.C. 1996. Proteolysis in the quiescent seed. *Physiol. Plant* 96: 519–525.

Dunaesky, Y.E. and Belozersky, M.A. 1989. Proteolysis of main storage protein of buckwheat seeds at the early stage of germination. *Physiol. Plant.* 75: 424–428.

Duranti, M., Gatehouse, J.A., Boulter, D. and Cerletti, P. 1987. *In vitro* proteolytic processing of pea and jackbean storage proteins by an endopeptidase from lupine seeds. *Phytochemistry* 26: 627–631.

Durgaprasad, K.M.R., Muthukumarasamy, M. and Pannerselvam, R. 1996. Changes in protein metabolism induced by NaCl salinity in soybean seedlings. *Indian J. Plant Physiol.* (new series) 1: 98–101.

Elpidina, E.N., Voskoboynikova, N.E., Belozersky, M.A. and Dunavaesky, Y.E. 1991. Localization of metalloproteinase and its inhibitor in the protein bodies of buckwheat seeds. *Planta* 185: 46–52.

Enari, T.M. and Saponen, T. 1986. Mobilization of endosperm reserves during the germination of barley. *J. Inst. Brew.* 92: 25–31.

Esen, A. 1986. Separation of alcohol soluble proteins (zeins) from maize into three fractions by differential solubility. *Plant Physiol.* 80: 623–627.

Ferreira, R.B., Melo, T.S. and Teixiera, A.N. 1995. Catabolism of seed storage proteins from *Lupinus albus*. Fate of globulins during germination and seedling growth. *Aust. J. Plant Physiol.* 22: 373–381.

Fincher, G.B. 1989. Molecular and cellular biology associated with endosperm mobilization in germinating cereal grains. *Ann. Rev. Plant Physiol. Plant Mol. Biol.* 40: 305-326.
Forest, J.C. and Wightman, F. 1971. Metabolism of amino acids in plants. I. Changes in the soluble amino acid fractions of bushbean seedlings (*Phaseolus vulgaris* L.) and the development of transaminase activity. *Can. J. Biochem.* 49: 709-720.
Galleschi, L. and Felicioli, F. 1994. Purification, characterization and activation by anions of an aspartic proteinase isolated from bran of soft wheat. *Plant Sci.* 98: 15-24.
Gopstein, S. and Ilan, I. 1980. Evidence for the involvement of cytokinins in the regulation of proteolytic activity in cotyledons of germinating beans. *Plant Cell Physiol.* 21: 57-63.
Hamabata, A., Garcia-Maya, M., Romero, T. and Bernal-Lugo, I. 1988. Kinetics of the acidification capacity of aleurone layer and its effects upon solubilization of reserve substances from starchy endosperm of wheat. *Plant Physiol.* 86: 643-644.
Hammerton, R. and Ho, T-H. D. 1986. Hormonal regulation of the development of protease and carboxypeptidase activities in barley aleurone layers. *Plant Physiol.* 80: 692-697.
Hara, I. and Matsubara, H. 1980. Pumpkin (*Cucurbita* sp.) seed globulin. V. Proteolytic activities involved in globulin degradation in germinating seeds. *Plant Cell Physiol.* 21: 219-232.
Hara-Nishimura, I., Hayashi, M., Nishimura, M. and Akazawa, T. 1987. Biogenesis of protein bodies by budding from vacuoles in developing pumpkin cotyledons. *Protoplasma* 136: 49-55.
Harvey, B.M.R. and Oaks, A. 1974. Characteristics of an acid protease from maize endosperm. *Plant Physiol.* 53: 449-452.
Hay, P.C., Ramasur, T., Smith, C. and Mierynk, J.A. 1991. Storage protein mobilization during germination and early seedling growth of *Zea mays*. *Physiol. Plant.* 81: 377-384.
Heimovaara-Dijkstra, S., Heistek, J.C. and Wang, M. 1994. Counteractive effects of ABA and GA_3 on extracellular and intracellular pH and malate in barley aleurone. *Plant Physiol.* 106: 359-365.
Higgins, C.F. and Payne, J.W. 1981. The peptide pool of germinating barley grains. Regulation to hydrolysis and transport of storage proteins. *Plant Physiol.* 67: 785-792.
Higgins, T.J.V. 1984. Synthesis and regulation of major proteins in seeds. *Ann. Rev. Plant Physiol.* 35: 191-221.
Hilhorst, H.W.M. and Karssen, C.M. 1989. Nitrate reductase independent stimulation of seed germination in *Sisymbrium officianale* L. (hedge mustard) by light and nitrate. *Ann. Bot.* 63: 131-137.
Hilton, J.R. and Thomas, J.A. 1986. Regulation of pregerminative roles of respiration in seeds of various species by potassium nitrate. *J. Exp. Bot.* 37: 1516-1524.
Holwedra, B.C. and Rogers, J.C. 1992. Purification and characterization of aleurain, a plant thiol protease functionally homologous to mammalian cathepsin H. *Plant Physiol.* 99: 848-855.
Jain, A. and Srivastava, H.S. 1981. Effect of salicylic acid on growth, nitrogen content and peroxidase activity in maize seedlings. *Proc. Nat. Acad. Sci.* (India) 51 B: 311-317.
Kent, N.L. and Evers, A.D. 1994. *Technology of Cereals*. Elsevier Science, Oxford.
Khademi, M., Koranski, D.S., Hannapel, D.J., Knapp, A.D. and Gladon, R.J. 1991. Water stress and storage protein degradation during germination of *Impatiens* seed. *J. Amer. Soc. Hort. Sci.* 116: 302-306.
Koehler, S. and Ho, T-H. D. 1988. Purification and characterization of gibberellic acid induced cysteine endoprotease in barley aleurone layer. *Plant Physiol.* 87: 95-103.
Kortt, A.L. and Caldwell, J.B. 1990a. Sunflower 11 S globulin susceptibility to proteolytic cleavage of the subunits of native helianthinin during isolation: HPLC fractionation of the subunits. *Phytochemistry* 29: 1389-1396.

Kortt, A.L. and Caldwell, J.B. 1990b. Low molecular weight albumins from sunflower seed: Identification of a methionine rich albumin. *Phytochemistry* 29: 2805–2810.

Kosanke, R., Muntz, K., Saalbach, G., Saalbach, I., Vogt, B. and Kohler, K.H. 1990. Globulins of *Agrostemma githago. Biochem. Physiol. Pflanzen.* 186: 243–250.

Krishnan, H.B., Franceshi, V.R. and Okita, T.W. 1986. Immunochemical studies on the role of Golgi complex in protein body formation in rice seeds. *Planta* 169: 471–482.

Mathur, A. Mathur, S. and Mathur, S.N. 1988. Measurement of protease activity and translocation of its hydrolytic products in germinating seeds of blackgram as affected by treatments of rogor, an organophosphate insecticide. *Indian J. Plant Physiol.* 31: 92–96.

McIntyre, G.I., Cessna, A.J. and Hsiao, A.I. 1996. Seed dormancy in *Avena fatua*: Interacting effects of nitrate, water and seed coat injury. *Physiol. Plant.* 97: 291–302.

Metivier, J.R., and Dale, J.E. 1977. The utilization of endosperm reserve during growth of barley cultivars and the effect of time of application of nitrogen. *Ann. Bot.* 41: 715–728.

Miege, M.N. 1982. Protein types and distribution. In: *Nucleic Acids and Proteins in Plants. I. Structure, Biochemistry and Physiology of Protein* (D. Boulter and B. Parthier, eds.). Encyclopaedia of Plant Physiology, New series Vol. 14A. Springer-Verlag Berlin, pp. 291–345.

Mikkonen, A., Begbie, R., Grant, G. and Pusztai, A. 1986. Intracellular localization of some peptidases and mannosidase in cotyledons of resting kidney bean, *Phaseolus vulgaris. Physiol. Plant.* 68: 75–80.

Munoz, J.L., Martin, L., Nicolas, G. and Villalobas, N. 1990. Influence of endogenous cytokinins on reserve mobilization in cotyledons of *Cicer arietenum* L. Reproduction of endogenous levels of total cytokinins, zeatin, zeatin riboside and their corresponding glucosides. *Plant Physiol.* 93: 1011–1016.

Muren, E., Ek, B. and Rask, L. 1995. Processing of the 2 S storage protein pronapin in *Brassica napus* and in transformed tobacco. *Eur. J. Biochem.* 227: 316–321.

Mycek, M.J., Clarke, D.D., Neidle, A. and Walesh, H. 1961. Transglutaminase. *Methods Enzymol.* 5: 833–838.

Nandi, S.K., Palni, L.M.S. and de Klerk, G.J.M. 1995. The influence of the embryonic axis and cotyledons on reserve mobilization of germinating lupin seeds. *J. Exp. Bot.* 46: 329–336.

Neto, G.C., Yunes, J.A., Da Silva, M.J., Vettore, A.L., Arruda, P. and Leite, A. 1995. The involvement of opaque-2 on *b*-prolamin gene regulation in maize and coix suggests a more general role for their transcriptional activator. *Plant Mol. Biol.* 27: 1015–1029.

Nikolova, A., Klisurska, D., Gajdarzhieva, K. and Gorchilova, M.G. 1993. Seasonal variation in the mobilization of easily soluble proteins isolated from germinating cereal and leguminous crops. *Bulgarian J. Plant Physiol.* 19: 83–93.

Ohtani, T., Galili, G. Wallace, J.C., Thompson, G.A. and Larkins, B.A. 1991. Normal and lysine containing zeins are unstable in transgenic tobacco seeds. *Plant Mol. Biol.* 16: 117–128.

Palmiano, E.P. and Juliano, B.O. 1972. Biochemical changes in the rice grain during germination. *Plant Physiol.* 49: 751–756.

Palni, L.M.S., Nandi, S.K., and Singh, S. 1990. The physiology and biochemistry of cytokinins. In: *Hormonal Regulation of Plant Growth and Development* (S.S. Purohit, ed.). Agro Bot. Publisher, Bikaner, India, pp., ed. 1–60.

Poulle, M. and Jones, B.L. 1988. A proteinase from germinating barley. Purification and some physical properties of a 30 kD cysteine endoproteinase from green malt. *Plant Physiol.* 88: 1454–1460.

Ranjan, R. and Lewak, S. 1995. Interaction of jasmonic acid and abscisic acid in the control of lipases and proteases in germinating apple embryos. *Physiol. Plant.* 93: 421–426.

Reddy, M.P. and Vora, A.B. 1985. Effect of salinity on protein metabolism in bajra (*Pennisetum typhoides*) leaves. *Indian J. Plant Physiol.* 28: 190–195.
Redinbaugh, M.G. and Campbell, W.H. 1991. Higher plant responses to environmental nitrate. *Physiol. Plant.* 82: 640–650.
Richardson, M. 1991. Seed storage proteins: The enzyme inhibitors. In: *Methods in Plant Biochemistry*, vol. 5 (P.M. Dey and J.B. Harbone, eds.). Acad. Press London, pp. 259–305.
Rogers, J.C., Dean, D. and Heck, G.R. 1985. Aleuroni: a barley thiol protease closely related to mammalian cathepsin H. *Proc. Nat. Acad. Sci. U.S.A.* 82: 6512–6516.
Saini, H.S., Bassi, P.K., Goudey, J.S. and Spencer, M.S. 1987. Breakage of seed dormancy of field penny cress (*Thlaspi awense*) by growth regulators, nitrate and environmental factors. *Weed Sci.* 35: 802–806.
Santino, A., Daminati, M.G., Vitale, A. and Bollini, R. 1992. The α-amylase inhibitor of bean seed: Two-step proteolytic maturation in the protein storage vacuoles of the developing cotyledons. *Physiol. Plant.* 85: 425–432.
Sarkkinen, P., Kalkinnen, P., Tilgmann, C., Siuro, J., Kervinen, J. and Mikola, L. 1992. Aspartic protease from barley grains is related to mammalian lysosomal cathepsin H. *Planta* 186: 317–323.
Scarafoni, A., Giani, D. and Cerletti, P. 1992. An endopeptidase in dormant lupine seeds. *Phytochemistry* 31: 3715–3723.
Schliephacke, M., Kremp, A., Schmid, H-P., Kohler, K. and Kull, U. 1991. Prosomes (proteosomes) of higher plants. *Eur. J. Cell. Biol.* 55: 114–121.
Sembdner, G. and Parthier, B. 1993. The biochemistry and the physiological and molecular actions of jasmonates. *Ann. Rev. Plant Physiol. Plant Mol. Biol.* 44: 569–589.
Shankar, N. and Srivastava, H.S. 1997. The response of NADH and NADPH specific nitrate reductase activities and nitrate and oxygen uptake to nitrate supply in maize seedlings. *Biol. Plant* 39: 583–589.
Sheoran, I.S. and Garg, O.P. 1978. Effect of salinity on the activities of RNase, DNase and protease during germination and early seedling growth of mung bean. *Physiol. Plant* 44: 171–174.
Shotwell, M.A. and Larkins, B.A. 1989. The molecular biology and biochemistry of seed storage proteins. In: *The Biochemistry of Plants*, vol. 15. *A. Comprehensive Treatise* (A. Marcus, ed.). Academic Press, San Diego CA., pp. 297–345.
Shutov, A.D. and Vaintraub, I. 1987. Degradation of storage proteins in germinating seeds. *Phytochemistry* 26: 1557–1566.
Shuttuck-Eidens, D.M. and Beachy, R.N. 1985. Degradation of *b*-conglycinin in early stage of soybean embryogenesis. *Plant Physiol.* 78: 895–898.
Skoda, B. and Malek, L. 1992. Dry pea seed proteosome. Purification and enzymic activities. *Plant Physiol.* 99: 1515–1519.
Srivastava, H.S. 1976. Some aspects of nitrate assimilation in the seedlings of normal and opaque-2 mutant of maize. *J. Exp. Bot.* 27: 1215–1222.
Srivastava, H.S., Oaks, A. and Bakyta, I.L. 1976. The effect of nitrate on early seedling growth in *Zea mays* L. *Can J. Bot.* 54: 923–929.
St. Angelo, A.J., Ory, R.L. and Hansen, H.J. 1969. Localization of an acid protease in hempseed. *Phytochemistry* 8: 1135–1141.
Staswick, P.E., Hermodson, M.A. and Nielsen, N.C. 1981. Identification of the acidic and basic subunit complexes of glycinin. *J. Biol. Chem.* 256: 8752–8755.
Takeba, G. 1980. Accumulation of free amino acids in the tips of non-dormant embryonic axis accounts for the increase in growth potential of New York lettuce seeds. *Plant Cell Physiol.* 21: 1639–1644.
Tanaka, K., Sugimoto, T., Ogawa, M. and Kasai, Z. 1980. Isolation and characterization of two types of protein bodies in the rice endosperm. *Agric. Biol. Chem.* 44: 1633–1639.
Tan-Wilson, A.L., Liu, X., Chen, R., Qi, X. and Wilson, K.A. 1996. An acidic amino acid-specific protease from germinating soybeans. *Phytochemistry* 42: 313–319.

Tomakangas, K., Keavinnen, J., Ostmann, A. and Teeri, T. 1994. Tissue specific localisation of aspartic proteinase in developing and germinating barley grains. *Planta* 195: 116–125.
Tsai, F-Y. and Kao, C.H. 1996. Methyl jasmonate inhibits endosperm reserve mobilization in germinating rice seeds. *Physiol. Mol. Biol. Plants.* 2: 67–70.
Vaintraub, I.A., Beltey, N.K. and Shutov, A.D. 1981. Deamidation of glutenin proteins during wheat grain germination. *Prikl. Biokhim. Mikrobiol.* 17: 166–169.
Vaintraub, I.A., Kotova, L.V. and Saha, R. 1992. Protein deamidase from germinating wheat grains. *FEBS Lett.* 302: 169–171.
Vaintraub, I.A., Kotova, L.V. and Saha, R. 1996. Protein deamidase from germinating seeds. *Physiol. Plant.* 96: 662–666.
Van Staden, J., 1983. Seeds and cytokinins. *Physiol. Plant.* 58: 340–346.
Vigil, E.L. and Fang, T.K. 1995. Comparative biochemical and morphological changes in imbibed cotton seed hypocotyls and radical *in situ* and *in vitro* protein breakdown and elongation growth. *Seed Sci. Res.* 5: 41–51.
Wang, M., Heimovaara-Dijkstra, S. and Van Duijn, B. 1995. Modulation of germination of embryos isolated from dormant and non-dormant grains by manipulation of endogenous abscisic acid. *Planta* 195: 586–589.
Willey, L. and Ashton, F.M. 1967. Influence of embryonic axis on protein hydrolysis in cotyledons of *Cucurbita maxima*. *Physiol. Plant.* 20: 688–696.
Wrobel, R. and Jones, B.L. 1992. Appearance of endoproteolytic enzymes during the germination of barley. *Plant Physiol.* 100: 1508–1516.
Yamagata, H., Sugimoto, T., Tanaka, K. and Kasai, Z. 1982. Biosynthesis of storage proteins in developing rice seeds. *Plant Physiol.* 70: 1094–1100.
Yamaoka, Y., Takeuchi, M. and Morohashi, Y. 1990. Purification and characterization of cysteine endopeptidase in cotyledons of germinating mungbean seeds. *Plant Physiol.* 94: 561–566.
Yomo, H. and Srinivasan, K. 1973. Protein breakdown and formation of protease in attached and detached cotyledons of *Phaseolus vulgaris* L. *Plant Physiol.* 52: 671–673.
Youle, R.J. and Huang, A.H.C. 1978. Albumin storage protein bodies of castor bean. *Plant Physiol.* 61: 13–16.
Zarain, M.H., Bernal, L.I. and Vazquez-Ramix, J.M. 1987. Effect of BA on the DNA synthesis during early germination of maize embryo axis. *Mutat. Res.* 181: 103–110.
Zheng, Z., Sumi, K., Tanaka, K. and Murai, N. 1995. The bean seed storage protein is synthesized, processed and accumulated in the vacuolar type II protein bodies of transgenic rice endosperm. *Plant Physiol.* 109: 777–786.

Chapter 2

ROOT ABSORPTION AND ASSIMILATION OF INORGANIC NITROGEN

D.J. Pilbeam and A.U. Jan

I. Introduction

II. Mechanism of Nitrate and Ammonium Uptake

III. Assimilation of Nitrate and Ammonium

IV. Control of Uptake and Assimilation of Nitrate and Ammonium by Roots

V. Conclusions and Future Prospects

Literature Cited

Abbreviations: HATS, high affinity transport system; LATS, low-affinity transport system; NR, nitrate reductase; NRA, nitrate reductase activity; NiR, nitrite reductase, NiRA, nitrite reductase activity; PEP, phosphoenolpyruvate; RA, relative addition (of nitrate); RGR, relative growth rate.

I. INTRODUCTION

Since the formulation of Sprengel's Law of Minimum, which states that plant growth is limited by whichever element is least available relative to the plant's demand for it, there has been considerable interest in the uptake and assimilation of nitrogen, the element taken up from soil by plants in the largest amounts.

Although it has long been assumed that the shoot is the major site for the assimilation of nitrogen taken up as nitrate, the important role of the roots in the acquisition of nitrogen has attracted the attention of plant physiologists for a considerable time, and more recently the role of the root system in assimilation has also been recognized.

Nitrogen freely circulates between the atmosphere, the lithosphere, and the hydrosphere, mainly as dinitrogen (N_2) gas, but in soils it is also present in organic molecules and as ammonium (NH_4^+), nitrite (NO_2^-) and nitrate (NO_3^-) ions. Of these forms dinitrogen is only taken up by those plants capable of forming symbiotic associations with bacteria, but nitrate and ammonium are capable of being taken up by most plant species. The conversion of dinitrogen to organic nitrogen in root nodules is a complex subject that has been extensively explored and shall not be covered here.

In an aerated, non-acidic agricultural soil the actions of the microorganisms that contribute to the nitrogen cycle ensure that nitrate is the major form of nitrogen present, and most crop plants preferentially take up nitrate. However, most plants also take up some ammonium, and those plants that are adapted to acidic soils, in which ammonium predominates, seem to have an exclusive requirement for that form of nitrogen.

The uptake mechanisms of the two ions vary considerably, not least because one is an anion and the other is a cation. Highly soluble nitrate is freely leached from the soil, but is made available to plants by the fact that the ongoing actions of the nitrogen cycle continually generate it from other forms of nitrogen and because it moves to plants in the mass flow of water to the roots. Ammonium is usually present in soils in lower concentration, but remains present because of ion exchange reactions with the negatively charged clay colloids. From here it is made available to plants by cation exchange reactions. Once at the root surface, nitrate and ammonium ions diffuse into the free space and enter the apoplastic cell walls. However, they must then pass across the plasmalemma to enter cells.

II. MECHANISMS OF NITRATE AND AMMONIUM UPTAKE

The plasmalemma represents a considerable barrier to the nitrate ion, both because of the hydrophobic nature of its interior and because of the negative charges associated with the inner surface. Since nitrate is present at a higher concentration inside the root than in the soil solution, uptake must be an active process, and the amount of energy required must partly depend on the concentration gradient between the outside and inside of the root. It has been calculated from an experiment on *Lolium perenne* L. that taking up one mole of nitrate from a 1.5 mmol.m^{-3} solution requires 28 kJ of energy whereas to take up one mole from the 10,000-fold higher concentration of 15 mol.m^{-3} requires only 9 kJ of energy (Clarkson and Hanson, 1980).

The energy is expended in the functioning of carriers. In the case of nitrate the carrier is almost entirely specific for the NO_3^- ion (although Cl$^-$ ions may compete for entry into roots via the same carrier (Dhugga et al., 1988) and the energy is provided by an H$^+$-ATPase. The recognition of nitrate may involve a planar, positively charged guanidinium group of an arginine residue in the carrier protein, since in experiments in which ketones and carbonyl compounds that bind to guanidinium groups were supplied to maize (*Zea mays* L.) roots, uptake of nitrate was inhibited (Ni and Beevers, 1990).

When plants are supplied with nitrate as an exclusive source of nitrogen, uptake of anions exceeds uptake of cations because of the high requirement for N (Kirkby and Mengel, 1967), and the excess anion uptake is exactly matched by an apparent excretion of OH$^-$ equivalents into the rooting medium (Dijkshoorn, 1962; van Beusichem et al., 1988). However, the mechanism by which this occurs is by the simultaneous uptake of H$^+$ and NO_3^- (NO_3^-/H^+ symport).

Earlier workers (Thibaud and Grignon, 1981; Deane-Drummond, 1984) failed to see any effect of the H$^+$-ATPase inhibitor diethylstilbestrol on nitrate uptake, but in experiments on soybeans (*Glycine max* (L.) Merr.), maize and oats (*Avena sativa* L.) (Ullrich, 1992) membrane potential was seen to become less negative immediately upon supply of nitrate, and then to go to negative hyperpolarization. This was taken to indicate a fall in membrane potential (and acidification of the cytoplasm) due to the influx of protons, and these effects caused stimulation of plasmalemma H$^+$-ATPases, which restored the negative charges on the membrane.

Factors that affect such H$^+$-ATPases exert a notable influence on the uptake of nitrate by plants. For example, in oats nitrate transport across membranes was increased by up to 50% by the action of low molecular weight humic substances. These stimulated H$^+$-ATPase activity (Varanini et al., 1993).

Like enzymes, carrier systems obey Michaelis-Menten kinetics and show a hyperbolic relationship between rate of uptake of nitrate and concentration of nitrate in the external medium. However, there appear to be at least two systems that operate for nitrate uptake in higher plants. One of these systems takes up nitrate at low concentration and shows a high affinity for the ion. In *Arabidopsis thaliana* the K_m value is 40 mmol m^{-3} and the maximum rate of uptake is 4.0 µmol g^{-1} h^{-1}. The second system takes up nitrate at the higher concentration more typical of soil solution, and has a K_m value of 25 mol m^{-3} and an I_{max} value of 75 µmol h^{-1} g^{-1} (Doddema and Telkamp, 1979). In a study on lettuce (*Lactuca sativa* L.) grown in solution culture saturation occurred at less than 15 mmol m^{-3}, and K_m values averaging 9.5 mmol m^{-3} were generally 10–20 times higher than the lowest concentration from which nitrate uptake could occur. The K_m values remained constant as the plants aged but the I_{max} values decreased (Swiader and Freiji, 1996).

Similar low-affinity and high-affinity transport systems (LATS and HATS) have been seen in a range of plant species. In barley (*Hordeum vulgare*) microelectrodes inserted in the root epidermis or cortical cells were used to show that the concentration of nitrate in the external medium that gives the K_m value for uptake gives the half-depolarization value of the membrane, indicating that each system operates as an H$^+$/NO$_3$-symport (Siddiqi et al., 1990; Glass et al., 1992). The activity of HATS is induced by nitrate in the medium, and in different experiments on one cultivar of barley the induction of uptake was exactly matched by the induction of the ability of nitrate to cause membrane depolarization (Siddiqi et al., 1989; Glass et al., 1992). The activity of LATS does not appear to be inducible, so it seems that LATS may serve to take up nitrate from flushes of the ion in the soil to provide the reduced nitrogen required for carrier synthesis in induction of HATS. Nitrate taken up by root tips appears to be important for induction of new carrier molecules in the root in general (Siebrecht et al., 1995).

It has been confirmed that it is the high-affinity uptake system itself that is induced, and not merely that induction of the enzymes of nitrate assimilation in the root make the uptake of nitrate thermodynamically easier. In experiments on membrane vesicles from maize roots the uptake of a radioactive analogue of nitrate, ^{36}ClO$_3^-$, was induced within 2 hours of supply of nitrate to the medium in which the plants were growing before the vesicles were prepared (Lu and Briskin, 1993). As the vesicles did not contain enzymes to assimilate nitrate, it must have been the carrier itself that was induced.

Induction of nitrate uptake occurs after a lag phase, which may typically last 30–90 minutes after exposure of uninduced cells to an external source of nitrate. A steady state uptake at 2 to 5 times the preinduced rate occur by 4–6 hours after induction (Clarkson, 1986). During this time *de novo* biosynthesis of polypeptides occurs, including synthesis of a membrane polypeptide of 30–31 kD (McClure et al., 1987). Ni and Beevers (1994) reported enhanced synthesis of 50, 49, 38 and 33 kD polypeptides after supply of nitrate to maize seedlings, of which the 49 and 33 kD polypeptides were definitely integrated into the plasma membrane. The 33 kD polypeptide seemed to be present in uninduced seedlings, so it could make up the constitutive system.

In experiments on *Brassica napus*, a plant with a fast rate of nitrate uptake, induction of nitrate uptake was inhibited by both a ribonucleic acid synthesis inhibitor and a protein synthesis inhibitor (Lainé et al., 1995). The inhibitors were supplied to plants in which half of a nitrate-starved root system was resupplied with nitrate and the other half was left without. The protein synthesis inhibitor caused 100% inhibition of nitrate uptake in the root system resupplied with nitrate. This indicated not only a complete inhibition of induction, but also a decrease in uninduced carrier activity. Continuous carrier activity must therefore depend on protein turnover. In a study of maize deprived of nitrogen for up to 72 h, the rate of nitrate influx on resupply increased up to 24 h of deprivation (probably because of a rapid flow into free space) and then decreased as carrier activity declined (Teyker et al., 1991).

The induction of HATS is inhibited by ammonium ions, probably through inhibition of carrier synthesis (Clarkson, 1986). However, Henriksen and Spanswick (1993) failed to show any inhibition of induction by metabolites downstream from nitrate.

The actual process of uptake of nitrate is definitely inhibited by ammonium (Ayling, 1993). Positively charged NH_4^+ ions depolarize the plasmalemma, which may limit nitrate uptake by affecting H^+-APTase activity. However, Ayling showed that removing NH_4^+ by washing led to hyperpolarization of the membrane, followed by gradual recovery. Inhibition of nitrate influx was not relieved in the same way. It seems likely that H^+ ions resulting from the assimilation of NH_4^+ affected membrane potential, but once NH_4^+ ions were removed the H^+ ions were used up by H^+-ATPase activity and the normal membrane potential was restored. The continuing inhibition of nitrate influx must have been due to repression of the carrier, and not to an effect on the proton motive force.

Both the rate of influx of nitrate and the induction of the carrier system seem to be dependent on the cytoplasmic nitrate concentration, or at least on the concentration gradient between the external medium and the root system. Therefore, the rate at which nitrate disappears from the cytoplasm is important here. This could be achieved by the rate of assimilation of nitrate in the cytoplasm, removal of nitrate into other cells and ultimately into the xylem, or removal of nitrate into the vacuole. Clarkson (1986) suggested that removal of nitrate into a storage pool in the vacuole was the key factor, and it has been seen in barley, maize and pea (*Pisum sativum*) roots that the cytoplasmic nitrate concentration is approximately 1/8th of the concentration in the vacuole (Belton et al., 1985). Internal nitrate concentrations in the root appear to have a considerable impact on the rate of nitrate efflux (Deane-Drummond and Glass, 1983), with efflux being greater when internal concentrations are higher. In an experiment in which maize plants were supplied $^{15}NO_3^-$ after nitrogen deprivation the efflux of $^{14}NO_3^-$ was linearly related to nitrate concentration in the root (Teyker et al., 1991).

This efflux is almost certainly a passive process, and data for net uptake of nitrate by *Lolium perenne* have been shown to give a diurnal pattern that matches a pump/leak model of active influx and passive efflux (Scaife, 1989). Although net uptake is less at night due to increased efflux, this is presumably a passive process linked to increased internal nitrate concentrations due to a lowered availability of carbon skeletons for N assimilation. When induction of nitrate uptake occurs, this is due more to an increase in influx component than to a decrease in the efflux component (Lainé et al., 1995).

Ammonium is taken up by higher plants by a constitutive carrier that shows Michaelis-Menten kinetics. The uptake occurs because of the attraction of NH_4^+ ions to the negative potential of the plasmalemma; this potential is restored by an H^+-ATPase (Ullrich, 1992). Some uncharged ammonia (NH_3) might enter roots by diffusion, but as the pK_a value of the dissociation of ammonia in water into ammonium is 9.25 it is not likely that there would be much NH_3 available for uptake in soils. In fact, the concentration of ammonia is likely to be higher inside a root than outside, and any diffusion that occurs might be outwards (Lee and Ayling, 1993).

It seems probable that there are two NH_4^+ uptake systems (Mäck and Tischner, 1994). One of these systems is a Low Affinity Transport System (LATS) which is inducible, but takes up NH_4^+ with no lag phase and thus also has a constitutive component. The other is a High Affinity Transport System (HATS), with high initial uptake rates (possibly into apparent free space in the root). The two systems have been observed in wheat seedlings when the external nitrate concentration was higher than 1 mol m^{-3} (Goldstein and Hunziker, 1985). In rice (*Oryza sativa*) both systems have been shown to be dependent on the supply of energy and of pH (Wang et al., 1993), further evidence that they operate by means of H⁺-ATPases. The stoichiometry of the reaction has apparently shown a 1 NH_4^+: 1.25 H⁺ relationship in sugar-beet (*Beta vulgaris*) (Breteler, 1973) and a 1:1 relationship in castor (*Ricinus communis*) (Allen and Raven, 1987). A 1:1 relationship would be a stable system over long time periods as the assimilation of one mole of ammonium ions in roots should release one mole of protons, so the link between proton extrusion and ammonium uptake would maintain constant pH within the root. A recent study of cranberry (*Vaccinium macrocarpon*) showed only 56% of the theoretical amount of protons released during assimilation of ammonium being excreted, presumably because the rest neutralized the OH⁻ equivalents released during the concurrent assimilation of sulphate and nitrate by the plants (Barak et al., 1996). The authors observed a rate of uptake of ammonium of up to 0.70 µmol g⁻¹ h⁻¹ from the highest concentration of ammonium in which they grew the plants.

III. ASSIMILATION OF NITRATE AND AMMONIUM

Assimilation of nitrate in the roots occurs through a reduction to nitrite and another reduction step giving rise to ammonium. These two steps are catalyzed by the enzymes nitrate reductase (NR) and nitrite reductase (NiR) acting in sequence (Fig. 2.1).

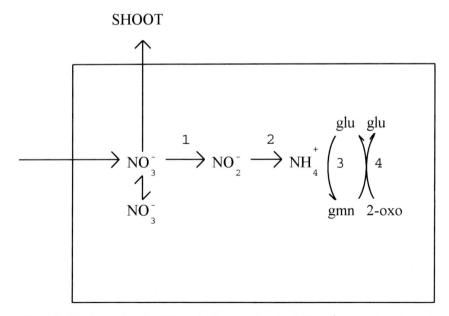

Fig. 2.1: Uptake and assimilation of nitrate and assimilation of ammonium, in roots. 1 — NR, 2 — NiR, 3 — glutamine synthetase, 4 — glutamate synthase, glu — glutamate, gmn — glutamine, 2-oxo — 2-oxoglutarate.

NR is considered a major enzyme of the photosynthetic tissues of plants. In roots it seems to be present in epidermal cells (Rufty et al., 1986), possibly more in the apical regions than in the basal (Wallace, 1973). However, there may be more activity of the enzyme in older parts of the root system than was once thought (Oaks and Long, 1992). In experiments on barley roots it was demonstrated that the enzyme has both a cytosolic form and a membrane-bound form (Ward et al., 1988), and like NR in the shoot is induced by nitrate (Oaks et al., 1972).

NiR is also thought to be primarily an enzyme of photosynthetic tissues, but it is found in roots relatively evenly distributed along the axis (Polisetty and Hageman, 1983). Its distribution inside root cells is restricted to the plastids (Dalling et al., 1972).

The activity of NR requires reducing power, either in the form of NADH or NADPH. The form of the enzyme identified first in roots requires NADH [EC 1.6.6.1], but there is also a bispecific form in roots [EC 1.6.6.2] that can use either NADH or NADPH (Redinbaugh and Campbell, 1981). It has been shown that in barley there are two separate genes that control the synthesis of these two forms of NR, and that the genes which code for each form do so for both shoots and roots (Warner et al., 1987). In maize roots the NADH-NR is the major form in the tips, but the bispecific form may be the major form in the more mature roots (Oaks and Long, 1992).

In roots the cofactors are supplied from respiratory reactions. Nitrite reductase activity in roots also seems to be dependent on NADH or NADPH as reductants, although the energy may actually be transferred to the enzyme via a ferredoxin-like substance or a pyridine nucleotide (Suzuki et al., 1985). NiR activity in roots appears to be linked to the activity of the oxidative pentose phosphate pathway (Emes and Fowler, 1983).

Ammonium produced by the reduction of nitrate in roots, or taken up directly from the rooting medium, is added to glutamate to give rise to glutamine. This reaction is catalyzed by the enzyme glutamine synthetase [EC 6.3.1.2], an enzyme also found in leaves. In roots there is one form of the enzyme (Mann et al., 1979), found in both the cytoplasm and plastids (Vezina et al., 1987), whereas in leaves there are different isozymes in the cytoplasm and chloroplasts (Mann et al., 1979; McNally et al., 1983). The main role of these isozymes in leaves is presumably in the reassimilation of ammonia released during photorespiration, as very little ammonium taken up from the rooting medium is translocated to the shoots.

Glutamine is deaminated in plants to give rise to glutamate, and the amino group released is transferred to α-oxoglutarate to generate another molecule of glutamate. In roots this is catalyzed by the enzyme glutamate synthase: NAD(P)H-dependent [EC 1.4.1.13]. This is actually two different forms, one of which uses NADH and the other NADPH. The NADH form is more common in plants, particularly in root nodules (Lea et al., 1992). There is a third form of the enzyme, glutamate synthase: ferredoxin-dependent [EC 1.4.7.1] that is found in leaves (Bottella et al., 1988).

The amino nitrogen is subsequently transferred to various amino acids by other transamination reactions and the amino acids incorporated into proteins. The amino acid products of assimilation of nitrate and ammonium in roots are either assembled into proteins in the roots themselves, or are exported to the shoot in the xylem. In many plants the major amino acid exported from the roots is glutamine (Pilbeam and Kirkby, 1992), although in others it is asparagine (McClure and Israel, 1979). It is therefore obvious that the synthesis of α-oxoglutarate and oxalacetate, the organic acid precursors of glutamine and asparagine, is essential for the assimilation of nitrogen by plants. This is generally thought to occur by the action of the enzyme PEP carboxylase [EC 4.1.1.31] and the tricarboxylic acid cycle (Fig. 2.2).

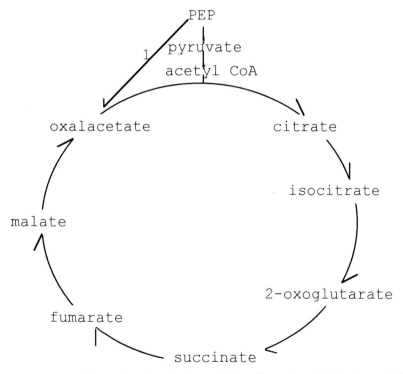

Fig. 2.2: Tricarboxylic acid cycle and the synthesis of organic acids in plants. PEP carboxylase.

The sites where these assimilatory reactions occur vary according to both the type of nitrogen supplied and the plant species concerned. When NH_4^+–N is the sole N source its assimilation occurs almost entirely in the roots (Findenegg et al., 1989 and others). There is some indication that plants supplied ammonium-nitrogen have higher PEP carboxylase activity in their roots than plants supplied nitrate-nitrogen (Schweizer and Erismann, 1985), which would be expected because of the demand for organic acids for amino acid synthesis in the roots of ammonium-grown plants.

For plants supplied nitrate-nitrogen the main site of the assimilation is the leaves, although there is considerable variability in the exact proportions of total assimilation that occur in roots and shoots between different species (Pate, 1973). However, these proportions may change as plants age, and in wheat and dwarf bean (*Phaseolus vulgaris*) for example it has been shown that the proportion reduced in the roots declines with ageing (Ashley et al., 1975; Breteler and Hanisch ten Cate, 1980; Talouizte et al., 1984).

The proportion of nitrate assimilated in different plant parts also varies with the rate of supply of nitrate to the plants. In maize seedlings transferred from 0.5 mol m^{-3} to 20 mol m^{-3} nitrate the amount of nitrate reduction in the roots remained constant, while the amount translocated to the shoots (and hence the proportion translocated to the shoots) increased (Morgan et al., 1985). A similar pattern was observed in a study on maize in which seedlings growing in 14 mmol m^{-3} nitrate were found to have retained 95% of ^{13}N-labelled nitrate 13 minutes after it was supplied, whereas seedlings in 1.4 mol m^{-3} nitrate retained only 75% of the label taken up in their roots (McNaughton and Presland, 1983). In intact plants of *Vicia faba* growing in different concentrations of nitrate the ratio of nitrate to total nitrogen in the xylem sap was found to be greatest for the highest concentration of nitrate in the medium (Sutherland et al., 1985). It appeared that high concentrations of nitrate in the growth medium gave rise to an increase in the reduction of nitrate in the roots, but above a certain limit of external concentration more nitrate was translocated to the shoot to be reduced.

A low rate of nitrogen supply also gives rise to differences in the partitioning of dry matter between root and shoot. With a low rate of nitrogen supply shoot growth decreases, so that root:shoot ratio increases (Brouwer, 1962). The roots keep growing (possibly fuelled by the nitrate taken up being assimilated in the roots themselves) so that their chances of growing into areas of soil where nitrogen supply is higher are maximized (Drew et al., 1973). When roots grow into areas of higher nitrogen content lateral branching occurs (Drew, 1975).

Increased root length relative to the overall plant size is important in nitrogen uptake as at high rates of supply there is a relatively constant relationship between nitrogen uptake and root length per plant (Fig. 2.3). However, as indicated earlier, nitrogen deficiency can cause enhanced uptake of inorganic nitrogen, at least before carrier turnover causes a decline in influx, and at lower rates of supply there is a higher rate of uptake.

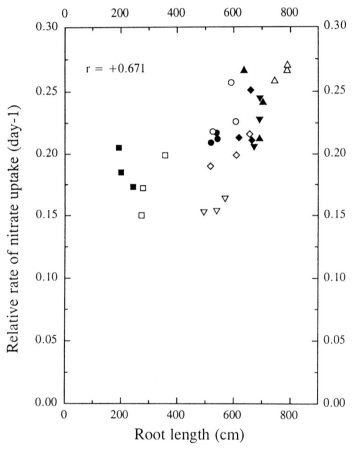

Fig. 2.3: Relationship between relative rate of nitrate uptake (natural logarithm of uptake per plant per day) and total root length in *Triticum* species. Unpublished data. Seedlings of *Triticum* species were grown in 2.0 mol m^{-3} nitrate in solution culture for 15 days in the conditions given in Jan and Pilbeam. (1993). ■ *T. monococcum* var. boeoticum; ¯ *T. monococcum* cv monococcum; ♦ *T. turgidum* cv turgidum; ▲ *T. turgidum* var. dicoccoides; ▼ *T. turgidum* cv dicoccon; ∆ *T. turgidum* cv durum; π *T. turgidum* cv polonicum; ∇ *T. turgidum* var. timopheevii; ○ *T. aestivum* group aestivum; ● *T. aestivum* group compactum.

The importance of both the increased proportion of plants represented by the root system and the enhanced capacity of roots to take up nitrogen when it is in short supply was shown by experiments on barley (Mattsson et al., 1991). In these experiments plants were given different rates of nitrate supply; increase in I_{max} of nitrate with increasing relative addition (RA) was observed up to an optimal value, followed by a decrease. An increase in RGR with an increase in RA across the range of RA values used was also recorded, but when the I_{max} values were replotted as relative I_{max} (rate of nitrogen uptake per amount of nitrogen already in the plant), a steady decrease with increasing RGR was found. This shows that the slower growing plants could have taken up nitrate faster relative to the amount of nitrogen they already contained because a much larger proportion of these plants was in their root systems.

In these experiments on barley the relationship between I_{max} and N concentration in the roots was found to be very similar to that between I_{max} and RA, i.e., I_{max} increased with N concentration up to an optimal value and then decreased. This indicates that the rate of uptake of nitrate is not entirely dependent on carriers, but also depends on the amounts (and activities) of root enzymes.

IV. CONTROL OF UPTAKE AND ASSIMILATION OF NITRATE AND AMMONIUM BY ROOTS

Because there is such an obvious link between the uptake of nitrate and its assimilation it was once thought that nitrate reductase was part of the carrier mechanism for nitrate. Treatment of barley root cells with anti-nitrate reductase immunoglobulin G fragments was found to inhibit nitrate uptake (Ward et al., 1988), but in many experiments in which NRA is inhibited by tungstate the uptake of nitrate still occurs, so there cannot be a direct physical link.

For a plant supplied nitrate-nitrogen the control of uptake and assimilation is a very complicated process. Carriers for nitrate are induced in the presence of nitrate and degraded in its absence; the enzymes of assimilation of nitrate and ammonium are induced both in leaves and roots by the presence of nitrate. It is hardly surprising that it has been difficult to distinguish between the control of uptake and the control of assimilation, but since assimilation in the shoot is physically separated from the site of uptake, it is easier to study there. Moreover, recent studies have shown that the controls of the two processes appear to be separated and distinguishable in the roots also.

Nitrate entering a root can be partitioned immediately into metabolic pools (probably the cytoplasm) and storage pools (probably the vacuoles), or at least in those cells far enough back from the root tips where vacuoles occur. From here it is either reduced to amino acids, or loaded into the xylem and transported to the leaves either in the transpiration flow or by root pressure. As might be expected from the inducible nature of carriers, the concentrations of nitrate in the cytoplasm vary according to the rates of nitrate supply to plants, and in barley roots cytoplasmic concentrations of nitrate from 0.66 mol m^{-3} and 3.9 mol m^{-3} for plants grown with no nitrate supplied or with high rates of supply respectively, have been observed (King et al., 1992). In this experiment the K_m value for NR was estimated to be 1.41 mol m^{-3} nitrate for the NADH-dependent form of NR and 0.16 mol m^{-3} for the NADPH-dependent form, so it can be seen that the range of cytoplasmic nitrate concentrations observed straddles the K_m value of one of the forms of root NR. If plants are exposed to increasing supplies of nitrate the increased cytoplasmic concentration of nitrate enables a much more rapid reduction to occur in the root irrespective of whether NR in the root is induced or not, assuming that supplies of NADH or NADPH are not limiting. However, the fact that cytoplasmic nitrate concentrations are higher in plants grown in high nitrate shows that the enhanced NRA that occurs in the roots of these plants is not sufficient to account for the extra nitrate that is taken up, and so more nitrate is available to be loaded into the xylem unreduced.

The exact proportions of nitrate assimilated in roots and shoots are probably determined by the availability of reduced carbon compounds to provide the skeletons for amino acid synthesis and to provide the substrates for the production of cofactors for nitrate reduction. Both NRA and mRNA for NR fluctuate in a diurnal pattern in maize roots, albeit to a lesser extent than in the shoots (Bowsher et al., 1991), so there would appear to be more assimilation when energy levels are high in a plant. In experiments on wheat seedlings, nitrogen deprivation caused an increase in the concentrations of soluble carbohydrates (Talouizte et al., 1984). In other experiments on wheat, competition for carbon skeletons between processes of carbon and nitrogen assimilation occurred when nitrate was supplied after its depletion for 3 weeks (Van Quy et al., 1991). In the latter series of experiments the synthesis of sucrose was inversely proportional to the rates of both uptake and assimilation of nitrate, illustrating the importance of energy supply in these processes. In determining the proportion of nitrate assimilation that occurs in roots the supply of photosynthates to the roots is apparently of paramount importance.

If this is so, there should be a big difference in the amount of energy supplied to roots when plants are supplied ammonium-nitrogen rather than nitrate-nitrogen as ammonium is assimilated almost entirely in the roots. However, in experiments on barley uptake and assimilation of ammonium were shown to account for 14% of root catabolism, whereas uptake of nitrate was shown to account for 5%, nitrate assimilation 15% and assimilation of the ammonium produced 3% (Bloom et al., 1992). Furthermore, there was an enhanced requirement for carbon skeletons in roots when ammonium was the nitrogen source, and when assimilation of nitrate increased. In the experiments of Van Quy et al. (1991) the main demand for sucrose was as carbon skeleton since the concentrations of sucrose increased when nitrate assimilation was inhibited by tungstate, even though the uptake of nitrate was still occurring.

In an experiment on young tobacco (*Nicotiana tabacum*) plants fed $^{15}NO_3^-$ a rapid decrease in reduction of nitrate was observed in whole plants immediately upon onset of the dark period, a decrease that corresponded to a rapid decrease in the amount of non-structural carbohydrate, particularly in the shoot and to a lesser extent in the roots (Rufty et al., 1992). Despite this overall decline in reduction the proportion of absorbed ^{15}N incorporated into root protein remained constant. Work on roots of maize kept in CO_2^- depleted air for 10 h showed there was an apparent excess activity of root NRA at the time of decrease in root reduction of nitrate (Pace et al., 1990). In the work of Rufty and coworkers a decrease in the rate of nitrate uptake occurred concomitant with a decrease in the rate of reduction, although much less severe, and the proportion of ^{15}N as nitrate in the roots actually increased during this time. It therefore appears likely that it is neither decline in NRA nor a lowered availability of substrate (nitrate) for the enzyme that causes a decline in the rate of reduction of nitrate under these circumstances. A decreased availability of energy and carbon skeletons is a much more likely possibility.

If the availability of soluble carbohydrates governs the rate of assimilation of nitrate in roots, there should be enhanced levels of these compounds in plants after short-term depletion of nitrate or ammonium, and resupply of either nitrogen source should enable uptake and reduction in roots to occur more quickly. Young barley plants grown with NO_3^- nitrogen and then kept without nitrate showed an enhanced ability to take up nitrate when it was resupplied (Lee and Rudge, 1986). This must be linked to the ability of the plants to assimilate the nitrate taken up due to an enhanced capacity for uptake of ammonium. In a study of wheat, barley and oats supplied NO_3^- nitrogen and then kept without nitrate an increase in uptake of $^{15}NH_4^+$ was noted (Morgan and Jackson, 1988). Since uptake and assimilation of nitrate have been shown to be not directly linked, there must be a mechanism that accounts for the decline in carrier activity.

This occurs before any obvious effect on growth, and at any one time the overall uptake of nitrogen appears to be subject to feedback repression. This could be caused by NO_3^- itself (Siddiqi et al., 1992), from NH_4^+ (Siddiqi et al., 1989) or from some products of NH_4^+ assimilation (Breteler and Siegerist, 1984). From the experiments of Lainé et al. (1995) on *Brassica napus* already referred to, starvation of one part of a root system led to enhanced uptake in another, so the regulation must operate at the whole plant level. Phloem-mobile amino acids appear to limit the uptake of nitrate by roots (Cooper and Clarkson, 1989; Muller and Touraine, 1992), and this is almost certainly due to an effect on influx rather than efflux (Muller et al., 1995). In experiments on soybean seedlings, pretreatment of the plants with alanine and arginine caused a decrease in NO_3^- influx across the plasmalemma. This may have been due to disruption of synthesis of a carrier protein involved in nitrate influx, which would also explain the increased I_{max} in NO_3^- depleted plants (Hole et al., 1990; Siddiqi et al., 1990).

Such a mechanism has also been implicated in the changes in net uptake of nitrate that occur during a day/night cycle (as well as changes in availability of carbohydrates). Although Scaife (1989) accounted for the diurnal rhythm in nitrate uptake by enhanced efflux of nitrate during the dark period, work on soybean has shown that it is the influx of nitrate that is most altered during the night (to give a net uptake rate of approximately 75% of that seen in light) (Delhon et al., 1995a). The assimilation of nitrate in the roots and the translocation of nitrate to the shoots also decreased strongly in darkness, so that nitrate accumulated in the roots in the dark. One of the products of nitrate assimilation, asparagine (the major reduced form of nitrogen normally present in the xylem sap of soybean), also accumulated in the roots, presumably as a result of the reduction of nitrate that was still occurring there and a limited translocation of reduced nitrogen as well as nitrate.

The decrease in nitrate translocation at night was not caused by lower loading of nitrate into the xylem, but by less water movement in the xylem at night (Delhon et al., 1995b). This did not seem to be the case for asparagine, which seemed to show lower levels of movement into the xylem in the dark. It seems that the effectors of nitrate uptake in the dark may well come from the shoot, and these results do not contradict the idea that this could be due to phloem-mobile amino acids. However, this could also be brought about by amino acids arising from the assimilation of nitrate or ammonium in the roots themselves; in an experiment on detopped maize seedlings a supply of exogenous glutamine enhanced xylem exudation (Barthes et al., 1996). This implies a possible double control wherein assimilation of inorganic nitrogen in the roots causes increased flow of the products of this assimilation to the shoots in the xylem, and the movement of amino-N to the roots in the phloem from ongoing shoot metabolism also increases xylem flow.

V. CONCLUSIONS AND FUTURE PROSPECTS

The control of uptake and assimilation of inorganic forms of nitrogen by roots, although complex, is becoming better understood. However, at the level of the whole plant understanding the integration of the various processes is still very difficult. It is generally thought that the rate of uptake of nitrate limits the rate of assimilation of nitrate (Imsande and Touraine, 1994) and although there are controls on the activities of the enzymes of nitrogen assimilation in both roots and shoots, the first effects of any controls are on the uptake mechanisms.

That plants should have evolved to maximize the uptake of inorganic nitrogen is self-evident. This is brought about by: 1) increased uptake capacity per unit root weight in nitrogen-deficient plants, 2) increased root:shoot ratio under conditions of nitrogen deficiency, 3) induction of carriers and assimilatory enzymes when the roots of deficient plants grow into areas of higher nitrogen availability and 4) increased root branching in areas of soil with high nitrogen content.

In root systems growing into areas of high nitrogen availability from areas of low nitrogen availability it is probable that the increased uptake per unit weight has a far bigger impact in enabling the rapid acquisition of this nitrogen than the increased branching (Robinson et al., 1994).

The increased root:shoot ratio under conditions of nitrogen deficiency is an obvious mechanism to increase the ability of a plant to obtain nitrogen from a nutrient-poor soil. There are considerable differences in root:shoot ratio between species, and it would be expected that the higher the proportion of shoot, the greater the plant productivity (because of greater photosynthetic capacity) as long as nutrient uptake is not limiting. Thus, for plants growing at high rates of nitrogen supply there is a negative correlation between root weight as a proportion of whole plant weight and RGR (Fig. 2.4). A larger root system is more useful for plants growing in nitrogen-deficient soils or soils in which the availability of inorganic nitrogen is patchy.

Why there should be big differences between plant species in the proportion of nitrate assimilation that occurs in the roots relative to the shoots is more difficult to understand.

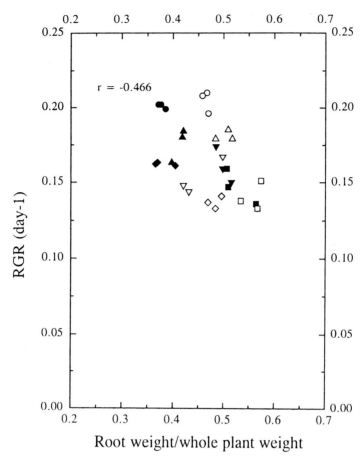

Fig. 2.4: Relationship between root/whole plant fresh weight and RGR of *Triticum* seedlings.
Unpublished data and data in Jan and Pilbeam (1993). Seedlings of *Triticum* species were grown in 2.0 mol m^{-3} nitrate for 15 days in the conditions given in Jan and Pilbeam (1993). Symbols as in Fig. 2.3.

LITERATURE CITED

Allen, S. and Raven, J.A. 1987. Intercellular pH regulation in *Ricinus communis* grown with ammonium and nitrate as N source: the role of long distance transport. *J. Exp. Bot.* 38: 580–596.

Ashley, D.A., Jackson, W.A. and Volk, R.J. 1975. Nitrate uptake and assimilation by wheat seedlings during initial exposure to nitrate. *Plant Physiol.* 55: 1102–1106.

Ayling, S.M. 1993. The effect of ammonium ions on membrane potential and anion flux in roots of barley and tomato. *Plant, Cell Environ.* 16: 297–303.

Barak, P., Smith, J.D., Krueger, A.R. and Peterson, L.A. 1996. Measurements of short-term nutrient uptake rates in cranberry by aeroponics. *Plant, Cell Environ.* 19: 237–242.

Barthes, L., Deléens, E., Bousser. A., Hoarau, J. and Prioul, J.-L. 1996. Xylem exudation is related to nitrate assimilation pathway in detopped maize seedlings: use of nitrate reductase and glutamine synthetase inhibitors as tools. *J. Exp. Bot.* 47: 485–495.
Belton, P.S., Lee, R.B. and Ratcliffe, R.G. 1985. A ^{14}N nuclear magnetic resonance study of inorganic nitrogen metabolism in barley, maize and pea roots. *J. Exp. Bot.* 36: 190–210.
Bloom, A.J., Sukrapanna, S.S. and Warner, R.L. 1992. Root respiration associated with ammonium and nitrate absorption and assimilation by barley. *Plant Physiol.* 99: 1294–1310.
Botella, J.R., Verbelen, J.P. and Valpuesta, V. 1988. Immunological localization of glutamine synthetase in green leaves and cotyledons of *Lycopersicon esculentum*. *Plant Physiol.* 87: 255–257.
Bowsher, C.G., Long, D.M., Oaks, A. and Rothstein, S.J. 1991. The effect of light/dark cycles on expression of nitrate assimilatory genes in maize shoots and roots. *Plant Physiol.* 95: 281–285.
Breteler, H. 1973. A comparison between ammonium and nitrate nutrition of young sugar beet plants grown in nutrient solution at constant acidity. 1. Production of dry matter, ionic balance and chemical composition. *Neth. J. Agric. Sci.* 21: 227–244.
Breteler, H. and Hanisch ten Cate, C.H. 1980. Fate of nitrate during initial nitrate utilization by nitrogen-depleted dwarf bean. *Physiol. Plant.* 48: 292–296.
Breteler, H. and Siegerist, M. 1984. Effect of ammonium on nitrate utilization by roots of dwarf bean. *Plant Physiol.* 24: 1099–1103.
Brouwer, R. 1962. Nutritive influences on the distribution of dry matter in the plant. *Neth. J. Agric. Sci.* 10: 399–408.
Clarkson, D.T. 1986. Regulation of the absorption and release of nitrate by plant cells: a review of current ideas and methodology. In: *Fundamental, Ecological and Agricultural Aspects of Nitrogen Metabolism in Higher Plants* (H. Lambers, J.J. Neeteson and I. Stulen, eds.) Martinus Nijhoff, Dordrecht, pp. 3–27.
Clarkson, D.T. and Hanson, J.B. 1980. The mineral nutrition of higher plants. *Ann. Rev. Plant Physiol.* 31: 239–298.
Cooper, H.D. and Clarkson, D.T. 1989. Cycling of amino-nitrogen and other nutrients between shoots and roots in cereals. A possible mechanism integrating shoot and root in the regulation of nutrient uptake. *J. Exp. Bot.* 40: 753–762.
Dalling, M.J., Tolbert, N.E. and Hageman, R.H. 1972. Intracellular location of nitrate reductase. II. Wheat roots. *Biochim. Biophys. Acta* 283: 513–519.
Deane-Drummond, C.E. 1984. Mechanism of nitrate uptake into *Chara corallina* cells; lack of evidence for obligatory coupling to proton pump and a new NO_3/NO_3 exchange model. *Plant, Cell Environ.* 7: 317–323.
Deane-Drummond, C.E. and Glass, A.D.M. 1983. Short-term studies of nitrate uptake into barley plants using ion-specific electrodes and $^{36}ClO_3^-$.I. Control of net uptake by NO_3^- efflux. *Plant Physiol.* 73: 100–104.
Delhon, P., Gojon, A., Tillard, P. and Passama, L. 1995a. Diurnal regulation of NO_3^- uptake in soybean plants. I. Changes in NO_3^- influx, efflux, and N utilization in the plant during the day/night cycle. *J. Exp. Bot.* 46: 1585–1594.
Delhon, P., Gojon, A., Tillard, P. and Passama, L. 1995b. Diurnal regulation of NO_3^- uptake in soybean plants. II. Relationship with accumulation of NO_3^- and asparagine in the roots. *J. Exp. Bot.* 46: 1595–1602.
Dhugga, K.S., Waines, J.G. and Leonard, R.T. 1988. Nitrate absorption by corn roots. Inhibition by phenylglyoxal. *Plant Physiol.* 86: 759–763.
Dijkshoorn, W. 1962. Metabolic regulation of the alkaline effect of nitrate utilization in plants. *Nature* 194: 165–167.
Doddema, H. and Telkamp, G.P. 1979. Uptake of nitrate by mutants of *Arabidopsis thaliana*, disturbed in uptake or reduction of nitrate. II. Kinetics. *Physiol. Plant.* 45: 332–338.

Drew, M.C. 1975. Comparison of the effects of a localized supply of phosphate, nitrate, ammonium and potassium on the growth of the seminal root system and the shoot of barley. *New Phytol.* 75: 479–490.

Drew, M.C., Saker, L.R. and Ashley, T.W. 1973. Nutrient supply and the growth of seminal root system in barley. I. The effect of nitrate concentration on the growth of axes and laterals. *J. Exp. Bot.* 24: 1189–1202.

Emes, M.J. and Fowler, M.W. 1983. The supply of reducing power for nitrite reduction in plastids of seedling pea roots (*Pisum sativum* L.). *Planta* 158: 97–102.

Findenegg, G.R., Nelemans, J.A. and Arnozis, P.A. 1989. Effect of external pH and Cl on the accumulation of NH_4^+ ions in the leaves of sugar beet. *J. Plant Nutr.* 12: 593–602.

Glass, A.D.M., Shaff, J.E. and Kochian, L.V. 1992. Studies on the uptake of nitrate in barley. IV. Electrophysiology. *Plant Physiol.* 99: 456–463.

Goldstein, A. and Hunziker, A.D. 1985. Induction of high affinity nitrate and ammonium uptake systems in wheat. *J. Plant Nutr.* 8: 721–730.

Henriksen, G.H. and Spanswick, R.M. 1993. Investigation of the apparent induction of nitrate uptake in barley (*Hordeum vulgare* L.) using NO_3^--selective microelectrodes. Modulation of coarse regulation of NO_3^- uptake by exogenous application of downstream metabolites in the NO_3^- assimilatory pathway. *Plant Physiol.* 103: 885–892.

Hole, D.J., Emran, A.M., Fares, Y. and Drew, M.C. 1990. Induction of nitrate transport in maize roots and kinetics of influx, measured with nitrogen-13. *Plant Physiol.* 93: 642–647.

Imsande, J. and Touraine, B. 1994. N demand and the regulation of nitrate uptake. *Plant Physiol.* 105: 3–7.

Jan, A.U. and Pilbeam, D.J. 1993. Nitrate assimilation by wheat species at low rates of nitrogen supply. In: *Optimization of Plant Nutrition* (M.A.C. Fragoso and M.L. van Beusichem, eds.). Kluwer Academic Publishers, Dordrecht, pp. 227–233.

King, B.J., Siddiqi, M.Y. and Glass, A.D.M. 1992. Studies of the uptake of nitrate in barley. V. Estimation of root cytoplasmic nitrate concentration using nitrate reductase activity — implications for nitrate influx. *Plant Physiol.* 99: 1582–1589.

Kirkby, E.A. and Mengel, K. 1967. Ionic balance in different tissues of the tomato plant in relation to nitrate, urea or ammonium nutrition. *Plant Physiol.* 42: 6–14.

Lainé, P., Ourry, A. and Boucaud, J. 1995. Shoot control of nitrate uptake by roots of *Brassica napus* L.: effects of localized nitrate supply. *Planta* 196: 77–83.

Lea, P.J., Blackwell, R.D. and Joy, K.W. 1992. Ammonia assimilation in higher plants. In: *Nitrogen Metabolism of Plants* (K. Mengel and D.J. Pilbeam, eds.). Oxford University Press, Oxford, pp. 153–186.

Lee, R.B. and Rudge, K.A. 1986. Effects of nitrogen deficiency on the absorption of nitrate and ammonium by barley plants. *Ann. Bot.* 57: 471–486.

Lee, R.B. and Ayling, S.M. 1993. The effect of methionine sulphoximine on the absorption of ammonium by maize and barley roots over short periods. *J. Exp. Bot.* 44: 53–63.

Lu, Q. and Briskin, D.P. 1993. Modulation of the maize plasma membrane H^+/NO_3^- symport carrier by NO_3^-. *Phytochemistry* 33: 1–8.

Mäck, G. and Tischner, R. 1994. Constitutive and inducible net NH_4^+ uptake of barley (*Hordeum vulgare* L.) seedlings. *J. Plant Physiol.* 144: 351–357.

Mann, A.F., Fenten, P.A. and Steward, G.R. 1979. Identification of two forms of glutamine synthetase in barley. *Biochem. Biophys. Res. Commun.* 85: 515–521.

Mattsson, M., Johansson, E., Lundborg, T., Larsson, M. and Larsson, C.-M. 1991. Nitrogen utilization in N-limited barley during vegetative and generative growth. I. Growth and uptake kinetics in vegetative cultures grown at different relative addition rates of nitrate-N. *J. Exp. Bot.* 42: 197–205.

McClure, P.R. and Israel, D.W. 1979. Transport of nitrogen in the xylem of soybean plants. *Plant Physiol.* 64: 411–416.

McClure, P.R., Omholt, T.E. Pace, G.M. and Bouthyette, P.-Y. 1987. Nitrate-induced changes in protein synthesis and translation of NRA in maize roots. *Plant Physiol.* 84: 52–57.

McNally, S.F., Hirel, B., Gadal, P., Mann, A.F. and Steward, G.R. 1983. Glutamine synthetase of higher plants. Evidence for a specific isoform content related to their possible physiological role and their compartmentation within the leaf. *Plant Physiol.* 72: 22–25.

McNaughton, G.S. and Presland, M.R. 1983. Whole plant studies using radioactive 13-nitrogen. I. Techniques for measuring the uptake and transport of nitrate and ammonium ions in hydroponically grown *Zea mays*. *J. Exp. Bot.* 34: 880–892.

Morgan, M.A. and Jackson, W.A. 1988. Suppression of ammonium uptake by nitrogen supply and its relief during nitrogen limitation. *Physiol. Plant.* 73: 38–45.

Morgan, M.A., Jackson, W.A. and Volk, R.J. 1985. Concentration-dependence of the nitrate assimilation pathway in maize roots. *Plant Sci.* 38: 185–191.

Muller, B. and Touraine, B. 1992. Inhibition of NO_3^- uptake by various phloem-translocated amino acids in soybean seedlings. *J. Exp. Bot.* 43: 616–623.

Muller, B., Tillard, P. and Touraine, B. 1995. Nitrate fluxes in soybean seedling roots and their response to amino acids: an approach using ^{15}N. *Plant, Cell Environ.* 18: 1267–1279.

Ni, M. and Beevers, L. 1990. The influence of arginine-specific reagents on nitrate uptake by corn seedlings. *J. Exp. Bot.* 41: 987–993.

Ni, M., and Beevers, L. 1994. Nitrate-induced polypeptides in membranes from corn seedling roots. *J. Exp. Bot.* 45: 355–365.

Oaks, A. and Long, D. 1992. NO_3^- assimilation in root systems: with special reference to *Zea mays* (cv. W64A X W182E). In: *Nitrogen Metabolism of Plants* (K. Mengel and D.J. Pilbeam, eds.). Oxford University Press, Oxford, pp. 91–102.

Oaks, A., Wallace, W. and Stevens, D. 1972. Synthesis and turnover of nitrate reductase in corn roots. *Plant Physiol.* 50: 649–654.

Pace, G.M., Volk, R.J. and Jackson, W.A. 1990. Nitrate reduction in response to CO_2 limited photosynthesis. Relationship to carbohydrate supply and nitrate reductase activity in maize seedlings. *Plant Physiol.* 92: 286–292.

Pate, J.S. 1973. Uptake, assimilation and transport of nitrogen compounds by plants. *Soil Biol. Biochem.* 5: 109–119.

Pilbeam, D.J. and Kirkby, E.A. 1992. Some aspects of the utilization of nitrate and ammonium by plants. In: *Nitrogen Metabolism of Plants* (K. Mengel and D.J. Pilbeam, eds.). Oxford University Press, Oxford, pp. 55–70.

Polisetty, R. and Hageman, R.H. 1983. Variation in nitrate accumulation, nitrate reductase activity and nitrite reductase activity along primary and nodal roots of corn (*Zea mays* L.) seedlings. *Plant Cell Physiol.* 24: 1163–1168.

Redinbaugh, M.G. and Campbell, W.H. 1981. Purification and characterization of NAD(P)H:nitrate reductase and NADH:nitrate reductase from corn roots. *Plant Physiol.* 68: 115–120.

Robinson, D., Linehan, D.J. and Gordon, D.C. 1994. Capture of nitrate from soil by wheat in relation to root length, nitrogen inflow and availability. *New Phytol.* 128: 297–306.

Rufty, T.W. Jr., Volk, R.J. and Glass, A.D.M. 1992. Relationship between carbohydrate availability and assimilation of nitrate. In: *Nitrogen metabolism of Plants* (K. Mengel and D.J. Pilbeam, eds.). Oxford University Press, Oxford, pp. 103–119.

Rufty, T.W. Jr., Thomas, J.F., Remmler, J.L., Campbell, W.H. and Volk, R.J. 1986. Intercellular localization of nitrate reductase in roots. *Plant Physiol.* 82: 675–680.

Scaife, A. 1989. A pump/leak/buffer model for plant nitrate uptake. *Plant and Soil* 114: 139–141.

Schweizer, P. and Erismann, K.H. 1985. Effect of nitrate and ammonium nutrition of nonnodulated *Phaseolus vulgaris* L. on phosphoenolpyruvate carboxylase and pyruvate kinase activity. *Plant Physiol.* 78: 455–458.

Siddiqi, M.Y., King, B.J. and Glass, A.D.M. 1992. Effects of nitrite, chlorate, and chlorite on nitrate uptake and nitrate reductase activity. *Plant Physiol.* 100: 644–650.

Siddiqi, M.Y., Glass, A.D.M., Ruth, T.J. and Fernando, M. 1989. Studies on the regulation of nitrate influx by barley seedlings using $^{13}NO_3^-$. *Plant Physiol.* 90: 806–813.
Siddiqi, M.Y., Glass, A.D.M., Ruth, T.J. and Rufty, T.W. Jr. 1990. Studies on the uptake of nitrate in barley. I. Kinetics of $^{13}NO_3^-$ influx. *Plant Physiol.* 93: 1426–1432.
Siebrecht, S., Mäck, G. and Tischner, R. 1995. Function and contribution of the root tip in the induction of NO_3^- uptake along the barley root axis. *J. Exp. Bot.* 46: 1669–1676.
Sutherland, J.M., Andrews, J., McInroy, A. and Sprent, J.I. 1985. The distribution of nitrate assimilation between root and shoot in *Vicia faba* L. *Ann. Bot.* 56: 259–263.
Suzuki, A., Oaks, A., Jaquot, J.-P., Vidal, J. and Gadal, P. 1985. An electron transport system in maize roots for reactions of glutamate synthase and nitrite reductase. Physiological and immunochemical properties of the electron carrier and pyridine nucleotide reductase. *Plant Physiol.* 78: 374–378.
Swiader, J.M. and Freiji, F.G. 1996. Characterizing nitrate uptake in lettuce using very-sensitive ion chromatography. *J. Plant Nutr.* 19: 15–27.
Talouizte, A., Giraud, G., Moyse, A., Marol, F. and Champigny, M.L. 1984. Effect of previous nitrate deprivation on ^{15}N-nitrate absorption and assimilation by wheat seedlings. *J. Plant Physiol.* 116: 113–122.
Teyker, R.H., Dallmeir, K.A., St. Aubin, G.R.S. and Lambert, R.J. 1991. Seedling nitrate reductase activity and nitrate partitioning in maize (*Zea mays* L.) strains divergently selected for post-anthesis leaf-lamina nitrate reductase activity. *J. Exp. Bot.* 42: 97–102.
Thibaud, J.B. and Grignon, C. 1981. Mechanism of nitrate uptake in corn roots. *Plant Sci. Lett.* 22: 279–289.
Ullrich, W.R. 1992. Transport of nitrate and ammonium through plant membranes. In: *Nitrogen Metabolism of Plants* (K. Mengel and D.J. Pilbeam, eds.). Oxford University Press, Oxford, pp. 121–137.
Van Beusichem, M.L., Kirkby, E.A. and Baas, R. 1988. Influence of nitrate and ammonium nutrition on the uptake, assimilation and distribution of nutrients in *Ricinus communis*. *Plant Physiol.* 86: 914–921.
Van Quy, L., Lamaze, T. and Champigny, M.-L. 1991. Short term effects of nitrate on sucrose synthesis in wheat leaves. *Planta* 185: 53–57.
Varanini, Z., Pinton, R., De Biasi, M.G., Astolfi, S. and Maggioni, A. 1993. Low molecular weight humic substances stimulate H^+-ATPase activity of plasma membrane vesicles isolated from oat (*Avena sativa* L.) roots. *Plant and Soil* 153: 61–69.
Vezina, L.P., Hope, H.J. and Joy, K.W. 1987. Iso-enzymes of glutamine synthetase in roots of pea (*Pisum sativum* L. cv Little Marvel) and alfalfa (*Medicago media* Pers. cv Saranac). *Plant Physiol.* 83: 58–62.
Wallace, W. 1973. The distribution and characteristics of nitrate reductase and glutamate dehydrogenase in the maize seedling. *Plant Physiol.* 52: 191–196.
Wang, M.Y., Siddiqi, M.Y., Ruth, T.J. and Glass, A.D.M. 1993. Ammonium uptake by rice roots. II. Kinetics of $^{13}NH_4^+$ influx across the plasmalemma. *Plant Physiol.* 103: 1259–1267.
Ward, M.R., Tischner, R. and Huffaker, R.C. 1988. Inhibition of nitrate transport by anti-nitrate reductase IgG fragments and the identification of plasma membrane associated nitrate reductase in roots of barley seedlings. *Plant Physiol.* 88: 1141–1145.
Warner, R.L., Narayanan, K.R. and Kleinhofs, A. 1987. Inheritance and expression of NAD(P)H nitrate reductase in barley. *Theor. Appl. Genet.* 74: 714–717.

Chapter 3

RHIZOBIUM-LEGUME ASSOCIATION

S.N. Mishra, P.K. Jaiwal, Rana P. Singh and H.S. Srivastava

I. Introduction
II. Rhizobium Biology
III. Nodule Formation
   A. Recognition of Symbionts and Signal Exchange between Plant and Bacterium
   B. Nod-Factor-Induced Response and Nodule Initiation
   C. Differentiation of Root Nodules
   D. Entry of Rhizobium in Host Cell
   E. Autoregulation of Nodule Organogenesis
IV. Factors Affecting Nodulation
   A. Edaphic Factor
   B. Chemical Factors
   C. Nodule Development Under Inorganic Nutrition
   D. Phytohormones and Nodulation
   E. Effect of Light/Dark Conditions
   F. Some other factors
V. Molecular Biology of Nodulation
   A. Rhizobium Nodulation Gene
   B. Host-Plant Genes Involved in Nodulation and N_2 Fixation
   C. Rhizobium Genes Involved in Nitrogen Fixation and their Regulation
VI. Biochemistry of Nitrogen Fixation
VII. Molecular Biology of Regulation of Nitrogen Fixation
VIII. Factors Affecting Nitrogen Fixation
   A. Effect of Inorganic Nitrogen
   B. Nitrogen-fixation Under Stress
   C. Phytohormones
IX. Genetic Engineering for Increasing Nodulation and Nitrogen Fixation
X. Conclusion and Future Prospects

Literature Cited

I. INTRODUCTION

Biological nitrogen fixation is an important aspect of nitrogen nutrition in crop production throughout the world. The process saves a huge sum of money, which otherwise would have to be pumped into agricultural productivity to consistently sustain an increasing world population. Among various symbiotic associations of biological nitrogen fixation, the *Rhizobium*-legume association is the most common in both temperate and tropical climates. Attempts have been made in recent years to increase its nitrogen fixation and also to induce non-leguminous species to develop symbiotic associations with rhizobia to inculcate the potential for atmospheric nitrogen fixation in them.

Many excellent reviews have been published in recent years describing the different aspects of the *Rhizobium*-legume associations (Burgess, 1990; Sindhu and Dadarwal, 1995; Geurts and Franssen, 1996). One of the most important aspects of this association is the recognition of the bacteria by the host legumes to develop an effective symbiosis. A molecular mechanism of this chemical conversation between the two partners through a signal exchange has recently been revealed (see Geurts and Franssen, 1996). Involvement of aerial tissues in regulating the number and activity of symbiotic root nodules has also been demonstrated recently at the physiological and molecular level (Sheng and Harper, 1997). Understanding the molecular regulation of nodule formation and nitrogen fixation is one of the most rapidly growing areas of knowledge. The importance of biological nitrogen fixation is now not limited to plant nutrition; increasing environmental concerns indicate that the high input of nitrogen fertilizers in agriculture may lead to many environmental problems, e.g. nitrate and nitrite pollution, volatilization of ammonia from the surface application of urea, eutrophication of streams and lakes, etc. This chapter covers the current status of various aspects of the *Rhizobium*-legume association, one of the major modes of nitrogen nutrition in plants.

II. *RHIZOBIUM* BIOLOGY

Beijerinck (1888) was the first to isolate and cultivate a micro-organism from nodules of legumes. He named it *Bacillus radicicola* which is now placed in Bergey's *Manual of Bacteriology* under the genus *Rhizobium*.

The genus *Rhizobium* (rhizo = root, bios = living) includes aerobic gram-negative soil organisms which have the ability to form symbiotic root nodules by infecting the roots of members of the family Leguminosae. Until recently, these bacteria were classified mainly on the basis of their host specificity, a concept first developed by Fred, et al. (1932). This classification assumed that the leguminous plants which are infected only by a particular species of *Rhizobium* belong to a so-called 'cross-inoculation group'. This classification was not perfect since there were many anomalies in cross-inoculation groups. It is now replaced by one which takes into consideration newer information generated by the use of modern methods, such as numerical taxonomy, nucleic acid hybridization and DNA sequencing (Jarvis et al., 1986; Dreyfus et al., 1988; Young and Johnston, 1989). Now the root nodule forming bacteria family (Rhizobiaceae) have been separated into 3 genera: *Rhizobium*, *Bradyrhizobium* and *Azorhizobium* (Table 3.1).

Members of the genus *Rhizobium* are relatively fast-growing (forming colonies within 2 days). They are closely related to *Agrobacterium*, the plant pathogens that form galls on plants by transferring bacterial DNA to plants. There are several species of *Rhizobium* and in general they have a fairly limited host range, i.e., each species can nodulate only a limited range of legumes. The species *R. leguminosarum* has been subdivided into three 'biovars' defined on the basis of the specific range of legumes that they nodulate.

The genus *Bradyrhizobium* is so called because the bacteria grow slowly (brady=slow), forming colonies after 4–5 days. In general, *Bradyrhizobium* species nodulate a broad range of tropical legumes whereas *Rhizobium* species nodulate a narrow range of temperate legumes. Although *Rhizobium* and *Bradyrhizobium* species reduce N_2 (nitrogen) in nodules, they cannot grow in culture using N_2 as a nitrogen source. Some *Bradyrhizobium* species (unlike *Rhizobium* species) can reduce N_2 under certain laboratory growth conditions outside the plant, but paradoxically they appear to be unable to utilize that reduced N_2 for growth. *Rhodopseudomonas palustris*, a soil phototroph is the closest relative of *Bradyrhizobium*.

A third genus, *Azorhizobium* (sp. *A. caulinodans*) (Azo = nitrogen, caulo=stem) can utilize N_2 as a nitrogen source during growth in free-living culture. These bacteria were first isolated from stem nodules on the tropical legume, *Sesbania rostrata*. *Azorhizobium* is closely related to *Xanthobacter* species.

Rhizobium species are widely distributed in the world's soil because of the natural distribution of Leguminosae and the cultivation of leguminous crops. With senescence and decay of nodular tissue, motile forms of *Rhizobium* are liberated in the soil which normally serve as a source of inoculum for the succeeding crop of a given species of legume. In the absence of a legume, the soil population of *Rhizobia* declines. However, *Rhizobia* are known to survive for many years despite the fact they are non-endospore formers. Many soils lack *Rhizobia* or contain

Table 3.1. Classification of rhizobia

Section	Gram negative, aerobic rods, cocci	
Family	Rhizobiaceae	
Genus I	*Rhizobium*: Fast-growing (symbiotic genes located on plasmid, comparatively restricted host range)	
Species	*R. meliloti* nodulate	Alfalfa (*Medicago sativa*)
		Sweet clover (*Melilotus alba*)
		Fenugreek (*Trigonella foenumgraecum*)
	R. leguminosarum	
	Biovar viciae nodulate	Pea (*Pisum sativum*)
		Vetch (*Vicia sativa*)
		Sweet pea (*Lathyrus sativa*)
	Biovar phaseoli nodulate	Bean (*Phaseolus vulgaris*)
	Biovar trifolii	Clover (*Trifolium* spp.)
	R. loti	Trefoil (*Lotus* spp.)
	R. fredii	Soybean (*Glycin max*)
	R. tropici	Bean (*Phaseolus vulgaris*)
	R. sp. NGR 234	26 legumes spp. including
		Cowpea (*Vigna unguiculata*)
		Siratro (*Macroptilium altropurpureum*)
		Soybean (*G. max*)
		Leucaena (*L. leucocephala*)
		Parasponia (non-legume)
Genus II	*Bradyrhizobium*	Slow-growing (symbiotic genes on chromosome, comparatively wide host range)
	B. japonicum	Soybean (*Glycine max*)
	B. sp. 'cowpea'	Cowpea (*V. unguiculata*)
		Mung bean (*Vigna radiata*)
		Chick-pea (*Cicer arietinum*)
		Pigeon-pea (*Cajanus cajan*)
		Peanut (*Arachic hypogaea*)
		Sunn hemp (*Crotolaria juncea*)
Genus III	*Azorhizobium*	
	A. caulinodans	Root and stem nodules on Sesbania (*S. rostrata, S. bispinosa*)

ineffective populations of rhizobia; under such circumstances, a beneficial response of seed inoculation with efficient rhizobium may be anticipated.

A *Rhizobium* strain should be introduced into soil at the time of sowing in a form that enables it to (1) remain viable until the host seedlings can be infected, (2) compete with any indigenous or naturalized rhizobia for infection sites and so form sufficient root nodules, (3) adequately nodulate its host(s) rapidly over a wide range of environmental conditions and (4) persist in soil in numbers sufficient to maintain nodulation of the perennial legume or to achieve prompt nodulation of regenerating annual species.

Considerable progress has been made in selecting efficient rhizobia that promptly form effective nodules with a specific legume host under a variety of environmental conditions. The criteria in *Rhizobia* strain selection for inoculants are (1) the ability to form nodules in competition with an indigenous or naturalized rhizobia population, (2) ability to survive in the absence of a host, (3) ability to grow in broth and peat culture, (4) the ability to survive in the inoculant carrier and on the seed, (5) ability to resist variation in soil pH, to tolerate macro- and micro nutrients and pesticides in close proximity and (6) genetic stability in culture.

Strain selection involves screening of several strains using plants grown in bottle jar assemblies, fully or partially closed tubes or growth porches under green house conditions for their excellent performance with particular hosts and/or environment.

Maintenance of Strain

The bacteria are normally maintained on yeast extract mannitol (YEM) agar medium either by subculturing at frequent intervals or as lyophilized cultures held at 5°C. This system maximizes the storage period (15 to 20 years) and minimizes genetic variations and contamination. The cultures can also be dried on porcelain beads over a desiccant in screw-cap bottles (storage period 3 to 4 years). Whichever system is employed, an authenticity check is carried out regularly.

Type of Inoculants

In early times, inoculation methods involved the transfer of soil from the roots of well-nodulated plants to the seeds at planting time. During the past few decades, different types of inoculants have been developed. The first commercial inoculant, 'Nitragin', was produced on gelatin and subsequently on agar nutrient medium. Such inoculants were directly applied to seeds. Due to high mortality rates during drying following seed inoculation, agar-culture inoculants were replaced by solid-based inoculants in the 1920s in the United States and in 1952 in Australia, a practice adopted by Asian (Subba Rao et al., 1987), African and Latin American (Nutman, 1976; Qusipel, 1974) countries with modifications to suit local needs and conditions. These solid-based inoculants (standard) consisted of a peat-based carrier inoculated with a large population of *Rhizobium* or *Bradyrhizobium* spp. Peat is an effective carrier because of its high water-holding capacity, good buffering capacity and high nutrient content. Recently, there have been significant advances in formulation technology and delivery of *Rhizobium* while maintaining user-friendly and cost-effective seed inoculation. These include the use of high-quality peats, some pre-sterilized and altered to promote growth and survival of *Rhizobia*. Because of the need to treat the seed with a striking agent to allow adhesion of the peat carrier, the inoculation procedure can be time consuming and inconvenient for farmers sowing a large quantity of seeds. Granulated peat inoculants are prepared by mixing high-count broth cultures of *Rhizobia* with dried peat. This inoculant is applied in the furrow along with seeds at the time of sowing and is particularly advantageous under these circumstances: (1) in soils which are difficult to colonize, (2) under conditions wherein seed inoculation is hazardous because of fragility of the seed-coat, (3) in small seed legumes, seed inoculation may not provide sufficient *Rhizobia* to effectively nodulate the plant and (4) circumstances wherein high levels of inoculant are needed to allow the inoculum strain to compete with naturally occurring *Rhizobia* and in species which have epigeal germination.

The use of seed auger has made the application of liquid inoculants (liquid culture of *Rhizobia*) to seeds very easy and row-cropped legumes permit in-furrow application of the liquid inoculant.

Methods of Inoculating Legumes

1. *Seed Inoculation*: The method used to inoculate seed must provide sufficient viable *Rhizobia* to ensure adequate nodulation of host plants. In some instances the number of *Rhizobia* per seed may be as low as 100 but in the case of severe environmental stresses as many as 10,000 or even 500,000 per seed may be necessary. Large quantities of inoculants can be attached to the seed by the use of such adhesives as 10% sucrose or 40% gum arabic. Inoculated seeds should be allowed to dry in a cool place (not in direct sunlight) and planted on the same day.

2. *Seed Pelleting*: Pelleting of seeds with finely ground coating materials (either limestone or bentonite or bauxite or talc or gypsum, or dolomite, calcium silicate) at near natural pH is used to protect rhizobia on the seed-coat against the deleterious effect of sunlight, low soil pH, desiccation, acidic fertilizers, fungicides and insecticides.

3. *Direct Implant Inoculation*: An alternative to pelleting and preinoculation has been the use of concentrated liquid or solid granular peat cultures which can be sprayed or drilled directly into the soil with the seed during planting. This method has been very successful wherever a very large population of competing naturally occurring *Rhizobia*

is available in the soil or in cases of adverse conditions such as hot dry soils.

In addition to rhizobial inoculants, there is interest in developing co-inoculants containing other micro-organisms which are able to improve legume growth (Nelson and Hynes, 1995). These include rhizobacteria which promote nodulation, nitrogen fixation, plant vigour and yield via such mechanisms as phytohormones, antibiotic or metal-binding compound production; bacteria or fungi which protect against specific root pathogens and others which aid in nutrient supply via phosphate solubilization. The challenges that still remain are microbial survival in the carrier or seed coating and consistency in beneficial effects under a variety of environmental conditions.

III. NODULE FORMATION

A unique mode of nitrogen nutrition in leguminous plants relates to their ability to develop symbiosis with bacteria belonging to the genera *Rhizobium*, *Bradyrhizobium* or *Azorhizobium*, which interact with the plant roots and produce N_2-fixing root nodules. The dinitrogen reduced to ammonia by the root nodules is utilized by the plant and in turn the plant provides energy, carbon substrates and reductants to the bacteria. The Leguminosae is a large family of about 750 genera and 2000 species. However, all the species or genera are not associated with *Rhizobia* to form the root nodules. Symbiotic interactions between the two organisms start from the recognition of each other in a specific manner. The attachment of rhizobia to the plant root hairs enable the bacteria to invade its roots by the formation of an infection thread, a tubular invading structure which increases successively within the infected root hairs. With the involvement of these two partners ultimately a specialized nodular structure is formed. The need for a specialized organ within the root system arises possibly because a major constraint of the N_2-fixation is the sensitivity of the process to oxygen. The confinement of the bacteria within the nodules does not allow them to integrate into the entire cell lineage of the plant, as has been achieved by chloroplasts and mitochondria (Verma, 1992).

Various steps in the process of nodule formation are discussed below.

A. Recognition of Symbionts and Signal Exchange between Plant and Bacterium

To prevent the entry of pathogenic micro-organisms plants usually possess some defence systems. So, a special mode for mutual recognition of the symbiotic partners is necessary to enable infection to occur and the association to develop. Further, due to the fine energy economy in plants, a system such as biological nitrogen fixation is established only if the soil nitrogen level is limited. The interaction between *Rhizobia* and leguminous plants is a host-specific process. This host-specific nature of the symbiosis was recognized long before any signal molecule was purified. It was observed that some *Rhizobium* strains can nodulate plant species belonging to a few genera, e.g. *Rhizobium leguminosarum* bv *viciae* can nodulate *Pisum, Vicia, Lathyrus* and *Lens*, whereas the closely related *R. leguminosarum* bv *trifolii* can only nodulate plants of the genus *Trifolium. Rhizobium* sp. strain NGR 234, on the other hand, nodulates a wide range of plant genera (Geurts and Franssen, 1996).

It was thought that the basis of recognition must be the molecules on the surfaces of the partners (Figure 3.1). The range of sugars in the surface polysaccharides of both plant and bacterium and a protein, lectin, which have several binding sites for sugars were the molecules considered initially as the recognition molecule on both symbionts (Bohlool and Schmidt, 1974; Barondes, 1981; Dazzo et al., 1984). Though very little was known earlier about the biochemical and molecular basis of this specificity of attachment of *Rhizobia* to the surface of legume roots, polysaccharide-polysaccharide and/or polysaccharide-lectin interactions were proposed as the mode of recognition between the two partners (see Robertson and Farnden, 1980).

A two-way molecular conversation of this recognition process was discovered in the early 1990s. It was realized that the host legume releases signal compounds that stimulate the co-ordinate expression of bacterial nodulation (nod) genes, which in turn encode the enzymes for the synthesis of nod factors (substituted oligosaccharides) (Verma, 1992; Sindhu and Dadarwal, 1995; Geurts and Franssen, 1996). The bacterial secreted nod factors play a major role in inducing nodule organogenesis and in determining host specificity (Lerouge et al., 1990; Verma, 1992; Sindhu and Dadarwal, 1995; Geurts and Franssen, 1996).

1. *Early Signals from Plant to Bacterium*: Plant-derived phenolic compounds, (iso) flavonoids (e.g. luteolin, naringenin, daidzein etc.), and betaines (e.g. strigonellina and stachydrine, which are structurally different from the flavonoids) are the specific plant molecules which have been identified as signal molecules in the initiation of the process of nodulation in N_2-fixing legume root nodules (see Sindhu and Dadarwal, 1995). These plant signal compounds are often exuded by the portion of the root with emerging root hairs, a region that is highly susceptible to infection by *Rhizobia*, and released in the rhizosphere as root exudates (Verma, 1992). The nature and amount of the compounds exuded

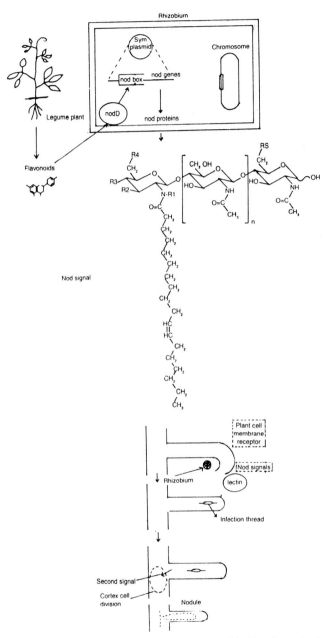

Fig. 3.1. Plant secreted (iso) flavonoids are recognized by *Rhizobium* bacteria via an interaction with the transcriptional activator Nod D protein. Upon binding of the flavonoid, Nod D is activated, leading to transcription of the bacterial nod genes. The nod proteins are involved in the production and secretion of nod factors.

depend on the plant cultivar and its developmental stages. A study of the flavonoids present in seed exudates of alfalfa, bean and soybean revealed, however that root exudates may show a spectrum of flavonoids that differs from seed exudates (Peters et al., 1986; Hartwig et al., 1989; Graham, 1991; Hungria et al., 1991a, b). One reason for this difference could be that gene(s) for some flavonoids may find expression only at specific developmental stages when the nodulation process is actually required by the plants.

Rhizobia respond only to specific signal molecules and not to all. For example luteolin and chalcone are most active inducers of *R. meliloti* nod gene whereas naringenin is active in *R. leguminosarum* bv viciae and diadzein is the nod gene inducer for *B. japonicum* (Sindhu and Dadarwal, 1995). On positive chemotaxis to plant root exudates the bacteria move towards localized sites on legume roots (Kape et al., 1991) and the plant derived flavonoids or other specific signal molecules induce the transcription of an important set of nodulation genes of *Rhizobium* and *Bradyrhizobium* (Peters et al., 1986; Kosslak et al., 1987; Long, 1989a; Sindhu and Dadarwal, 1995). These nod genes are not expressed or are poorly expressed in free-living bacteria lying beyond the plant rhizosphere, however. It appears that the flavonoids or other plant signal molecules actually regulate the 'nod' factor synthesis and 'nod' gene expression in a complex manner since some flavonoids present in exudates can interfere with each other's ability to induce 'nod' gene expression and some secreted flavonoids can act as competitive inhibitors (anti-inducers) rather than transcriptional activators (see Sindhu and Dadarwal, 1995).

The attachment of rhizobia to root hairs in response to the plant signals is observed throughout the growth area but the hairs located just behind the apical meristem are most responsive to infection (Bhuvaneswari et al., 1981). Rhizobial attachment to the root hairs has been proposed as a two-step process, first a loose attachment to a plant receptor through an acidic extracellular polysaccharide or via a calcium-binding protein on the bacterial surface known as rhicadesin and then a tighter adherence by means of cellulose fibrils or fimbriae (see Sindhu and Dadarwal, 1995 and references therein).

2. *Nod-Factor synthesis and Signal Exchange from Rhizobium to Plant*: The low molecular weight diffusible rhizobial signalling substances are synthesized in recognition to plant secreted (iso) flavonoids through the transcriptional activator nod-D genes in bacteria. The newly synthesized nod proteins are involved in the production and secretion of nod factors (see Verma, 1992; Geurts and Franssen, 1996). However, in the overall induction and production of nod factors identified as lipo-oligosaccha-

ride (glycolipid) both common and host specific nodulation genes are involved at various steps.

Nod factors have recently been isolated and characterized from several rhizobial strains. They consist of β-1, 4-linked N-acyl-D-glucosamine backbone of three to six units and a variable fatty acid substituted on the CO_2 position of the non-reducing terminal sugar moiety. Nod factors that vary in length of the glucosamine backbone, in structure of fatty acid and in decoration of terminal sugar residues are produced by all the *Rhizobium* spp. studied thus far (Geurts and Franssen, 1996).

The nod signal compound from *Rhizobium meliloti* (nod Rm-1) has been identified as a sulphated and acylated oligo-N-acetylglucosamine (Lerouge et al., 1990). The common nod ABC genes are considered to activate the synthesis of the N-acyl-D-glucosamine backbone structure and the host range-determining specific gene(s), e.g., nod H and nod Q in this case, participate in the addition of sulphur to the nod factor (Verma, 1992). The signal molecule from *Bradyrhizobium japonicum* is very similar to nod Rm-1, having a backbone of pentaglucosamine, but is not sulphated. *R. leguminosarum* bv *viciae* also produces non-sulphated nod signals, but these molecules contain a unique, highly unsaturated fatty acid and differ in number of N-acetylglucosamine units (Spaink et al., 1991). Both the fatty acid moiety (nod E-dependent) and O-acetyl substitution (nod L-dependent) are essential for initiation of the meristematic activity that gives rise to modular structures on the host plant (Verma, 1992). For a more detailed account of the chemical nature of various Nod-factors excellent reviews by Verma (1992), Carlson et al. (1994), Demont et al. (1994), Sindhu and Dadarwal (1995), Geurts and Franssen (1996) and others are available.

Purified nod factors are able to induce responses similar to those induced by rhizobial strains. The initial responses, which may lead to root hair deformations, are provoked at spatially separated sites, viz., the epidermis, cortical cells, and pericycle. However, it is not yet clear how nod factors are perceived by the host plant and transduced. Initiation of the meristematic activity that gives rise to the nodule occurs before the infection thread is formed and can occur prior to root hair curling, which is an early host response to rhizobia (see Verma, 1992 and references therein). These studies suggest that diffusible nod factors produced by rhizobia are responsible for initiating the concatenation of events in the host that lead to nodule organogenesis.

It has been suggested that a specific receptor(s) exists in the epidermal cells as they are the first site to encounter nod factors and to perceive them at very low concentration and in a host-specific manner. It is unclear, however, whether nod factors are perceived directly in the inner cortical tissue or some secondary signals are generated in the epidermis to provoke the responses in these inner cells (see Geurts and Franssen, 1996).

Studies on the bacterial-induced plant genes, the ENOD genes, may be the useful markers to understand this gap. As yet investigations into the formation of effective root nodules in the absence of bacteria involving the nod factor alone are not possible and the constraints on the bacterium still need further elucidation. Some recent studies indicate that the activity or the recognition of nod factors is not limited to legumes, nod factors were involved in rescuing the lost ability to form somatic embryos in a mutated carrot cell line (De Jong et al., 1993), triggering alkalinization of a tomato suspension culture medium (Staehelin et al., 1994) and like phytohormones stimulated division of tobacco protoplast (Röhrig et al., 1995). Further, a chitin oligosaccharide similar to the nod-factor backbone could be synthesized by the *Xenopus* developmental protein DG 42, which has a homology with the *Rhizobium* nod-C encoded protein (Semino and Robbins, 1995).

On the basis of these observations, Geurts and Franssen (1996) have suggested that nod factors can be considered as (plant) growth regulators and that nod-factor-like molecules normally occur in plants and even in animals. Hence, it seems likely that *Rhizobium* has evolved a genetic pool that enables it to synthesize molecules by which it becomes possible to redirect (plant) developmental programmes and to manipulate them for its own benefit. For example, homology can be seen in the formation of an infection site and a typical cell-tip growth for initiation of root hairs, trichomes and the extending pollen tube.

B. Nod-Factor-Induced Responses and Nodule Initiation

Nod-factor-induced initiation of cell division occurs in the root cortex of the infectible zone of roots, prior to the entry of bacteria into the host. Very small (picomolar) concentrations of the nod factor are sufficient for the induction of responses in the epidermis which may be provided as purified chemical or diffused out from the *Rhizobia* (see Verma, 1992; Sindhu and Dadarwal, 1995; Geurts and Franssen, 1996). A receptor(s)-mediated nod-factor recognition model at the root hair membrane has been proposed (Ardourel et al., 1994; Geurts and Franssen, 1996). The model describes the existence of two nod factor receptors: a signalling receptor, which is less selective for nod factor structure and an entry receptor, which is highly selective and recognizes nod factor(s) only with the specific decorations leading to root infection (Geurts and Franssen, 1996). These receptors are considered to be located on different sites at the root-hair membrane. Upon binding to the non-specific signalling receptor, the nod factor triggers the initiation of cortical cell division as well as root-hair deformation.

The nod factor initiates several types of effects even at the host epidermis e.g. membrane depolarization of the epidermal cells (Ehrhardt et al., 1992; Felle et al., 1995; Kurkdjian, 1995), calcium spiking in the root hairs (Ehrhardt et al., 1996), deformation of root hairs (Heidstra et al., 1994) and induction of early noduline gene(s) expression in the nodule primordia such as ENOD 12, ENOD 40, Mt RiPil and PrP4 (Scheres et al., 1990a; Yang et al., 1993; Wilson et al., 1994; Journet et al., 1994; Cook et al., 1995). These responses occur in a host-specific manner and take place in a narrow zone of the root where young root hairs are present (Geurts and Franssen, 1996). Further, the nod factor responses usually follow a definite order in relation to the time factor. Depolarization of the membrane in trichoblasts as well as in atrichoblasts has been detected within minutes after nod-factor application and seems to be the one of the earliest responses of the signal molecule (Felle et al., 1995). Root-hair deformation, the first detectable morphological change, takes only up to 10 min after contact of the root with the nod factor. Swelling in the tip of a mature root hair and new tip growth are initiated within about 3 h (Heidstra, et al., 1994). It has been proposed that the gene products of early nodulins, e.g. ENOD 12 (encodes a pro-rich protein) and Mt Rip 1 (encodes peroxidase), are involved in the infection by altering root-hair wall structure at sites of incipient penetration and expression of these gene(s) occurs in the root epidermis within 2–3 h after nod-factor application (Salzwedel and Dazzo, 1993; Journet et al., 1994; Cook et al., 1995; Geurts and Franssen, 1996).

Development of the root nodule primordium requires the dedifferentiation of fully differentiated cortical cells. Nod-factor-induced mitotic activity started within 12–24 h after inoculation with rhizobia in the cortical tissues of soybean root (Calvert et al., 1984). The plane of cell division was initially oriented to keep the axis of the new wall deposition perpendicular to the longitudinal axis of the root. Subsequently, this primordium gave rise to an organized meristem consisting of a mass of small cells dividing in all planes.

On the basis of similarity between nod-factor-induced initiation of cell division and the effects caused by the cytokinin or auxin transport inhibitors (e.g. triodobenzoic acid naphthyl phthalamic acid), it has been suggested that nod factors either act by altering phytohormones in some way or act as phytohormones themselves (see Verma, 1992; Geurts and Franssen, 1996; and references therein).

In the case of determinate (spherical) nodules, e.g. nodules of soybean, phaseolus bean, lotus, etc., hypodermal cells of the cortex are the first to respond to infection (Verma, 1992). In indeterminate nodules, e.g. of alfalfa, clover, vetch, pea, etc., on the other hand, cells of the inner cortex appear to be the primary target for initiation of meristematic activity (Libbenga and Harkes, 1973; Newcomb, et al., 1979). It has been suggested by Geurts and Franssen (1996) that nod-factor-induced cell divisions might be mediated via regulation of the cell cycle. During infection of the root hairs, the bacteria trigger the cell cycle machinery to arrest the cells in the G_2 phase, which leads to the formation of a preinfection thread, whereas the cells that proceed with the cell cycles lead to the formation of a nodule primordium.

Though many infections occur in the root surface, only a small fraction of them give rise to cortical cell division. This suggests the involvement of some other factor(s) along with the path of infection process for eventual success in generating nodule primordia. Many bacterial (Long, 1989; Sindhu and Dadarwal, 1995) and plant (Greeshoff and Delves, 1986; Guinel and LaRue, 1991) mutants are known that abort this process by perturbing the signal transduction process. However, the precise steps in the process are not understood (Verma, 1992; Sindhu and Dadarwal, 1995). The expression of ENOD 40 in the pericycle precedes the first cortical cell division; thus it has been hypothesized that ENOD 40 may encode a regulator of cell division whose induction after the *Rhizobium* infection starts the process of root cortical cell division leading to nodule formation (Geurts and Franssen, 1996). Nod-factor-induced changes via the expression of ENOD 40 to switch on the cortical cell division might be mediated by changes in the phytohormones balance in the roots. Recently, uridine was identified as a factor of the stele from pea that in combination with phytohormones is capable of inducing cell division in the inner cortex of pea root explant (Smit et al., 1995). A possible similarity in the signal transduction pathways involved in the symbiosis and in the other plant process needs to be elucidated for better understanding of the existing knowledge.

In spite of the different ontogeny of nodule meristem (originates from cortical cells) and lateral root primordia (originates from pericycle), they share a part of their genetically controlled developmental programme (see Verma, 1992). *Rhizobia* invade the dividing cells in the cortex, indicating a different nature of these cells. An extensive need for endoplasmic reticulum (ER) and Golgi activity to generate large amount of membrane required to enclose the bacteria invading the dividing cortical cells. In determinate nodules cell division ceases before the commencement of N_2-fixation. In indeterminate nodules, however, cell division ceases in the infected cells while the nodule meristem persists, generating new cells that continue to be infected. The type of nodule depends on the type of host as the same bacteria can form different types of nodules in different hosts. After the cessation of cell division, nodular growth continues by cell enlargement. The factors controlling the continuity of the indeterminate nodule formation in the legume root nodules are, however, not known.

C. Differentiation of Root Nodules

Root nodules are highly differentiated structures consisting of enlarged cells, which normally contain the bacteroid, and small interstitial cells containing more starch. It has been indicated that nodule development is largely under the control of the nod-factor-triggered organogenesis programme and that differentiation of the nodule can occur even without the entry of rhizobia (see Verma, 1992; Sidhu and Dadarwal, 1995; Geurts and Franssen, 1996). The entry of bacteria and their release from the infection thread completes the differentiation process and commences nitrogen fixation. At the time of N_2-fixation not only nodule differentiation, but bacterial divisions also cease. It has been noticed that the expression of many early nodulin genes related to nodule differentiation is independent of the presence of bacteria in the nodules (Nap and Bisseling 1990).

Well-differentiated root nodules consist of both infected and uninfected cells and a symbiotic zone is formed in which these cells perform different metabolic roles. This zone is surrounded by the nodule parenchyma and endodermis possibly to prevent diffusion of the fixed ammonia (Miao et al., 1991). A vital regulation of oxygen availability is also essential to protect the nitrogenase activity in the N_2-fixing root nodules and thus such arrangements are also maintained. It is believed that the nodule parenchyma controls the diffusion of oxygen and the infected cells produce leghaemoglobin to regulate the oxygen flux in the nodules (Verma, 1992).

D. Entry of *Rhizobium* in Host Cell

One of the most crucial events which has not been adequately understood at the molecular level is the endocytosis of bacteria inside the host cell without provoking the host defence mechanisms. Formation of a membrane compartment, in the form of an infection thread, is essential for successful nodulation (Werner et al., 1985). Studies related to the entry of bacterial parasites in animals and plants and an analogy of the *Rhizobia* to those bacteria have indicated that a receptor-mediated endocytosis may occur in this symbiotic entry (Morrison and Verma, 1987). No information is available as yet on the transmembrane signalling system that triggers endocytotic or phagocytotic events (see Verma, 1992). Another significant event triggered with the infection of *Rhizobia* is the generation and continued proliferation of the peribacteroid membrane (PBM). This membrane is essential for enclosing the *Rhizobia* so that contact of the bacteria with the host cytoplasm is avoided. In soybean root nodules almost 30 times more PBM was generated compared to that of plasma membrane during the process (Verma et al., 1978). Proliferation of the PBM begins concomitant with the release of bacteria from the infection thread.

Encapsulation of the *Rhizobia* within the PBM leave a peribacterioid space between the bacteria and PBM which is equilibrated with certain metabolites, including dicarboxylic acid (used as C-source by the bacterioid). This arrangement eliminates the concentration gradient between the host and the *Rhizobia* (Verma, 1992). However, equilibration of the peribacterioid space is apparently accomplished by opening specific channels in the PBM. Nodulin-26 has been suggested to be operating for such channels (Verma, 1992).

To entrap the bacteria in the contact of the root hair, the curled root hair tip facilitates encircling of the *Rhizobium*. A hyaline spot is usually the first sign of infection thread penetration. Root-hair cytoplasmic streaming increases in response to the bacterial attachment and the nucleus of the host cell migrates towards the refractile spot. Following dissolution of the host cell wall by the bacterial hydrolytic enzymes, the plasma membrane of the root hair invaginates and a cell wall is deposited around it (Verma and Long, 1983). Invagination with the newly formed cell wall forms the infection thread (Kijne, 1992).

E. Autoregulation of Nodule Organogenesis

The extent of nodulation is generally restricted by a plant-mediated, feedback-regulated process, termed autoregulation, upon successful nodulation. This involves suppression of nodule emergence from ontogenetically younger root tissues by the previously formed nodules on older parts of the root system (Bhuvaneswari et al., 1981; Pierce and Bauer, 1983; Kosslak and Bohlool, 1984). Feedback suppression of symbiotic root nodulation (autoregulation) was discovered as early as 1952 by Nutman in red clover and has since been demonstrated in several other legumes (Rolfe and Gresshoff, 1988). However, the precise molecular mechanism of this regulation is not yet known.

Studies with the supernodulating mutants in legumes have shown that the autoregulatory mechanism exists both in the leaf/shoot as well as in the root (Sheng and Harper, 1997). It has been proposed that once a critical number of the subepidermal cell divisions (SCDs) in the root cortex are initiated, a precursor molecule from the root is transported to the shoot where it is converted into the shoot-derived inhibitor (SDI), and transported back to the roots to supress the SCDs formed later from the developing emergent nodules (Caetano-Anolles and Gresshoff, 1990, 1991). Takats (1990) showed an early autoregulation of symbiotic root nodulation in soybean. Autoregulation in soybean cv Williams 82 and its hypernodulating mutants does not always maintain a constant number of nodules. Rather, the nodule number in Williams 82 can be affected by an available infection site (root site) at the time of inoculation; thus a delayed inoculation increased the nodule number (Francisco and Harper, 1995). It appears that new nodule primordia are arrested during early nodule ontogeny by the previously formed SCDs in the root through a shoot-mediated feedback process and signal communication must occur in the root-shoot interactions (Delves et al., 1986; Francisco and Harper, 1995).

Our understanding of the plant-translocatable signals is poor because the biochemical nature, the biosynthetic site(s), and the pathway by which signals are transported and function are not known. Francisco and Harper (1995) demonstrated that the leaf autoregulates the number of nodules in soybean plants, which confirmed earlier studies by Delves et al. (1988). It appears, however, that the root also affects the nodulation pattern to some extent. On the basis of their most recent studies in soybean genotypes, Sheng and Harper (1997) concluded that (a) soybean leaves are the dominant site of the autoregulatory signal production, which controls the nodule number; (b) *Glycine max* and *Lablab purpureus* genotypes have a common, translocatable, autoregulatory control signal; (c) seedling vegetative growth and nodule number are independently controlled; and (d) two signals, inhibitor and promoter, may be involved in controlling nodule numbers.

IV. FACTORS AFFECTING NODULATION

Nodule formation is a response to a series of specific interactions between symbiotic bacteria and host plants. Thus, it is regulated by many factors other than genetical factors of host and bacteroids. Therefore, all the factors, either environmental or internal which affect the physiological and biochemical processes of the host plant and *Rhizobium* are prerequisites for modification of nodulation. In fact, the incoming solutes and metabolites in the nodule from the host and vice versa are the great legacy of symbiosis. Regulation by internal factors is called autoregulation (Pierce and Baur, 1983; Caeteno-Annolles and Gresshoff, 1990), already described in section III. The effect of some external factors is described here.

A. Edaphic Factors

Initiation of infection, the first step towards nodulation, is affected by soil factors. The plant root secretes sugars, amino acids and organic acids etc. in the rhizosphere which enhances *Rhizobium* multiplication. The bacterial number may be very high under these rhizospheric conditions. As described earlier, the host secretions around the rhizosphere are liable to specificity also. Inhibition in nodulation may also be due to change in soil pH and chemical composition (Yoneyama et al., 1979). They may limit *Rhizobium* multiplication and infection of roots.

1. *Water Stress*: The water status of the soil and also of the plants controls various developmental processes (Hanson and Hitz, 1982). The optimum soil moisture required for nodulation is around 65–70%. Decline in this critical moisture level reduces nodule development. Decline in moisture leading to soil desiccation may also act as a physical barrier to *Rhizobium* infection, either by inhibiting *Rhizobium* movement, dependent on water tension (Hamidi, 1970), or encouraging adherence of sand/silica particles to root hairs, thereby reducing the proximity of *Rhizobium* with the host. Reduced water potential checks *Rhizobium* infection (Worrall and Roughley, 1976) and nodule development (Galiacher and Sprent, 1978). Drought may induce senescence in the developed nodule.

Water stress may show a differential response depending on nodular type, i.e. difference in their anatomy. An indeterminate type of nodule exhibits considerable resistance to water stress (Engin and Sprent, 1973; Wahab and Zahran, 1979) while spherical or determinate type nodules are less resistant (Sprent, 1976). In tree legumes *Albizia stipulata* and *Ougeinia dalbergioides*, nodulation and nodular growth decreased under mild stress. However, nodulation was found to be more sensitive in *A. stipulata* than in *O. dalbergioides* (Purohit et al., 1996). In fact, growth and development of the two nodular types seem to be entirely different. Indeterminate nodules have a complex anatomy with more structural compartments. Indeterminate nodular development is alongwith the nodule axis, from the distal region (meristematic zone I), passing consecutively through the prefixing zone II (infection zone), the amyloplast-rich interzone II–III region, the nitrogen-fixing zone III, the inefficient zone III and finally the senescence zone IV (Hirsch and Smith, 1987; Truchet et al., 1989; Vasse et al., 1990). These zonal differences have been distinguished at the ultrastructure level in alfalfa (Vasse et al., 1990) and may be one of the factors contributing to resistance to water stress. It has been demonstrated that nodular development may be reversible until the moisture content of the nodule falls below 80% under water stress (Swaraj et al., 1984). Water-stress inhibition in nodulation has been attributed to disruption of interconnections between plant cell and nodule due to the breakdown of plasmodesmata, which leads to shrinkage and desiccation of the cytoplasm of the cortical cells surrounding the bacteroidal tissue (Sprent, 1976). Under water stress, direct ion excess toxicity may be another factor in nodular growth inhibition. Ammonium accumulation under stress contribute to reduced nodulation (Marinda-Ham and Loyola-Vargas, 1994). A change in sink-source relationship may also disrupt nodulation and nodular growth (Singleton, 1983). Alterations in protein content and configuration under water stress may also inhibit nodular growth. Swaraj (1988) demonstrated decrease in leghaemoglobin protein during water stress.

2. *Salinity Effects*: It is generally considered that *Rhizobia* can withstand a higher salinity level compared to their host (Sprent, 1984). Information regarding the sustainability of various *Rhizobium* strains under salinity is meagre. Hua and coworkers (1982) observed that *Rhizobium* sp. strain WR 1001 from *Prosopis* was able to maintain osmoticum. However, salinity caused reduction in nodulation in cowpea and broad bean (Ibrahim et al., 1970), soybean and lupin (Sprent, 1972) and growth in cluster bean and moth bean (Garg et al., 1984). Salinity may affect various steps of nodulation and growth. Inhibition in nodulation (Subba Rao et al., 1987), and decrease in mucilaginous sheath (Lakshmi Kumari et al., 1974) have been reported.

Nodular growth inhibition under salinity may also depend on the host and microsymbiont. *R. melioloti* can tolerate 0.3% NaCl but loses the potential for nodule formation at salinity beyond 0.6% NaCl. It has been also observed that salinity effects may be more severe for nodular initiation than for nodular growth (Balasubramanian and Sinha, 1976a, b). It has been further observed that some crop plants try to alleviate the stress effect by producing large nodules, as reported in *Glycine wightii* and *Medicago sativa* (Yousef and Sprent, 1983). Large-size nodules may play some role in maintaining the osmoticum, but their nitrogen-fixation potential has to be estimated. Maintenance of osmoticum in larger nodules may be due to increased partitioning of photosynthates and other metabolites, although this may be at the expense of upper parts of the host plant. This may ultimately be detrimental to the growth of root hairs, reducing a new site for new nodule formation and causing a net decline in nitrogen-fixation potential.

In nature, the chlorides and sulphates of K^+, Ca^{2+}, Mg^{2+} and Na^+ also present along with NaCl could restrict nodulation and nodular growth.

3. *Environmental Pollutants (Heavy Metals or Gaseous)*: It is widely known that pollutants such as toxic metals (Pb, Cd, Al, Hg being non-essential elements for plant growth) and gaseous pollutants (especially

oxides of nitrogen, and sulphur and ozone) are phytotoxic. Reduction in early seedling growth in legumes such as soybean (Huang et al., 1974) and pea (Sinha et al., 1988) in the presence of different levels of Pb has been recorded. Soybean growth was found to be more sensitive to Cd than Pb (Dubey and Dwivedi, 1987). Plant growth, especially root hair, is a prerequisite for nodulation and further nodular development; hence pollutants, especially heavy metals, may cause reduction in nodulation. However, extensive studies on the effects of heavy metal on nodulation and nodular growth are scarce. Pb inhibition of nodular development but little effect on nodulation (increased number of nodules noted) was recorded in *Vigna radiata* (Dabas et al., 1995).

Gaseous pollutants also restrict nodular development. Inhibition of nodulation by NO_2 was observed in soybean (Srivastava and Ormrod, 1986; Trinchant and Rigaud, 1988) and by O_3 in landino clover and soybean (Tingey and Blum, 1973; Letchworth and Blum, 1977).

B. Chemical Factors

Stress-induced formation of free radicals, potential inhibitor of membrane permeability, may also influence membrane organization of the root cells, which in turn may disturb the secretions of leachates required for interaction with micro-symbionts. Moreover, under stress condition the free radicals may also damage directly the nodule membrane and induce senescence. Free radical scavengers such as ascorbic acid application enhanced nodulation in chick-pea (Swaraj and Garg, 1970b), clover, ground nut and cowpea (Prabha and Bharti, 1979).

From these studies it may be inferred that such compounds might be used to overcome the stress effect on nodulation up to some extent. A study of the phytohormone level under stress would also provide a clue to the specific hormone role in nodulation and nodular growth and thus a means for ameliorating the process. A decrease in hormonal level has been observed in plants under stress conditions (Sanchez-Roldan et al., 1990, Mishra et al., 1994). It was recently demonstrated that salinity and toxic metal stress increase IAA oxidase activity (Singh, 1996) which in turn may decrease the auxin level.

C. Nodule Development Under Inorganic Nitrogen Nutrition

Nitrate supplementation in the nutrient medium usually decreases nodulation (Munns, 1977; Franco et al., 1979; Carroll and Mathews, 1990). Reduction in nodular mass per plant has been observed in legumes and actinorhizal plants receiving high levels of combined nitrogen also (Granhall et al., 1983; Roberts et al., 1983). However, Vogel and Dawson (1991) found no significant difference in nodule dry weight in nitrate-treated *Alnus glutinosa*. It seems that the response of symbiotic activity to nitrate supply varies in crop and tree legumes.

By way of mechanism, it is generally believed that nitrate nutrition induces nodular senescence, marked by decrease in macromolecules. Singh et al. (1994) demonstrated decrease in total soluble carbohydrate and leghaemoglobin in chick-pea in response to nitrate. This may change the physiological activity of the nodules (Becana et al., 1988). Marques et al. (1983) demonstrated that NO_3-assimilating *P. vulgaris* had higher photosynthesis while nitrate had no effect on distribution of photosynthetic products compared with those of N-free nutrition. Whereas in NH_4-fed plants CO_2-fixation in carbohydrate decreased, C-incorporation into organic acids increased (Larsen et al., 1981).

Nitrogen assimilation in general results in synthesis of protein and other macromolecules. Though, Becana et al. (1985) found a decrease in nodular mass and protein in alfalfa with nitrate. Generally, nodules do not fix N_2 until the haemoglobin has formed because the Fe-protein is involved in regulation of the ATP/NADPH required for N_2-fixation. There has been considerable decrease in leghaemoglobin-mRNA in nodules of alfalfa with 5 mM KNO_3 (De-Billy et al., 1991). These workers found that short period (24 h) exposure to nitrate led to the disappearance of Lb-transcripts from the proximal region of the hybridizing zone whereas longer exposure (72 h) led to considerable decrease in the intensity of hybridization signal with the amyloplast-rich interzonal region. It indicates that nitrate mediated effect on nodule growth might be through leghaemoglobin synthesis.

Some authors have suggested that nitrate inhibition of nodulation might be due to inhibition in initial cortical cell division and infection thread formation also (Malik et al., 1987). However, studies on the interaction between NO_3 and autoregulation signals (Malik et al., 1987, Day et al., 1989) may well throw some light on the inhibition process. An enhanced ethylene production during inoculation in alfalfa root in the presence of NO_3 has been reported (Ligero et al., 1987), which correlated negatively with nodulation. It was recently documented that ethylene inhibitor aminoethoxy vinylglycine (AVG) treatment eliminates the NO_3 inhibitory effect on nodulation (Ligero et al., 1991). This indicates that the nitrate inhibitory effect on nodulation might be mediated through ethylene. Since ABA has also been implicated in nodular inhibition, studies are required to ascertain the possible mode of action of NO_3 on nodulation through ABA.

D. Phytohormones and Nodulation

Studies on the exchange of signals between host and *Rhizobia* to establish the relationship culminating in nodulation have been the main focus for understanding the mechanism. Phytohormones have been considered to be the potential mediators of the nodulation process (Grobbelaar et al., 1971; Phillips and Torrey, 1972). But it has yet to be conclusively demonstrated which contributes the most, host or nodule, to this process, since both are potential sources for phytohormones.

Exogenously applied phytohormones showed responses depending on host species. Exogenous auxin or naphthylic acetic acid supplementation showed stimulation in nodulation and nodular growth in *Medicago sativa* and *Arachis hypogaea* (Srinivasan and Gopalkrishna, 1977), while 1-Naphthoxy acetic acids and (p-chlorophenoxy)-isobutyric acid increased the nodulation process in excised bean roots (Cartwright, 1967). Kefford and associates (1960) contrarily observed increase in nodular growth but little effect on nodulation with naphthylic acetic acid. Induction of nodulation by cytokinins (Nandwal and Bharti, 1982) and GAs (Swaraj and Garg, 1970a; Kandaswamy and Prasad, 1976; Bishnoi and Krishnamoorty, 1990) has also been recorded. GA application, on the other hand, reduced nodulation in bean (Thurbur et al., 1958), alfalfa (Kefford et al., 1960), clover (Prakash, 1966), black gram and mung bean (Kandaswamy and Prasad, 1976).

Hormonal regulation of nodulation and nodular growth appear to be circumstantial. A higher level of nodular auxin (Dullaart, 1970; Badenoch-Jones et al., 1984), cytokinin (Badenoch-Jones et al., 1987) and GA (Dullaart and Duba, 1970) has been observed. It was suggested that there is preferential transport of auxin (Badenoch-Jones et al., 1984), GA (Evensen and Blevins, 1981) and cytokinin (Henson and Wheeler, 1976) to the nodules. These observations led to the proposal that the higher levels of phytohormones in nodules are involved in the process of nodulation and nodular growth (Libbenga and Bogers, 1974; Nutman, 1977; Bhattacharya and Basu, 1995). Induction of infection and nodule development by auxin and cytokinin have also been observed (Libbenga et al., 1973; Syono et al., 1976). Auxins or auxin-like compounds induced initiation without affecting curling of root hairs in clover (Dart, 1977).

Some compounds other than the growth promoters also affect nodulation. Phenols such as salicylic acid (SA) in combination with chlorogenic acid induced nodulation (Garg et al., 1989). Nodulation and nodular growth were retarded by ABA (Phillips, 1971) and ethylene (Drennan and Nortton, 1972; Goodlass and Smith, 1979). Surprisingly, Charbonneau and Newcomb (1985) found ABA promotion of nodules in pea, which Jaiwal and Gulati (1989) suggested were pseudonodules. But whether these nodules were able to fix dinitrogen was not determined.

E. Effect of Light/Dark Conditions

Light has an indirect effect on nodule development in legumes. Light is essential for normal plant development and hence also for nodule formation and development. However, exposure of the root system to light caused a reduction in nodulation (Grobbelaar et al., 1971). In darkness or in low light the nodules might be inhibited because of reduction in leaf photosynthesis. Prolonged darkness can thus induce nodular senescence (Wong and Evans, 1971; Pfeiffer et al., 1983). Nodular growth could be recovered in optimum light in soybean but not in cowpea (Garg and Swaraj, 1984). Light stress also caused reduction in leghaemoglobin content and in bacteroids in the nodules of pea (Roponen, 1970), chickpea and cowpea (Swaraj et al., 1986).

F. Some other factors

In the past two decades the influence of vesicular-arbuscular mycorrhizae (VAM) on plant growth and development has attracted the attention of plant scientists throughout the world. VAM increased nodulation in soybean (Kawai and Yamamoto, 1986), lentils (Singh et al., 1984) and chick-pea (Thiagrajan et al., 1992; Rawat et al., 1991). Since VAM do not enter into the plant or nodule, they might be mediating their effects by modifying the rhizosphere (Blevins et al., 1977). Kawai and Yamamoto (1986) suggested that VAM channelizes nutrients such as P, Mg^{2+} and Ca^{2+} to the nodule. Penicillin, a fungal antibiotic, was found to promote nodulation in *Vigna radiata* (Mukhopadhyay et al., 1990).

V. MOLECULAR BIOLOGY OF NODULATION

A. Rhizobium Nodulation Gene

The bacterial nodulation (nod) genes in several *Rhizobia* are located on sym (symbiotic) plasmid (of about 200 to 1500 kb size) whereas in *Bradyrhizobium* such symbiotic genes are located on the chromosome. Recent molecular techniques such as transposon mutagenesis, complementation analysis, DNA sequences and *in-vitro* transcription/translation have made it possible to identify and characterize as many as 44 bacterial nod genes in various *Rhizobium* species. The nod genes may be divided into regulatory, common and host-specific genes.

The nod D is a constitutively expressed regulatory gene whose product acts as a transcriptional activator for all other nod genes after activation due to binding to plant-secreted signals, as described in section IV. *R. leguminosarum* and *R. trifolii* contain a single copy of nod D (Kondorosi, 1992) whereas some strains of *Rhizobium* such as *R. meliloti* possess multiple copies of nod D. The different nod D genes are conserved at the nucleotide sequence level and share homology with the Lys R family of prokaryote regulatory genes. The function of nod Ds may be diverse. In *R. meliloti*, two inducer requiring nod Ds (nod D_1 and nod D_2) are important for optimal responses to different hosts. The inducer independent circuit controlled by nod D_3 may be important for late nod gene expression (Kondorosi et al., 1991) and/or may respond to nitrogen, growth and/or inducers (Wang and Stacey, 1991).

Upstream of all the inducible nod operons, there is a highly conserved DNA sequence of about 47 bp nucleotides, the nod boxes, to which the nod D gene product binds directly. In *R. meliloti*, nol R regulates expression of the nod gene by inhibiting expression of nod D_1 and nod D_2 (Kondorosi, 1992).

1. *Common Nod Genes*: Besides nod D, the other common nod ABC genes have been found in all *Rhizobium, Bradyrhizobium* and *Azorhizobium* species (Stacey, 1990). These genes are structurally very similar and functionally interchangeable between *Rhizobium* and *Bradyrhizobium* species without altering the host range (Barbour et al., 1992). These highly conserved nodulation genes are required to produce lipid-linked chito-oligosaccharide signal molecules (nod factor) which elicit nodule morphogenesis (Spaink et al., 1991). It was recently shown that nod B gene product is an oligosaccharide-modifying enzyme which deacetylates the non-reducing terminus N-acetylglucosamine residue of the chito-oligosaccharides (John et al., 1993). The deacetylated nod B together with nod A protein are involved in generating small heat-stable compounds that stimulate mitosis in protoplasts of legumes and nonlegumes (Schmidt et al., 1988). Nod IJ genes are present in *R. leguminosarum, R. trifolii* and *B. japonicum* (Gottfert et al., 1990a, b). The presence of nod ABC IJ genes in the genetically distant *Rhizobium, Bradyrhizobium* and *Azorhizobium* suggests a common origin of these genes. Recently, other common nod genes, the nod M and nod N genes have been found in *R. leguminosarum* and *R. meliloti*.

The DNA sequence of nod M is homologous to the glm S gene in *E. coli* which encodes D-glucosamine synthetase (Baev et al., 1991). Nod C with a sequence similarity to yeast chitin synthetase (Hirsch, 1992) may be involved with linking individual glucosamine units synthesized by the activity of nod M. The nod L gene product is thought to acetylate the glucosamine residues (Downie, 1989) while nod FE-encoded proteins are involved in synthesizing the fatty acid side-chain (Shearnman et al., 1986).

2. *Host-specific Nod Genes*: The host-specific nod genes determine the specificity of nodulation on a particular host and are present in one

species and not in others. The nod PQ and H genes which determine alfalfa specificity are present in R. *meliloti* but not in R. *leguminosarum*, R. *trifolii* and B. *japonicum*. Similarly, nod SU genes are present in B. *japonicum* and the broad-host range tropical *Rhizobium* sp. NGR 234 and nod T in R. *leguminosarum*. Some host-specific nod genes can be present only in some biovars, for example, the nod O gene is found in R. *leguminosarum* bv viciae but not in closely related biovar *trifolii*. Between these closely related biovars, host specificity is determined by the structure of the acyl chain of the nod factor. R. *leguminosarum* bv *viciae* produces mainly nod factors that contain cis-vaccenyl (C18:1) or highly unsaturated (18:4) fatty acid that is determined by nod E (Spaink et al., 1991a, b). The biovar *trifolii* does not produce nod factors that contain a C18:4 but instead produce a complex mixture of nod factors carrying unsaturated fatty acyl groups containing either two, three or four double bonds. The loss of nod E in R. *leguminosarum* bv *trifolii* enables this bacterium to nodulate the non-host pea. Strikingly, the nod O of R. *leguminosarum* bv *viciae* can partially complement the nod factor structure deficiency due to a nod E mutation, although nod O is not involved in nod-factor production or secretion. Nod O encodes a protein that forms calcium-regulated monovalent cation channels when added to planar lipid bilayers (Sutton et al., 1994).

Upon mutation, some of the host-specific nod genes may result in alteration of the host range. In R. *meliloti*, the nod H, nod P and nod Q genes encode for enzymes involved in sulphation. R. *meliloti* nod H mutants produce nod factors lacking the sulphate group at the reducing end (Roche et al., 1991). These mutants lost their ability to nodulate on their host plant, alfalfa.

B. Host-Plant Genes Involved in Nodulation and N_2 Fixation

Several nodule-specific plant genes that encode proteins called *nodulins* are activated during *Rhizobium* infection, nodule morphogenesis and establishment of the appropriate environment for nitrogen fixation (Kondorosi and Kondorosi, 1986; Nap and Bisseling 1990; Verma, 1992). Plant nodulin genes have been classified into early and late nodulin genes according to their timing of appearance during nodule development. Gene-encoding early nodulins are induced during rhizobial infection, growth of infection thread and stimulation of cortical cell divisions. The late nodulin genes are activated in developing and mature nodules concomitant with the onset of nitrogen fixation activity (Verma and Dalauney, 1988).

1. *Early Nodulin Genes*: The first described early noduline gene was ENOD2 (nodulin-75) whose expression in the nodule parenchyma

results in the formation of plant cell-wall proteins which help in establishing and maintaining the oxygen diffusion barrier in the nodule (Van de Wiel et al., 1990a, b). An extracellular glycoprotein of 95 kD was also identified recently in cell walls and intracellular spaces of nodule parenchyma cells (Rae et al., 1991). The 95 kD protein had earlier been found within the infection thread matrix (Vanden Bosch et al., 1989), a site where ENOD2 transcripts have not been detected thus far.

The expression of two early nodulin genes, PsENOD5 and PsENOD12, was induced during the infection process in pea. ENOD5 expression was restricted to those cells containing the tip of the growing infection thread whereas ENOD12 expression was detected in cells containing infection threads as well as in cells preparing for infection thread passage (Nap and Bisseling, 1990). Both of these early nodulins are proline-rich proteins. The PsENOD12 early nodulin is for the most part composed of two repeated peptapeptides, each of which contains two prolines. This structure suggests that this early nodulin is a hydroxyproline-rich cell-wall protein (Scheres et al., 1990 a, b). The putative cell-wall protein PsENOD12 may thus be that part of the additional cell wall formed in the cortex cells which prepares for infection thread passage and may also be a component of the infection thread itself. PsENOD5 lacks the repetitive structure of PsENOD12 but is rich in proline and also in alanine, glycine and serine, suggesting it is related to arabinogalactan-like proteins, known to be components of the plasma membrane. The PsENOD5 protein may be a component of the plasma membrane of the infection thread.

The expression of some nodulin genes such as ENOD12 can be induced by rhizobial lipoligosaccharides as a known nod factor (Journet et al., 1994). However, others such as ENOD2 cannot be induced by a nod factor and appear to be under hormonal regulation (Dehlo and Bruijn, 1992). Expression of ENOD12 alongwith the peroxidase-encoding MtRip1 (Journet et al., 1994; Cook et al., 1995) was found to occur in the root epidermal cells, especially in the area of root-hair differentiation, within 2 to 3 h following the nod factor. Based on the timing and the site of expression, it is not likely that these genes are involved in root-hair deformation, but they might play a role in the next step to occur, infection.

The expression of several nodule-specific genes has been detected in nodule primordia such as ENOD12, ENOD40 and PrP_4 (Scheres et al., 1990; Yang et al., 1993; Wilson et al., 1994) but has not yet been observed in other primordia. Activation of ENOD40 by nod factors causes changes in phytohormones which stimulate divisions in inner cortical cells located opposite to protoxylem poles (Smith et al., 1995).

The other early nodulin genes, ENOD3 and ENOD14, are expressed in the infected cells of the central tissue and encode for metal-binding proteins involved in, for example, transport of Fe and Mo to the bacteroids for a functional nitrogenase/hydrogenase.

2. *Late Nodulin Genes*: Leghaemoglobin (Lb) is the most abundant late nodulin; it constitutes up to 25% of the total soluble protein in a mature nodule. It is a 16 kD myoglobin-like haemoprotein with a high oxygen affinity and controls the concentration of free oxygen supply to the bacterioids that balance protection of nitrogenase against oxygen damage with support of respiration. Lb is a true 'symbiotic protein' because the haem group is presumed to be a bacterioid product and the globin proteins are encoded by the host-plant genome. Lb is localized in the cytoplasm of the infected cells.

In soybean, there are six Lb genes in two independent clusters, one containing 4 genes in the order 5'-Lba-LbC_1-ψ_1Lb-LbC_3-3' and the other cluster two genes in the order 5'ψ_2Lb, LbC_2 3' (Marcker et al., 1984) (Figure 3.2).

A study by Stougaard et al. (1986) showed that the promoter region of a soybean Lb gene directed nodule-specific and developmentally correct expression of a reporter gene when transformed into other leguminous plants. Apparently, the Lb promoter region not only carries all required cis-regulatory sequences, but also conserves all relevant transacting factors among various leguminous species. Detection analysis of soybean Lb promoter defined a relatively small region as responsible for the correct expression pattern. Jacobsen et al. (1990) also found nodule-specific transacting factors interacting with the Lb promoter in soybean. The factors responsible for the expression of Lb genes are not presently known. Molecular analysis of promoters and their transacting factors will give insight into the signals presumed to trigger the expression of Lb genes.

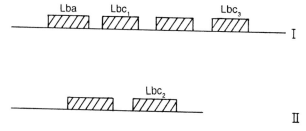

Fig. 3.2. Organization of Lb genes in soybean.

Other late nodulins have been identified as enzymes or subunits of enzymes involved in the assimilation of nitrogen (glutamine synthase, uricase) or carbon (sucrose synthetase).

The second most abundant nodulin is N-35, a 33 kD subunit of a nodule-specific uricase (uricase II, a key enzyme in the ureide biosynthesis, Verma et al., 1996), which is exclusively found in the uninfected nodule cells. A nodule-specific form of glutamine synthetase (GS) has now been identified in several legumes (Sengupta-Gopalan and Pitas, 1986; Dunn et al., 1988; Forde and Cullimore, 1989). This new form is required for assimilating the large quantities of ammonium produced by bacteroids. Nodulin-100 is the subunit of sucrose synthetase, an enzyme involved in the breakdown of sucrose translocated to the nodule from leaves to support the carbon requirement of the nodules (Thummler and Verma, 1987).

The above-mentioned and other late nodulins are not unique to nodules but occur in other parts of the plant as well. This supports the viewpoint that nodule formation evolved from relatively minor alternations in the pathway of root differentiation in which common plant genes became adapted to fit physiological regulatory constraints of symbiosis (Nap and Bisseling, 1990). Several nodule enzymes have been shown to differ in physical, kinetic or immunological characteristics from the forms found in the root (phosphoenol pyruvate carboxylase, cholinekinase, xanthine dehydrogenase, purine nucleosidase and malate dehydrogenase). It is not known whether the nodule-specific forms of these enzymes originate from the expression of nodulin genes or result from nodule-specific, post-translational modifications of gene products also found elsewhere in the plant (Nap and Bisseling, 1990).

Late nodulins are also associated with the formation of peribacteroid membrane (PBM) around the bacteria within the plant cytoplasm. One such best-studied PBM nodulin is nodulin 26 (N 26) which is an intrinsic membrane protein. Its amino acid sequence is homologous to the *E. coli* glycerol facilitator protein, which is the only known pore-type protein in *E. coli* cytoplasmic membrane (Sweet et al., 1990) and functions in the transport of small molecules. Expression of N 26 occurs prior to and independent of nitrogen fixation; this protein functions as an ion channel for the translocation of small compounds, such as the carbon source, succinate, across the peribacterioid membrane (Verma, 1992).

Nodulin 24 (N 24), a surface PBM protein, showed homology with nodulin 16 which was isolated from soybean nodules (Nirunsuksiri and Sengupta-Gopalan, 1990). A highly expressed late nodulin gene family of unknown function, the nodulin-A family, was found in soybean (Jacobs et al., 1987). The nodulin-A family comprises at least six members: Ngm

20, Ngm 23, Ngm 26b, Ngm 22/Ngm 27, Ngm 44 and 15-9-A. The members of this gene family show two common characteristics: (a) two domains arranged in paired cys-X7-Cys motifs, resembling zinc-finger sequences, and (b) a putative signal peptide.

In *P. vulgaris*, a group of late nodulin gene (Npu 30) encode a 30 kD protein detected in infected cells. This protein has (a) a putative signal sequence at the deduced amino-terminal region, (b) a proline-rich stretch at the carboxy terminus and (c) a characteristic domain of 4 cysteines that resemble metal-binding sites (Campos et al., 1995). Further work to elucidate the functions of these nodulins and the nature of the signal triggering their induction is required.

3. *Lectin*: Lectins, the sugar-binding plant proteins, are involved in the *Rhizobium*-legume symbiosis. These proteins are localized at two different sites during the symbiotic interaction. First, they are found at the tip of growing root hair (Diaz et al., 1986) suggesting their role in the attachment and/or infection process (Ho and Kijne, 1991; Kijne et al., 1992). Second, they are localized in both infected and peripheral tissues of mature nitrogen-fixing nodules (Law and Van Tonder, 1992; Vanden Bosch et al., 1994) suggesting their role as storage proteins for organic nitrogen. Three lectin genes, Mtlec 1, Mtlec 2 and Mtlec 3, have been identified in *Medicago* species. Mtlec 1 and Mtlec 3 encode functional lectin monomer precursors of 277 and 265 amino acids, respectively. On the other hand, Mtlec 2 appears to a pseudogene with a single two-nucleotide frame shift mutation in the coding region (Bauchrowitz et al., 1992). Transgenic plants of *Medicago* expressing gene fusions between each of the three lectin promoters and GUS reporter gene showed that Mtlec 1 and Mtlec 3 lectin genes are expressing independently of symbiosis in root tissues. Further, Mtlec1 and Mtlec3 are transcribed in dividing cells of the nodule primordia whether elicited by *Rhizobium* or purified nod factors. In contrast, Mtlec2 is not expressed in uninoculated roots or during early stages of the symbiotic interaction. At later stages, all three lectins are expressed in non-infected cells of the central and peripheral tissues in the mature nodule (Bauchrowitz et al., 1996). These findings favour roles for lectins throughout both early and late stages of nodule ontogeny. However, their precise roles during nodulation remain to be elucidated. What is nevertheless clear is that lectin genes can serve as tissue-specific molecular markers related to the nodulation process and on this basis such genes might therefore be included within a broader definition of nodulin genes (Bauchrowitz et al., 1996).

C. Rhizobium Genes Involved in Nitrogen Fixation and their Regulation

The *Rhizobium* genes involved in the fixation of N_2 are known as 'nif' genes and 'fix' genes. *Rhizobium* nif genes are homologous in structure and/or function to nif genes found in *Klebsiella pneumoniae*, as determined by comparing the nif DNA sequences of rhizobia with Klebsiella, whereas 'fix' genes are additional nitrogen genes that have been found only in rhizobia. In *R. meliloti* the 'nif' genes are present on a symbiotic megaplasmid and some other genes required for symbiotic nitrogen fixation may be located on the chromosome and second megaplasmid, unlike in *Bradyrhizobium*, where both 'nif' and 'fix' genes are present on the chromosome. The organisation of nif genes in *R. meliloti* and *B. japonicum* is shown in figure 3.3.

In *B. japonicum* nif genes are clustered on two unlinked sites on the chromosome in the region that does not contain genes essential for growth (Kaluza et al., 1995). The nif H and nif DK in cluster I are

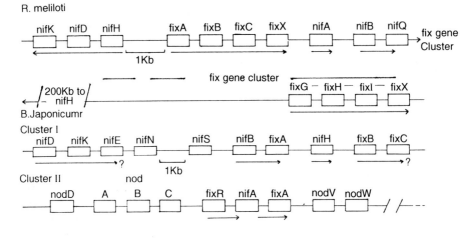

Fig. 3.3. Organization of genes involved in symbiotic nitrogen fixation in *Rhizobium meliloti* and *Bradyrhizobium japonicum*. The arrows indicate the direction of gene transcription, while the broken vertical lines in *R. meliloti* designate large interruption in the map (adopted from Sindhu and Dadarwal, 1995).

separated by 17 kb of intervening DNA, unlike in R. meliloti in which the nif HDK region is one operon. The other nif genes present in cluster I are nif E, nif N, nif S and nif B. The nif E gene is located immediately downstream from the nif K gene, nif S is located approx. 6 kb 3' to nif N and nif B is found about 1 kb downstream from it (Ebeling et al., 1987). Mutation in nif E, nif N and nif B genes resulted in nif-, fix-phenotypes.

In R. meliloti nif A is present approx. 5 kb upstream of the nif HDK gene cluster, while in B. japonicum the nif A gene is located in the cluster II region that contains the fix R nif A operon (Hennecke et al., 1987). In slow-growing Rhizobia (B. japonicum etc.) the number of copies of the nif H and nif DK appears to be one per genome while in fast-growing rhizobia (R. meliloti, etc.), some nif genes are reiterated. At least four copies of nif H appear to exist in R. phaseoli on two different plasmids. Whether these additional genes are also expressed in the cell is presently not known but their presence should be of significance for those who wish to construct strains with additional nif genes.

R. meliloti: Another cluster of genes required for nitrogen fixation is present in R. meliloti between the nif HDK and nif AB gene clusters and is called the fix ABC (Puhler et al., 1984). In B. japonicum the fix B and fix C genes are located adjacent to each other approximately 2.5 kb downstream from the nif H gene on symbiotic gene cluster I. The fix A is located apart from fix BC genes in cluster II. Transcription of both the operons fix BC and fix A is mainly dependent on the presence of nif A and low concentrations of O_2. The fix ABC genes have functions that involve the transport of electrons to nitrogenase and are unique to organisms that fix N_2 in the presence of O_2.

A small open reading frame (ORF) which encodes a polypeptide of 74 amino acid and shows homology with the ferrodoxin gene, is designated 'frx A'; it is present immediately downstream of nif B. Frx A is cotranscribed with nif B from the nif B promoter. Another gene, fix X, whose product contains a cluster of cysteine characteristic of ferrodoxins, is found immediately downstream of the fix C gene; it is probably involved in the transfer of electron to nitrogenase.

A cluster of four fix genes, namely fix G, H, I and X, lies in a single transcribed region of 12 kb; these encode membrane integrated proteins that may be involved in a redox process which is specific to symbiotic nitrogen fixations (Kahn et al., 1987). The products of most of the nitrogen-fixing genes are known but the nature of the role of some of these products is yet to be ascertained.

VI. BIOCHEMISTRY OF NITROGEN FIXATION

The site of N_2-fixation in the symbiotic root nodules is the bacterioid, which has all the necessary enzymes for the reduction of dinitrogen (N≡N) to ammonia (see Robertson and Farnden, 1980). This process requires the enzyme nitrogenase which catalyzes the reaction of the biological nitrogen fixation (BNF) and energy obtained by the carbohydrate metabolism in the plants, in the form of a reductant, e.g. Fdred and ATP.

Bacterioid nitrogenase has been studied less extensively than the enzyme from the free-living prokaryotes, e.g. *Klebsiella pneumoniae*. However, it has been indicated that the rhizobial nitrogenase closely resembles the enzyme from other N_2-fixing bacteria (see Yates, 1980 for review). It has been shown that the enzyme from *Rhizobium japonicum* and *R. lupini* bacterioids contains the two proteins characteristics of almost all nitrogenase—an Mo-Fe protein (MW 200,000) and an Fe protein (MW 50,000–65,000) containing acid-labile sulphur at their active sites (ses Yates, 1980; Robertson and Farnden, 1980; Shah et al., 1984; Sindhu and Dadarwal, 1995). The structure and reactivity of nitrogenase has been reviewed by various workers (Burris, 1984; Smith, 1990; Shah et al., 1990). Both the Fe-protein (dinitrogenase reductase; component II) and the MoFe protein (dinitrogenase; component I) are necessary for the nitrogenase activity and both are irreversibly inactivated in the presence of oxygen. The basic reaction involves an MgATP activated electron transfer from the Fe-protein to the Mo-Fe protein followed by substrate reduction on the Mo-Fe protein. Two moles of ATP are required for every electron transfer in which ATP is hydrolyzed to ADP and Pi during the protein-protein electron transfer. For every mole of nitrogen reduced, a pair of electrons are eliminated in proton reduction to H_2 through an ATP-dependent reaction. The current stoichiometry for nitrogen fixation is:

$$N_2 + 8H^+ + 8e^- + 16ATP \rightarrow 2NH_3 + H_2 + 16ADP + 16Pi$$

The Mo-Fe protein of the nitrogenase enzyme exists as a tetramer of α_2 and β_2 subunits with a total molecular weight of about 180 (*Rhizobia*) to 220 kD (*Klebsiella* and *Clostridium*), while the Fe protein exists as a homodimer of two identical subunits (d_2) with molecular weight of 32 kD each (see Sindhu and Dadarwal, 1995). The Mo-Fe protein is known to contain redox centres of two distinct types: two iron molybdenum cofactors called FeMoco or M centre and four Fe-S (4Fe-4S) centres called 'P'clusters. The Fe protein containing two identical subunits with one 4 Fe-4S centre, donates low potential e^- to the Mo-Fe protein. For each e⁻ transferred to the MoFe protein, the Fe protein must bind and hydrolyze 2 MgATP.

The iron-molybdenum cofactor (Fe-Moco) is composed of Fe, Mo, S and was successfully isolated from the Mo-Fe protein in 1977 (Shah and Brill, 1977; Shah et al., 1984; Burgess, 1990). It has been suggested to be the actual site of nitrogen reduction on the enzyme (Hawkes et al., 1984; Burgess, 1990; Sindhu and Dadarwal, 1995).

Miller et al. (1993) demonstrated that hydrolysis of MgATP to MgADP is essential for nitrogenase action in *Klebsiella pneumoniae*. There is good evidence for binding both nucleotides to the Fe protein of nitrogenase, but data indicating their binding to the MoFe protein has been controversial. It has been proposed that the ADP binding sites are transiently filled during the enzyme turnover by the hydrolysis of ATP originally bound to the Fe protein, and that hydrolysis occurs on a bridging site on the MoFe-Fe protein complex.

The chemical nature of leghaemoglobin (Lb) and conditions inside nodules, such as slightly acid pH and the presence of metal ions, chelators and toxic metabolites, e.g. nitrite, superoxide radical, peroxides etc., are conducive for oxidation of ferrous Lb (Lb^{2+}) to nonfunctional ferric Lb (Lb^{3+}) and ferryl Lb (see Becana and Klucas, 1992 for review).

Lb^{2+} is a hemoprotein of about 16 kD. Though it is an essential component of nodule N_2-fixation in the legumes, its role seems to be indirect as the isolated bacterioids fix N_2 even in the absence of Lb (Becana and Klucas, 1992). Bacterioid microaerobes require O_2 for respiration and N_2 fixation but simultaneously nitrogenase is to be protected from O_2. Thus O_2 supply to the central zone of nodules, Lb facilitates diffusion of O_2 from the plasmalemma of infected cells to the peribacterioid membrane and the free O_2 then diffuses through the peribacterioid space, which lacks Lb, to ultimately reach to the efficient terminal oxidases of the bacterioids. The enzyme ferric leghaemoglobin reductase has been shown to play a physiological role in maintaining Lb in the functional ferrous state in the nodules (Ji et al., 1991; Becana and Klucas, 1992).

VII. MOLECULAR BIOLOGY OF REGULATION OF NITROGEN FIXATION

The regulation of nif and fix genes in *Rhizobium* and *Bradyrhizobium* is more or less similar to the nif gene regulation of *K. pneumoniae* except for two important differences. In *Klebsiella*, nif gene regulation is at two levels: the first involves the centralized nitrogen regulation system involving the ntr system mediated by the products of ntr A, ntr B and ntr

C genes and the second involves the specific control of operons (except nif LA operon) by nif A and nif L gene products. Under conditions of nitrogen starvation, the products of nif A, ntr A and ntr C interact to activate the nif LA promoter. The nif A product in turn activates (positively regulates) the transcription of all other nif transcriptional units.

Nif L and nif A activation requires the ntr A gene product. The nif L product effects a negative control on nif regulation in the presence of ammonia, high oxygen and amino acids.

The exact mechanism of regulation of nif by the ntr system is not fully understood.

The nif gene expression in *Rhizobia* and *Klebsiella* is almost similar due to the similarity in properties of *Rhizobial* nitrogenase to those of *Klebsiella*. In *Rhizobium* and *Bradyrhizobium* activation of nif genes is dependent on the nif A gene product but nif A is not activated by ntr C and there is no counterpart of nif L (oxygen-sensing repressor) unlike in *Klebsiella* (Beynon et al., 1988).

Nif and fix genes are under the control of a concatenating regulatory system controlled by the fix LJ proteins. These two proteins are homologous to the prokaryotic regulatory system known as the 'sensor-kinase/response-regulator elements' which control many developmental processes in bacteria. The sensor protein responds to an environmental stimulus and then phosphorylates a second protein, the response-regulator, thereby modulating the biological activity of the regulator (Downie and Brewin, 1992).

In the fix LJ system the principal effector appears to be the low concentration of oxygen which is sensed by haemoprotein, the transmembrane sensor-kinase fix L. This protein then phosphorylates the response-regulatory protein fix J, such that it now activates the transcription of fix K and nif A. The nif A gene product then activates transcription of several nif and fix operons via an interaction with an upstream activator site about 100 nucleotides upstream of the transcription start site. The nif A protein appears to interact with a specialized RNA polymerase sigma factor encoded by ntr A to stimulate the transcription of nif genes. In parallel fix K protein can also activate transcription of another fix operon, fix NOQP, while it negatively regulates the expression of nif specific activator nif A as well as its own expression by autoregulation (figure 3.4). This complex control system ensures that the bacteria do not make the commitment to the process of nitrogen fixation unless the appropriate conditions prevail and such an environment is found within the nodule (Downie and Brewin, 1992).

The nitrogenase reducing dinitrogen to ammonia is functional in developed nodules only. Thus enzyme biosynthesis is regulated by its

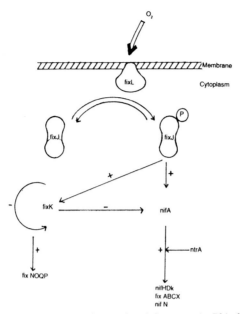

Fig. 3.4. Model for regulation of the nif and fix genes in *Rhizobium meliloti*.

own and host plant genes. Therefore, the environmental factors which affect plant and nodular growth would affect the nitrogenase activity and dinitrogen fixation also.

VIII. FACTORS AFFECTING NITROGEN FIXATION

A. Effect of Inorganic Nitrogen

As mentioned, nodule development is inhibited by exogenous nitrogen nutrition (Carrol and Mathews, 1990; Singh et al., 1994). Studies have shown that the supply of nitrogen inhibits nitrogenase activity in leguminous and actinorhizal nodules to various degrees, which is dependent on the form of nitrogen, its concentration, duration of supply and plant age (Cote and Dawson, 1989; Mackay et al., 1987; Streeter, 1985). It is assumed that the change in nitrogenase activity would be parallel to the change in respiration because the enzyme is highly energy dependent for its catalytic role. However, respiration increased while nitrogenase activity declined in pea under ammonium nitrate nutrition (Mahon, 1977a). On the other hand, it was found that little nitrate and nitrite reduction occurred in root or nodules and respiration per unit of acetylene reduction was not affected by either form of nitrogen added

(Mahon, 1977b). A decrease in nitrogenase activity has been found in many cases with nitrate supply also (Munns, 1977; Franco et al., 1979; Srivastava and Ormrod, 1986; Streeter, 1988; Carrol and Mathews, 1990). The lack of significant difference in nitrogenase activity in *Robinia pseudoacacia* (Roberts et al., 1983) and *Alnus glutinosa* (Vogel and Dawson, 1991) in the presence of either nitrate or combined nitrogen has also been noticed. Variable effects due to difference in nitrogen treatment time have been observed in other cases. Gallusci et al. (1991) (referenced in De-Billy et al., 1991) showed that 5 mM KNO_3 dropped the specific acetylene reduction activity by 20–30% after 24 h and 70–80% after 3–4 days in *Medicago truncata* nodules with concomitant decrease in leghaemoglobin mRNA; increase in treatment time did not affect either further. Various hypotheses have been put forward regarding the inhibition of nitrogenase activity by nitrate supply. Prominent among these is the toxicity due to the accumulation of nitrite in nodules (Kennedy et al., 1975; Rigaud and Puppo, 1977; Pagan et al., 1977; Trinchant and Rigaud, 1980; Manhart and Wong, 1980; Streeter, 1982). However, Gibson and Pagan (1977), and Manhart and Wong (1980) demonstrated that NO_2^- produced due to NRA may not play a role in inhibition in acetylene reduction. These studies of nitrogen effects were done with whole plants. Therefore it was visualized that a clear picture of NO_3^- effects might not be obtained because of complicated limitations imposed by the process of uptake, transport, storage, compartmentation and reduction of NO_3^-. To overcome these problems, experiments were performed with excised nodules having mutant NR in strains of *R. japonicum* (NR^- 108 and 303) (Vasconcelos et al., 1980) and USDA 110 (NR^+) mutant (Stephens and Neyra, 1983). The addition of either KNO_3 or KNO_2 to detached nodules in the wild type caused inhibition of nitrogenase activity up to 50 and 65% respectively compared to control (minus nitrogen treatment). The minus nitrate reductase strains (NR^- mutant) were insensitive to either nitrate or nitrite supply. Thus they provided direct evidence for localized effect of nitrate or nitrite at the nodular level. Further treatment of nodules with 100 µM sucrose stimulated nitrogenase activity probably by decreasing the level of nitrate or nitrite.

The availability of photosynthates has been considered to be another factor in reduction of nitrogen fixation in nitrogen supplied plants. Bean plants with NO_3^- had higher photosynthates, although their distribution pattern did not differ compared to minus nitrogen plants (Marques et al., 1983). Though, a lower CO_2-fixation rate in NH_4-fed plants was noticed, there was no limitation of C-skeleton for the formation of organic acids even in these plants (Larsen et al., 1981). McClure et al. (1983) found that acetylene and carbon dioxide reduction is considerably inhibited on feeding 20 mM NO_3 to *A. glutinosa*. De-Jong and Phillips (1981) showed that N_2-fixation interactions with leaf photosynthetic efficiency and plant growth in a manner dependent on the allocation of symbiotically fixed nitrogen.

It was recently demonstrated that aminoethoxy vinylglycine (ethylene inhibitor) increases nitrogenase activity in nitrate-fed plants (Ligero et al., 1991). It is implicated here that nitrate inhibition of nitrogenase might be through production of ethylene, a known inhibitor of plant growth. However, it would be very relevant to ascertain the level of auxins and other hormones in the presence of nitrate. Nitrate induction of auxin in non-leguminous plants in known (Mishra et al., 1994; Singh, 1996).

Ammonium toxicity at above a critical concentration is known to be an inhibitor for various plant processes. Ammonium suppression of nitrogenase gene expression is known in free-living bacteria (*Klebsiella pneumoniae* (Kaluza and Henneck, 1981). Other nutrients may also play role in modification of nitrogenase activity. Israel (1987) suggested that nitrogenase activity determines the level of phosphorus in legumes. A sufficient supply of phosphorus stimulates N_2-fixation specifically (Bethlenfalvay and Yoder, 1981; Jacobsen, 1985; Israel, 1987). It was also observed that specific activity of nitrogenase decreased in soybean nodules under P-deficient conditions (Sa and Israel, 1991). But these authors found no significant decrease in ATP and energy charge in bacteroids, although these two decreased in the plant fraction of the nodule under phosphorus-deficient conditions. Induction of nitrogenase activity by a few kreb cycle metabolites, such as succinate or malate (Bergersen and Turner, 1975) and oxalate (Trinchant et al., 1994) has also been demonstrated. These compounds probably provide a C-skeleton to support the N_2-fixation (Day and Copeland 1991; Streeter, 1991).

B. Nitrogen-fixation Under Stress

Retardation in nitrogen fixation as a reduced function of nitrogenase under water stress; (Sprent, 1981; Aparicio-Tejo and Sancheg-Diaz, 1982; Sheoran et al., 1988; Kaur et al., 1985), salinity (Garg et al., 1984; Lauchli, 1984), metal and gaseous pollutant stresses (Huang et al., 1974; Chugh et al., 1991; Dabas et al., 1995; Srivastava and Ormrod, 1986) has been observed. While water stress might have a generalized effect on nodule development and function, a few studies have shown it to have a direct effect on nitrogenase activity (Pankhurst and Sprent, 1975; Tu and Hietkamp, 1977; Chorpa et al., 1984; Swaraj et al., 1986; Sheron et al., 1988). Nitrogenase activity decreased with increase in water stress with concomitant decrease in respiration also (Pankhurst and Sprent, 1975). Streeter (1993) proposed that availability of water for export of nitrogenous products from legume nodules would be a major factor limiting the efficiency of symbiotic N_2-fixation. It is understood that water for export of solutes in the xylem probably depends largely on the import of water and reduced carbon in the phloem; one function of respiration might be to dispose of reduced carbon in order to increase the supply of water. Secondly, control of gas diffusion in soybean nodules is largely restricted to the cortex nearby the vascular bundles, thus making possible the linkage of solute balances in the xylem and phloem with resistance to diffusion. Streeter (1993) further suggested that solute concentration in bundles could influence gas diffusion by controlling the movement of water in and out of the intercellular spaces near bundles in the inner cortex. This could be affected by loading and unloading of the xylem and phloem, which is dependent on water.

It has been suggested that internal O_2 may regulate nitrogenase physiologically as well as genetically (Hennecke et al., 1990; Hunt and Layzell, 1991). It is interesting to note that nitrogenase regulation by internal oxygen is the same in pea, lupin and soybean (Castillo et al., 1992), albeit the nodules in each differ in morphology, physiology and biochemistry. For example, soybean nodules are determinate type nodules and N≡N fixed in the ammonium gets converted to ureids before being exported to the xylem. Pea nodules on the other hand are, indeterminate type and lupin produces a collar-type nodule; both export fixed nitrogen to the aerial parts as asparagine. Structural studies have indicated that the nodular cortex regulates O_2 transport by controlling the degree of occlusions of intercellular spaces with glycoproteins (James et al., 1991; Iannetta et al., 1993a, b; Lorenzo et al., 1993). Alternatively, it has been suggested that the cortical barrier exerts osmotic control of water in intercellular spaces (Hunt and Layzell, 1993). Expansion of the inner cortex has been observed in stressed nodules of soybean (Iannetta et al., 1993a, b) and lupin (Lorenzo et al., 1993). Osmotic control of water movement may be involved in any increase in glycoprotein occlusions in intercellular spaces (Witty and Minchin, 1990). Webb and Sheehy (1991) suggested that glycoproteins, which may have hydrophobic surfaces, could hold water in intercellular spaces.

The water holding capacity of glycoproteins may itself be governed by its association with ions and inorganic/organic compounds (Witty and Minchin, 1990). X-ray microanalysis has shown the presence of Mg^{++}, S, Ca^{++} and K^+ in a distinct ring in glycoproteins. There may be differences in these ions depending on species type and stress. This might possibly be one reason for variable effects of drought or salinity or pollutant stresses on N_2-fixation. Salinity-induced inhibition of nitrogen fixation has been reported in quite a few species, including mung bean (Balasubramanian and Sinha, 1976a, b). In some cases, it might be due to reduced water availability but in others might be the effect of specific ions.

Temperature stress also causes reduction in nitrogenase activity. High temperature (35°C to 44°C) caused a 30–50% reduction in nitrogen fixation in *Vigna radiata* cv PS 16 and Pusa Biasakhi (Panwar et al., 1988). Chilling temperature (–1 to 4°C) also decreased nitrogenase activity in *Alnus glutinosa* (Vogel and Dawson, 1991), *Alnus incana* (Huss-Denell et al., 1987) and in black alder (Cote and Dawson, 1989). Complete inhibition in nitrogenase may occur at nodular temperatures of 5 to 8°C; the reasons assigned to this inhibition are loss of enzyme activity and resistance to gas diffusion (Huss-Denell et al., 1987; Vogel and Dawson, 1991). The free radicals responsible for protein hydrolysis are also generated under stresses (Kennis et al., 1994). Possibly free radicals may inactivate nitrogenase, leghaemoglobin and change the membrane permeability of root and nodular cells, which in turn disturb the O_2 and N_2 diffusion leading to reduced N_2-fixation. This mechanism needs to be explored in further studies.

C. Phytohormones

Growth promoter auxins, gibberellic acids and cytokinins induce N_2-fixation. The effect on N_2-fixation of these phytohormones might be attributed to promotion in general growth and development of the host as well as of the nodule by inducing cell division and differentiation (Zeroni and Hall, 1980), in turn enhancing nitrogenase activity. The interdependency of N_2-fixation and photosynthesis is known (De-Jong and Phillips, 1981). Phytohormones might be mediating its effects on N_2-fixation through either synthesis and/or translocation of photosynthates or exchange of metabolites between host and microsymbionts. Cytokinin induction of nitrogenase has been noticed in many cases (Nandwal and Bharti, 1982). Bishnoi and Krishnamoorty (1990) recently demonstrated GA_3 (10μM) induction in nitrogenase in groundnut. Salicylic acid induction of N_2-fixation has been reported (Garg et al., 1989). Growth inhibitors ABA (Phillips, 1971) and ethylene (Goodlass and Smith, 1979; Drennan and Nortton, 1972) are considered repressors of nitrogenase. Ligero et al. (1991) suggested that ethylene inhibition may be mediated through change in nodulation. Ethylene checks the auxin and cytokinin effects on nodulation (Apelbaum et al., 1974). Phytohormonal interaction with the process of senescence in nodule/host may also change nitrogenase activity. Moreover, phytohormonal interaction with nitrogenase might well be elucidated by use of ^{15}N to precisely ascertain N_2-fixation in legumes.

IX. GENETIC ENGINEERING FOR INCREASING NODULATION AND NITROGEN FIXATION

There has been great interest in recent years, to increase nodulation and nitrogen fixation within symbiotic relations and to extend this ability in plants outside the symbiotic relationship in important crop plants such as rice, wheat, oil-seed rape, etc.

Attempts to improve nitrogen fixation and legume productivity by inoculation with superior genetically manipulated rhizobia have often failed under field conditions because indigenous strains occupied the root nodules rather than the inoculants. Therefore, it is essential that constructed superior strains be able to compete successfully for establishment in nodules with wild strains. For example, R. leguminosarum bv trifoli strain T24 which produces the antirhizobial bacteriocin, trifolitoxin, a small peptide, was found to be highly competitive for nodulation against a bacteriocin-sensitive strain in a co-inoculation experiment (Triplett and Barta, 1987). The recombinant strain developed by introduction of genes for trifolitoxin production (tfx) (Triplett, 1988) or competitiveness (PF2) (Dowling et al., 1987, 1989) showed increased nodulation competitiveness with the wild type strain.

Non-motile spontaneous mutants of R. meliloti (Caetano-Anolles et al., 1988), Tn5 mutants of R. leguminosarum bv phaseoli, R. tropica and B. japonicum deficient in exopolysaccharide (EPS) production and nif A mutant of R. meliloti were found to be less competitive for nodule occupancy than the corresponding wild strains (Bhagwat et al., 1991).

Addition of multiple copies of nif A from Klebsiella pneumoniae to R. meliloti strains increased nodulation and competitiveness of R. meliloti strains (Sanjuan and Olivares, 1991). Thus strains with an additional copy of nif A may have the ability to occupy nodules to the exclusion of indigenous rhizobia.

Application of nitrogen fertilizers of nitrate and ammonia can inhibit legume nodulation and nitrogen fixation through multiple and complex effects (see reviews, Streeter, 1988; Carroll and Mathews, 1990). The mutants of B. japonicum selected as resistant to chlorate and that of Azotobacter chrococcum resistant to azide showed higher nitrogenase activity in the presence of nitrate and ammonium ions respectively (Hua et al., 1981; Gordon and Brill, 1972). Similarly, plant mutants formerly resistant to nitrate have been developed with the ability to form a large number of nodules and to increase nitrogen fixation in the presence of nitrate (Hansen et al., 1989). Two such independent nitrate-tolerant symbiotic(nts) mutants were characterized using complementation analysis to define a single supernodulation locus (Delves et al., 1988).

On the other hand, strains with constitutive or derepressed nif genes have been developed to improve soil fertility. These strains are able to excrete large amounts of fixed nitrogen as NH_4^+ (up to 20.2 μ mole of NH_4^+ are produced per mg^{-1} of cell protein 24 h^{-1}). Strains with a derepressed system continue to synthesize nitrogenase, in the presence of NH_4^+, in amounts up to 65% of that produced in the absence of NH_4^+.

Several environmental factors, such as pH, temperature or moisture, affect nodulation and nitrogen fixation. In acid soil a low number of very poor nitrogen-fixing bacteria are present (Holding and King, 1963). Further, the seedlings release flavonoids in a low amount at low pH. Therefore, modification of the host to produce higher levels of flavonoids and/or modification of nod genes to be readily induced at low pH, would overcome nodulation problems in acid soils (Richardson et al., 1988).

One of the most promising approaches for improving symbiotic nitrogen fixation is to increase the number of nitrogen-fixing genes in the microsymbiont with concomitant increase in number of active nitrogenase molecules. Multiple copies and overexpression of nif A gene, which positively regulate other nif genes, may result in increased nitrogen fixation (Cannon et al., 1988; Ronson et al., 1990). Similarly, an increase in copy number and expression of nif H gene whose product dinitrogenase reductase (reduced) binds to dinitrogenase, followed by one electron transfer, could result in increased turnover rate of nitrogenase. An excess of the nif H gene product is required for maximal nitrogen fixation activity in *K. pneumoniae* (Thorneley and Lowe, 1984). This may be the basis of more than one copy of the nif H gene in some diazotrophs, such as *Azotobacter vinelandi* (Jacobson et al., 1986), *Rhizobium phaseoli* (Quinto et al., 1985) and *Azorhizobium sesbaniae* (Norel and Elmerich, 1987).

The supply of photosynthates to bacterioids is thought to limit the rate of nitrogen fixation in *Rhizobium*-legume symbiosis; therefore, increasing the ability of the endosymbiont to effectively utilize photosynthates in nodules might lead to increased nitrogen fixation rates (Ryle et al., 1979). Transfer of the dct gene (for dicarboxylate transport) from *R. meliloti* to *B. japonicum* (Birkenhead et al., 1988) and its high expression in alfalfa and soybean (Ronson et al., 1990) resulted in increased uptake of decarboxylic acids and a corresponding increase in levels of nitrogen fixation.

In many legumes, an increase in C_2H_2 reduction has been observed in nodules/bacterioids formed by inoculation with HUP + rhizobium strains. Thus, yield can be improved by increasing the activity of hydrogenase uptake in a strain that already possesses it. This has been achieved by mutagenizing a HUP + *B. japonicum* strain and selecting for growth on higher than normal levels of oxygen. Similarly mutants of cowpea miscellany *Rhizobium* strains S25 and GR4 were isolated for increased hydrogenase activity; these mutants showed increase in dry matter yield when inoculated onto green gram and black gram (Sindhu and Dadarwal, 1992, 1995).

To achieve induction of nodules in non-legumes, researchers in a number of labs have attempted to transfer the set of about 20 genes responsible for nitrogen fixation from *Rhizobium* or other nitrogen-fixing bacteria directly into plants.

One serious problem is the necessity to provide protection to the oxygen-labile nitrogenase enzyme from oxygen in an aerobic organism. Moreover, nitrogen fixation consumes vast amounts of energy which most non-legumes cannot provide. Cocking and coworkers (1990) achieved the formation of nodules on roots of rice and oil-seed rape by exposing the roots to rhizobia in the presence of polyethylene glycol after treating them with an enzyme that breaks down the cellulose walls surrounding the root cells. This method suffers from a disadvantage in that the seedlings would have to be treated every year before planting.

Jing et al., (1990) developed a bacterial strain that nodulated rice by exposing *Rhizobium sesbania*, which normally infects an Asian bush called *Sesbania cannabina*, to a mutagenic chemical. Two Australians Rolfe and Chan reported small nodules on rice and wheat respectively (see Moffat, 1990). Tohan and Kennedy (1989) induced paranodules on wheat following treatment with the herbicide, 2,4-D or with auxins, IAA and NAA and by inoculation with *Rhizobia* or *Azospirillum*. Despite such promising results with nodulation in non-legumes, no one has yet documented actual nitrogen fixation in these plants.

An understanding of the various physiological and genetic processes of *Rhizobium*-legume symbiosis and the essential characters present in legumes is required to develop the symbiotic capacity in agriculturally important non-legumes.

X. CONCLUSION AND FUTURE PROSPECTS

In the last decades, given advancement in molecular techniques, significant progress has been made in understanding the signal transduction between legume plants and *Rhizobium*, their cell-surface interaction, induction of *Rhizobium* and host genes during their symbiotic association. The molecular biology of the bacterial partner had, until recently, progressed faster than that of the higher plant. But now, due to the development of advance molecular techniques, genes concerned with symbiosis have been identified, cloned, sequenced and transferred to regenerable plant tissues to probe their tissue-specific expression, which in turn analyse the sequence of signals and responses which give rise to nodule. However, many aspects of *Rhizobium*-legume symbiosis have still to be investigated. A better understanding of the molecular mechanism of nod gene action, perception and responses of nod factors, role of several nodule-specific proteins/transcripts and factors switching 'nif' and 'fix' genes on or off would not only help in enhancing nodulation and nitrogen fixation, but also enable extension of the symbiosis to non-leguminous plants.

LITERATURE CITED

Aparicio-Tejo, P.M. and Sanchez-Diaz, M.F. 1982. Nodule and leaf nitrate reductases and nitrogen fixation in *Medicago sativa* L. under water stress. *Plant Physiol.* 69: 479–482.

Apelbaum, A., Sfakiotakis, E. and Dilley, D.R. 1974. Reduction in extractable deoxyribonucleic acid polymerase activity in *Pisum sativum* seedlings by ethylene. *Plant Physiol.* 54: 125–128.

Appleby, C.A. 1984. Leghemoglobin and *Rhizobium* respiration. *Ann. Rev. Plant Physiol.* 35: 443–478.

Ardourel, M., Demont, N., Debelte, F., Maillet, F., De Billy, F., Prome, J.C., Denarie, J. and Truchet, G. 1994. *Rhizobium meliloti* lipooligosaccharide nodulation factors: different structural requirements for bacterial entry into target root hair cells and induction of plant symbiotic developmental responses. Plant Cell 6: 1357–1374.

Badenoch-Jones, J., Parker, C.W. and Letham, D.S. 1987. Phytohormones, *Rhizobium* mutants and nodulation in legumes. VII. Identification and quantification of cytokinins in effective and ineffective pea root nodules using radio-immunoassay. *J. Plant Growth Regul.* 6: 97–111.

Badenoch-Jones, J., Summons, R.E., Rolfe, B.G. and Letham, D.S. 1984. Phytohormones; *Rhizobium* mutants and nodulation in legumes. IV. Auxin metabolites in pea root nodules. *J. Plant Growth Regul.* 3: 23–29.

Baev, N., Endre, G., Petrovies G., Banfalvi, Z. and Kondorosi, A. 1991. Six nodulation genes of nod box locus-4 in *Rhizobium meliloti* are involved in nodulation signal production-nod M codes for D-glucosamine synthetase. *Mol. Gen. Genet.* 228: 113–124.

Balasubramanian, V. and Sinha, S.K. 1976a. Nodulation and nitrogen fixation in chick pea (*Cicer aeritinum* L.) under salt stress. *J. Agric. Sci.* 87: 465–466.

Balasubramanian, V. and Sinha, S.K. 1976b. Effect of salt stress on nodulation and nitrogen fixation in cowpea and mungbean. *Plant Physiol.* 36: 197–200.

Barbour, W.M., Wang, S.P. and Stacey, G. 1992. Molecular genetics of *Bradyrhizobium* symbiosis. In: *Biological Nitrogen Fixation* (G. Stracey, H.J. Evans and R.H. Burris, eds.). Chapman and Hall, New York, pp. 645–681.

Barondes, S.H. 1981. Lectins: their multiple endogenous cellular functions. *Ann. Rev. Biochem.* 50: 207–231.

Bauchrowitz, M.A., Barker, D.G. and Truchet, G. 1996. Lectin genes are expressed throughout root nodule development and nitrogen fixation in *Rhizobium-Medicago* symbiosis. *Plant J.* 9: 31–43.

Bauchrowitz, M.A., Barker, D.G., Nadaud, I., Rouge, P. and Lescure, B. 1992. Lectin genes from the legume *Medicago truncatula*. *Plant Mol. Biol.* 19: 1011–1017.

Becana, M. and Klucas, R. 1992. Oxidation and reduction of leghaemoglobin in root nodules of leguminous plants. *Plant Physiol.* 98: 1217–1221.
Becana, M., Aparicio-Tejo, P.M. and Sanchez-Diaz, M. 1985. Nitrate and nitrite reduction by alfalfa root nodules: Accumulation of nitrite in *Rhizobium meliloti* bacterioids and senescence of nodules. *Physiol. Plant.* 64: 353–358.
Becana, M., Aparicio-Tejo, P.M. and Sanchez-Diaz, M. 1988. Nitrate and hydrogen peroxide metabolism in *Medicago sativa* nodules and possible effect on leghaemoglobin function. *Physiol. Plant.* 72: 755–761.
Bergersen, F.J. and Turner, G.L. 1975. Leghaemoglobin and supply of O_2 to nitrogen fixing root nodule bacterioids, presence of two oxidase system and ATP production at low free O_2 concentration. *J. Gen. Microbiol.* 91: 343–354.
Bethlenfalvay, G.J. and Yoder, J.F. 1981. The *Glycine Glomus-Rhizobium* symbiosis. I. Phosphorous effects on nitrogen fixation and mycorrhizal infection. *Physiol. Plant* 52: 141–145.
Beynon, J.L., Williams, M.K. and Cannon, F.C. 1988. Expression and functional analysis of the *Rhizobium melioti* nif A gene *EMBO J.* 7: 7–14.
Bhagwat, A.A., Tully, R.E. and Keister, D.L. 1991. Isolation and characterization of a competition-defective *Bradyrhizobium japonicum* mutant. *Appl. Environ. Microbiol.* 57: 3496–3501.
Bhattacharya, R.N. and Basu, P.S. 1995. Phytohormones in root nodules of leguminous climber (*Psophocarpus tetragonolobus* DC) and its rhizobial symbiont. *Indian J. Plant Physiol.* 38: 18–21.
Bhuvaneswari, T.V., Bhagwat, A. and Bauer, W.D. 1981. Transient susceptibility of root cells in four common legumes to nodulation by rhizobia. *Plant Physiol.* 68: 1144–1149.
Birkenhead, K., Manian, S.S. and O'Gara, F. 1988. Dicarboxylic acid transport in *Bradyrhizobium japonicum*: Use of *Rhizobium meliloti* dct gene(s) to enhance nitrogen fixation. *J. Bacteriol.* 170: 184–189.
Bishnoi, N.R. and Krishnamoorty, H.N. 1990. Effect of water logging and gibberellic acid on nodulation and nitrogen fixation on peanut. *Plant Physiol. Biochem.* 28: 663–666.
Blevins, D.G., Barnett, N.M. and Bottino, P.J. 1977. The effect of Ca^{++} and ionophoric A23187 on nodulation, nitrogen fixation and growth of soybean. *Physiol. Plant.* 41: 235–238.
Bohlool, B.B. and Schmidt, E.L. 1974. Lectins: A possible basis for specificity in the *Rhizobium*-legume root nodule symbiosis. *Science.* 185: 269–271.
Burgess, B.K. 1990. The iron-molybdenum cofactor of nitrogenase. *Chem. Rev.* 90: 1377–1406.
Burris, R.H. 1984. Enzymology of nitrogenase. In: Advances in *Nitrogen Fixation Research* (C. Veeger and W.E. Newton eds.). Martinus Nijhoff, The Hague, pp. 143–144.
Caetano-Anolles, G. and Gresshoff, P.M. 1990. Early induction of feedback regulatory responses governing nodulation in soybean. *Plant Sci.* 71: 69–81.
Caetano-Anolles, G. and Gresshoff, P.M. 1991. Plant genetic control of nodulation. *Ann. Rev. Microbiol.* 45: 345–382.
Caetano-Anolles, G., Wall, L.G., De Micheli, A.T., Macchi, E.M., Bauer, W.D. and Favelukes, G. 1988. Role of motility and chemotaxis in efficiency of nodulation by *Rhizobium meliloti. Plant Physiol.* 86: 1228–1235.
Calvert, H.E., Pence, M.K., Pierce, M., Malik, N.S.A. and Bauer, W.D. 1984. Anatomical analysis of the development and distribution of *Rhizobium* infection in soybean roots. *Can. J. Bot.* 62: 2375–2384.
Campos, F., Carsolio, C., Kuin, H., Bisseling, T., Rocha-Sosa, M. and Sanchez, F. 1995. Characterization and gene expression of nodulin (Npv 30) from common bean. *Plant Physiol.* 109: 363–370.
Cannon, F.C., Beynon, J., Hankinson, T., Kwiatkowski, R. and Legock, R.P. 1988. Increasing biological nitrogen fixation by genetic manipulation. In: *Nitrogen Fixation:*

Hundred Years After (H. Bothe, F.J. de Bruijn and W.E. Newton, eds.). Gustan Fischer, Stuttgart, pp. 735-740.

Carlson, R.W., Price, N.P.J. and Stacey, G. 1994. The biosynthesis of rhizobial lipo-oligosaccharide nodulation signal molecules. *Mol. Plant Microbe Interact.* 7: 684-695.

Carroll, B.J. and Mathews, A. 1990. Nitrate inhibition of nodulation in legumes. In: *Molecular Biology of Symbiotic Nitrogen Fixation* (P.M. Gresshoff, ed.). CRC Press, Boca Raton, Fla., pp. 159-180.

Cartwright, P.M. 1967. The effect of growth regulators in the growth and nodulation of excised roots of *Phaseolus vulgaris*. L. *Wiss. Z. Univ. Rostock, Math. Naturwiss Reith* 16: 537-538.

Castillo, L. D-del., Hunt, S. and Layzell, D.B. 1992. O_2 regulation and O_2 limitation of nitrogenase activity in root nodules of pea and lupin. *Physiol. Plant.* 86: 269-278.

Charbonneau, G.A. and Newcomb, W. 1985. Growth regulators in developing effective root nodules of the garden pea (*Pisum sativum* L.). *Biochem. Physiol. Pflanzen* 180: 667-681.

Chopra, R.K., Kaundal, K.R. and Sinha, S.K. 1984. A simple technique of studying water deficit effects on nitrogen fixation in nodules without influencing the whole plant. *Plant Physiol.* 76: 254-256.

Chugh, L.K., Gupta, G.K. and Sawhney, S.K. 1991. Effect of cadmium on enzymes of nitrogen assimilation in pea seedlings. *Phyto Chemistry* 31: 395-400.

Cocking, E.C., Al-Mallah, M.K., Bensen, E. and Davey, M.R. 1990. Nodulation of Non-legumes by rhizobia. In: *Nitrogen Fixation: Achievements and Objectives* (P.M. Gresshoff, L.E. Roth, G. Stacey and W.E. Newton, eds.). Chapman and Hall, New York, pp. 813-823.

Cook, D., Dreyer, D., Bonnet, D., Howell, M., Nony, E. and Van den Bosch, K. 1995. Transient induction of a peroxidase gene in *Medicago truncatula* precedes infection by *Rhizobium meliloti, I. Plant Cell,* 6: 215-225.

Cook, D., Dreyer, D., Bonnet, D., Howell, M., Nony, E. and Van den Bosch, K. 1995. Transient induction of a peroxidase gene in *Medicago truncatula* precedes infection by *Rhizobium meliloti II. Plant Cell* 7: 43-55.

Cote, B. and Dawson, J.O. 1989. Effect of regime and fertilization on nitrogenase activity of black alder seedlings during autumn in Illinois USA. *Can. J. For. Res.* 19: 1644-1647.

Dabas, S., Singh, R.P. and Sawhney, V. 1995. Nitrogen fixation and ammonia assimilation in mungbean nodules during lead contamination. *Physiol. Mol. Biol. Plants* 1: 135-140.

Dart, P. 1977. Infection and development of leguminous nodules. In: *A Treatise on Dinitrogen Fixation*. Section III. Biology. (R.W.F. Hardy and W.S. Silver, eds.). John Wiley and Sons, New York, pp. 367-477.

Day, D.A., Carrol, B.J., Delves, A.C. and Gresshoff, P.M. 1989. Relationship between autoregulation and nitrate inhibition of nodulation in soybean. *Physiol. Plant.* 75: 37-42.

Day, D.A. and Copeland, L. 1991. Carbon metabolism and compartmentation in nitrogen fixing legume nodules. *Plant Physiol. Biochem.* 29: 185-201.

Dazzo, F.B., Truchet, G.L., Sherwood, J.E., Hrabak, E.M., Abe, M. and Pankratz, S.H. 1984. Specific phases of root hair attachment in the *Rhizobium* trifolii-clover symbiosis. *Appl. Environ. Microbiol.* 48: 1140-1150.

De-Billy, F., Barker, D.G., Gallusci, P. and Truchet, G. 1991. Leghaemoglobin gene transcription triggered in a single cell layer in the indeterminate nitrogen fixing root nodule of alfalfa. *The Plant J.* 1: 27-35.

De Jong, A.J., Heidstra, R., Spaink, H.P., Haertog, M.V., Meijer, E.A., Hendrinks, T., Lo Schiavo, F., Terzi, M., Bisseling, T., Vankammen, A., and De Vries, S.C. 1993. *Rhizobium* lipooligosaccharides rescue a carrot somatic embryo mutant. *Plant Cell* 5: 615-620.

De-Jong, T.M. and Phillips, D.A. 1981. Nitrogen stress and apparent photosynthesis in symbiotically grown *Pisum sativum* L. *Plant Physiol.* 68: 309–313.

Dehlo, C. and de Bruijn F.J. 1992. The early nodulin gene Sr ENOD 2 from *Sesbania rostrata* in inducible by cytokinin. *Plant J.* 2: 117–128.

Delves, A.C., Carroll, B.J. and Gresshoff, P.M. 1988. Genetic analysis and complementation studies on a number of mutant supernodulating soybeans. *J. Genet.* 67: 1–8.

Delves, A.C., Mathews, A., Day, D.A., Carter, A.S., Carroll, B.J. and Gresshoff, P.M. 1986. Regulation of the soybean *Rhizobium* nodule symbiosis by shoot and root factors. *Plant Physiol.* 82: 588–590.

Demont, N., Ardourel, M., Maillet, F., Prome, J.C., Ferro, M., Prome, J.C. and Denarie, J. 1994. The *Rhizobium* regulatory nod D_3 and syr M genes control the synthesis of a particular class of nodulation factors N-acetylated by (W-1)-hydrolated fatty acids. *EMBO J.* 13: 2139–2149.

Diaz, C.L., Van Sponsen, P.C., Bakhuizen, R., Logman, G.J.J., Lugtenberg, E.J.J. and Kijne, J.W. 1986. Correlation between infection by *Rhizobium leguminosarum* and lectin on the surface of *Pisum sativum* L. roots. *Planta* 168: 350–359.

Dowling, D.N., Stanley, J. and Broughton, W.J. 1989. Competitive nodulation blocking of Afghanistan pea is determined by nod ABC and nod FE alleles in *Rhizobium leguminosarum*. *Mol. Gen. Genet.* 216: 170–174.

Dowling, D.N., Samrey, U., Stanley, J. and Broughton, W.J. 1987. Cloning of *Rhizobium leguminosarum* genes for competitive nodulation blocking on peas. *J. Bacteriol.* 169: 1345–1348.

Downie, J.A. 1989. The nod L gene from *Rhizobium leguminosarum* is homologous to the acetyl transferase encoded by lac A and cys E. *Mol. Microbiol.* 3: 1649–1651.

Downie, A., and Brewin, N. 1992. The *Rhizobium*-legume symbiosis. In: *Development: The molecular Genetic Approach* (V.E.A. Russo, ed.). Springer-Verlag, Berlin, pp. 257–270.

Drennan, D.S., and Nortton, C. 1972. The effect of ethrel on nodulation in *Pisum sativum* L. *Plant Soil* 36: 57–67.

Dreyfus, B., Garcia, J.L. and Gillis, M. 1988. Characterization of *Azorhizobium caulinodans* gen. nov. sp. nov., a stem nodulating nitrogen fixing bacterium isolated from *Sesbania rostrata*, *Int. J. Syst. Bacteriol.* 38: 89–98.

Dubey, R.C. and Dwivedi, R.S. 1987. Effect of heavy metals on seed germination and seedling growth of soybean. *Natl. Acad. Sci. Lett.* 10: 121–124.

Dullaart, J. 1970. The auxin contents of root nodules of *Alnus glutinosa* L. Vill. *J. Exp. Bot.* 21: 975–984.

Dullaart, J. and Duba, L.I. 1970. Presence of gibberellin-like substances and their possible role in auxin biproduction in root nodules and roots of *Lupinus luteus*. *Acta Bot. Neerl.* 19: 877–881.

Dunn, K., Dickstein, R. Feinbaum, R., Burnett, B.K., Peterman, T.K., Thoidis, G., Goodman, H.M. and Ausubel, F.M. 1988. Developmental regulation of nodule specific genes in alfalfa root nodule. *Mol. Plant-Microbe Interact.* 1: 66–74.

Ebeling, S., Hahn, M., Fisher, H.M. and Hennecke, H. 1987. Identification of nif F-, nif N- and nif S-like genes in *Bradyrhizobium japonicum*. *Mol. Gen. Genet.* 207: 503–508.

Engin, M. and Sprent, J.I. 1973. Effects of water stress on growth and nitrogen fixing activity of *Trifolium repense*. *New Phytol.* 72: 117–126.

Erhardt, D.W., Atkinson, E.M. and Long, S.R. 1992. Depolarization of alfalfa root hair membrane potential by *Rhizobium meliloti* nod factors. *Science* 256: 998–1000.

Erhardt, D.W., Wais, R. and Long, S.R. 1996. Calcium spiking in plant root hairs responding to *Rhizobium* nodulation signals. *Cell* 85: 673–681.

Evensen, K.B. and Blevins, D.G. 1981. Differences in endogenous levels of gibberellins like in nodules of *Phaseolus lunatus* L. Plants inoculated with two *Rhizobium* strains. *Plant Physiol.* 68: 195–198.

Felle, H.H., Kandorosi, E., Kondorosi, A. and Schultze, M. 1995. Nod signal-induced plasma membrane potential changes in alfalfa root hairs are differentially sensitive to structural modifications of the lipochito-oligosaccharide. *Plant J.* 7: 939–947

Forde, B.G. and Cullimore, J.V. 1989. The molecular biology of glutamine synthetase in higher plants. *Oxford Surv. Plant Mol. Cell Biol.* 6: 247–296.

Francisco, Jr., P.B. and Harper, J.E. 1995a. Autoregulation of soybean nodulation: delayed inoculation increases nodule number. *Physiol. Plant.* 93: 411–420.

Franco, A.A., Peripira, J.C. and Neyra, C.A. 1979. Seasonal patterns of nitrate reductase and nitrogenase activities in *P. vulgaris* L. *Plant Physiol.* 63: 421–424.

Fred, E.B., Baldwin, I.L. and McCoy, E. 1932. *Root Nodule Bacteria and Leguminous Plants.* University of Wisconsin, Madison, Wisconsin.

Galiacher, A.E. and Sprent, J.I. 1978. Effect of different water regimes on growth and nodule development of greenhouse grown *Vicia faba*. *J. Exp. Bot.* 29: 413–423.

Garg, B.K., Venkates Warlu, B., Vyas, S.P. and Lahiri, A.N. 1984. Nodulation and nitrogenase activity in clusterbean and mothbean under salt stress. *Indian J. Exp. Biol.* 22: 511–512.

Garg, N., Garg, O.P. and Dua, I.S. 1989. Symbiotic nitrogen fixation in relation to mono-, di- or polyphenols in *Cajanus cajan*. *Indian J. Plant Physiol.* 32: 86–89.

Garg, O.P. and Swaraj, K. 1984. Effect of prolongation of dark period on the rate of nitrogen fixation and some associated metabolic factors in cow pea. In: *Advances in Nitrogen Fixation Research* (C. Veeger and W.E. Newton, eds.). Martinus Nijhoff/Dr. W. Junk Publ., The Hague-Wugenin, pp. 503–509.

Geurts, R. and Franssen, H. 1996. Signal transduction in *Rhizobium* induced nodule formation. *Plant Physiol.* 112: 447–453.

Gibson, A.H. and Pagan, J.D. 1977. Nitrate effect on nodulation of legumes inoculated with nitrate reductase-deficient mutants of *Rhizobium*. *Planta* 134: 17–22.

Goodlass, G. and Smith, K.A. 1979. Effects of ethylene on root extension and nodulation of pea (*Pisum sativum* L.) and white clover (*Trifolium repense* L.). *Plant Soil* 51: 387–395.

Gordon, J.K. and Brill, W.J. 1972. Mutants that produce nitrogenase in presence of ammonia. *Proc. Natl. Acad. Sci. USA* 69: 3501–3503.

Gottfert, M., Grob, P. and Hennecke, H. 1990a. Proposed regulatory pathway encoded by the nod V and nod W genes, determinants of host specificity in *Bradyrhizobium japonicum*. *Proc. Natl. Acad. Sci. USA* 87: 2680–2684.

Gottfert, M., Hitz, S. and Hennecke, H. 1990b. Identification of nod S and nod U, two inducible genes inserted between *Bradyrhizobium japonicum* nod YABC and nod IJ genes. *Mol. Plant-Microbe Interact.* 3: 308–316.

Graham, T.L. 1991. Flavonoid and isoflavonoid distribution in developing soybean seedling tissues and in seed and root exudates. *Plant Physiol.* 95: 594–603.

Granhall, U., Ericsson, T. and Clarholm, M. 1983. Dinitrogen fixation in nodulation by *Frankia* and *Alnus incana* as affected by inorganic nitrogen in pot experiments with peat. *Can. J. Bot.* 61: 2956–2963.

Gresshoff, P.M. and Delves, A.C. 1986. Plant genetic control to symbiotic nodulation and nitrogen fixation in legumes. In: *A Genetic Approach to Plant Biochemistry* (A.D. Blonstein and P.J. King, eds.). Springer-Verlag, New York, pp. 159–206.

Grobbelaar, N., Clark, B. and Hough, M.C. 1971. The nodulation and nitrogen fixation of isolated roots of *Phaseolus vulgaris* L. II. The influence of light on nodulation. *Plant Soil* special volume, pp. 203–214.

Guinel, F.C. and Larue, T.A. 1991. Light microscopy study of nodule initiation in *Pisum sativum* L. cv sparkle and its low nodulating mutant E 2 (Sym 5). *Plant Physiol.* 97: 1206–1211.

Hamidi, Y.A. 1970. Soil water tension and the movement of rhizobia. *Soil Biol. Chem.* 3: 121–126.

Hansen, A.P., Peoples, M.B. Gresshoff, P.M., Atkins, C.A., Pate, J.S. and Carroll, B.J. 1989. Symbiotic performance of supernodulating soybean (*Glycine max* (L.) Merr) mutants during development on different nitrogen regimes. *J. Exp. Bot.* 40: 715–724.

Hanson, A.D. and Hitzt, W.D. 1982. Metabolic responses of mesophytes to plant water deficits. *Ann. Rev. Plant Physiol.* 33: 163–203.

Hartwig, U.A., Maxwell, C.A., Joseph, C.M. and Phillips, D.A. 1989. Interactions among flavonoid nod gene inducers released from alfalfa seeds and roots. *Plant Physiol.* 91: 1138–1142.

Hawkers, T.R., McLean, P.A. and Smith, B.E. 1984. Nitrogenase from nif. V. mutants of *Klebsiella pneumoniae* contain an altered form of the iron-molybdenum cofactor. *Biochem. J.* 217: 317–321.

Heidstra, R., Geurts, R., Franssen, H., Spaink, H.P., Van Kammen, A. and Bisseling, T. 1994. Root hair deformation activity of nodulation factors and their fate on *Vicia sativa*. *Plant Physiol.* 105: 787–797.

Hennecke, H., Bott, M. Ramseier, T., Thony-Meyer, L., Fischer, H.M., Anthematten, D., Kullik, I. and Thony, B. 1990. A genetic approach to analyse the critical role of oxygen in bacterioid metabolism. In: *Nitrogen Fixation: Achievements and Objectives* (P.M. Gresshoff, L.E. Roth, T. Stacey and W.E. Newton eds.). Chapman and Hall, New York, New York, pp. 293–300.

Hennecke, H., Fischer, H.M., Ebeling, S., Gubler, M., Thony, B., Gottfert, M., Lamb, J., Hahn, M., Ramseier, T., Regensberger, B., Alvarez-Morales, A. and Studer, D. 1987. Nif, fix and nod gene cluster in *Bradyrhizobium japonicum* and nif A-mediated control of symbiotic nitrogen fixation. In: *Molecular Genetics of Plant-Microbe Interactions* (D.P.S. Verma and N. Brisson, eds.). Martinus Nijhoff, Dordrecht, Netherlands, pp. 191–196.

Henson, T.J.V. and Wheeler, G.T. 1976. Hormones in plants bearing nitrogen fixing nodules. The distribution of cytokinins in *Vicia faba* L. *New Phytol.* 76: 433–438.

Hirsch, A.M. 1992. Development biology of legume nodulation. *New Phytol.* 122: 211–237.

Hirsch, A.M. and Smith, C.A. 1987. Effects of *Rhizobium meliloti nif* and *fix* mutants on alfalfa root nodule development. *J. Bacteriol.* 169: 1137–1146.

Ho, S.C. and Kijne, J.W. 1991. Lectins in *Rhizobium*-legume symbiosis. In: *Lectin Reviews*, vol. I. (D.C. Kilpatrick, E. van Driessche and T.C. Bog-Hansen, eds.). Sigma Chemical Co. St. Louis, MO, pp. 171–181.

Holding, A.J. and King, J. 1963. The effectiveness of indigenous populations of *Rhizobium trifolii* in relation to soil factors. *Plant Soil* 18: 191–198.

Hua, S.S.T., Scot, D.B. and Lim, S.T. 1981. A mutant of *Rhizobium japonicum* 110 with elevated nif activity in free living culture. In: *Genetic Engineering of symbiotic Nitrogen Fixation* (J.M. Lyons, R.C. Valentine, D.A. Phillips, D.W. Rains and R.C. Huftaker, eds.). Plenum Press, New York, pp. 95–105.

Hua, S.T., Tsai, V.Y., Lichens, G.M. and Noma, A.T. 1982. Accumulation of amino acids in *Rhizobium* sp. strain WR 1001 in response to sodium chloride salinity. *Appl. Environ. Microbiol.* 44: 135–140.

Huang, C.Y., Bazzaz, F.A. and Vanderhoef, L.N. 1974. The inhibition of soybean metabolism by Cd and Pb. *Plant Physiol.* 54: 122–124.

Hungria, M., Joseph, C.M. and Phillips, D.A. 1991a. Anthocyanidins and flavonols, major nod gene inducers from seeds of a black-seeded common bean (*Phaseolus vulgaris* L.). *Plant Physiol.* 97: 751–758.

Hungria, M., Joseph, C.M. and Phillips, D.A. 1991b. *Rhizobium nod* gene inducers exuded naturally from roots of common bean (*Phaseolus vulgaris* L.). *Plant Physiol.* 97: 759–764.

Hunt, S. and Layzell, D.M. 1991. The role of oxygen in nitrogen fixation and carbon metabolism in legume nodules. In: *Active Oxygen/Oxidative Stress and Plant Metabolism* (E.J. Pell, and K.L. Steffen, eds.). American Society of Plant Physiologists, Rockville, MD pp. 26–39.

Hunt, S. and Layzell, D.B. 1993. Gas exchange of legume nodules and the regulation of nitrogenase activity. *Ann. Rev. Plant Physiology & Mol. Biol.* 44: 483–511.
Huss-Denell, K., Winship, L.J. and Hahlin, A.H. 1987. Loss and recovery of nitrogenase in *Alnus incana* nodules exposed to low oxygen and low temperature. *Physiol. Plant.* 46: 31–35.
Iannetta, P.P.M., James, E.K., McHardy, P.D., Sprent, J.I. and Minchin, F.R. 1993a. An ELISA procedure for quantification of relative amounts of intracellular glycoprotein in legume nodules. *Ann. Bot.* 71: 85–90.
Iannetta, P.P.M., de Lorenzo, C., James E.K., Fernandez-Pascaul, M., Sprent, J.I., Lucab, M.M., Witty, J.F., de-Felip, M.R. and Minchiw, F.R. 1993b. Oxygen diffusion in lupin nodules. I. Visualization of diffusion barrier operation. *J. Exp. Bot.* 44: 1461–1467.
Ibrahim, A.N., Kamel, M. and Khadr, M.S. 1970. Nodule formation and growth of legumes as influenced by certain sodium salts. *Agrokem Talajaton* 19: 164–172.
Israel, D.W. 1987. Investigation of the role of phosphorus in symbiotic dinitrogen fixation. *Plant Physiol.* 84: 835–840.
Jacobs, F.A., Zhang, M., Fortin, M.G. and Verma, D.P.S. 1987. Several nodulins of soybean share structural domains but differ in their subcellular location. *Nucleic Acids Res.* 15: 1271–1280.
Jacobsen, I. 1985. The role of phosphorus in nitrogen fixation by young pea plants (*Pisum sativum*). *Physiol. Plant.* 64: 190–196.
Jacobsen, K., Laursen, N.B., Jensen, E.O., Marcker, A., Poulson, C. and Marcker, K.A. 1990. HMG-like proteins from leaf and nodule nuclei interact with different AT-motif in soybean nodule promoters. *Plant Cell* 2: 85–94.
Jacobsen, M.R., Premakumar, R. and Bishop, P.E. 1986. Transcriptional regulation of nitrogen fixation by molybdenum in *Azotobacter vinelandii*, *J. Bacteriol.* 167: 480–486.
Jaiwal, P.K. and Gulati, A. 1989. Morphactin and abscisic acid induced pseudonodules formation in the roots of *Cicer arietinum* L. *Proc. Nat. Acad. Sci. India.* 59: 463–477.
James, E.K., Sprent, J.I., Minchin, F.R. and Brewin, J.J. 1991. Intercellular location of glycoprotein in soybean nodules: effect of altered rhizosphere oxygen concentration. *Plant Cell Environ.* 14: 467–476.
Jarvis, B.D.W., Gills, M. and Deley, J. 1986. Intra and intergeneric similarities between the ribosomal ribonucleic acid cistrons of *Rhizobium* and *Bradyrhizobium* species and some related bacteria. *Int. J. Syst. Bacteriol.* 36: 129–138.
Ji, L., Wood, S., Becana, M. and Klucas, R.V. 1991. Purification and characterization of soybean root nodule ferric leghaemoglobin reductase. *Plant Physiol.* 96: 32–37.
Jing, Y., Zhang, B.T. and Shan, X.O. 1990. Pseudonodule formation on barley roots induced by *Rhizobium astragali*. *FEMS Microbiol. Lett.* 69: 123–128.
John, M., Rohrig, H., Schmidt, J., Wieneke, U. and Schell, J. 1993. Rhizobium Nod B protein involved in nodulation signal synthesis is a Chitooligosaccharide deacetylase. *Proc. Natl. Acad. Sci. USA* 90: 625–629.
Journet, E.P., Pichon, M., Dedieu, A., De-Billy, F., Truchet, G. and Barker, D.G. 1994. *Rhizobium meliloti* Nod-factors elicit cell-specific transcription of the ENOD 12 gene in transgenic alfalfa. *Plant J.* 6: 241–249.
Kahn, D., Batut, J., Biostard, P., Daveran, M.L., David, M., Domergue, O., Gamerone, A.M., Ghai, J., Hertig, C., Infante, D. and Renalier, M.H. 1987. Molecular analysis of a fix cluster from *Rhizobium meliloti*. In: *Molecular Genetics of Plant-Microbe Interactions* (D.P.S. Verma and N. Brisson, eds.). Martinus Nijhoff, Dordrecht, Netherlands, pp. 258–268.
Kaluza, K. and Hennecke, H. 1981. Regulation of nitrogenase messenger RNA synthesis and stability in *Klebsiella pneumonia*. *Arch. Microbiol.* 130: 38–43.
Kaluza, K., Hahn, M. and Hennecke, H. 1985. Repeated sequences similar to insertion elements clustered around the nif region of the *Rhizobium japonicum* genome. *J. Bacteriol.* 162: 535–542.

Kandaswamy, D. and Prasad, N.N. 1976. Foliar spray of growth regulators on nodulation in three plants and its relation to *Rhizobium* population. *Indian J. Exp. Biol.* 14: 737–739.

Kape, R., Parniske, M. and Werner, D. 1991. Chemotaxis and *nod* gene activity of *Bradyrhizobium japonicum* in response to hydroxycinasmic acids and isoflavonoids. *Appl. Environ. Microbiol.* 57: 316–319.

Kaur, A., Sheoran, I.S. and Singh, R. 1985. Effect of water stress on the enzymes of nitrogen metabolism in mung bean (*Vigna radiata* Wilczek) nodules. *Plant Cell Environ.* 8: 195–200.

Kawai, Y. and Yamamoto, Y. 1986. Increase in the formation of soybean nodules by V.A. mycorrhizal *Plant Cell Physiol.* 27: 399–405.

Kefford, N.P., Zwar, J.A. and Bruce, M.I. 1960. The symbiotic synthesis of auxin by legume and nodule bacteria and its role in nodule development. *Aust. J. Biol. Sci.* 13: 456–467.

Kennedy, I.R., Rigaud, J. and Trinchant, J.C. 1975. Nitrate reductase from bacterioids of *Rhizobium japonicum*: enzyme characteristics and possible interaction with nitrogen fixation. *Biochem. Biophys. Acta.* 397: 24–35.

Kennis, J., Nuana, D., Rouby, M.B. Edelman, M.O. and Silvente, S.T. 1994. Inhibition of nitrate reductase by water stress and oxygen in detached oat leaves. A possible mechanism of action. *J. Plant Physiol.* 149: 735–739.

Kijne, J.W. 1992. The *Rhizobium* infection process. In: *Biological Nitrogen Fixation* (G. Stacey, R.H. Burnis and H.J. Evans, eds.). Chapman and Hall, New York, pp. 349–398.

Kijne, J.W., Lugenberg, B.J.J. and Smit, G. 1992. Attachment, lectin and initiation of infection in (Brady) *Rhizobium*-legume interactions. In: *Molecular Signals in Plant-Microbe Communications* (D.P.S. Verma, ed.). CRC Press, Boca Raton, FL, pp. 281–294.

Kondorosi, A. 1992. Regulation of nodulation genes in rhizobia. In: *Molecular Signals in Plant-Microbe Communications* (D.P.S. Verma, ed.). CRC Press, Boca Raton, FL, pp. 325–340.

Kondorosi, E. and Kondorosi, A. 1986. Nodule induction on plant roots by *Rhizobium*. *Trends Biochem. Sci.* 11: 296–299.

Kondorosi, E., Burie, M., Cren, M., Iyer, N. Hoffman, B. and Kondorosi, A. 1991. Involvement of the syr M and nod D_3 genes of *Rhizobium meliloti* in nod gene activation and in optimal nodulation of the plant host. *Mol. Microbiol.* 5: 3035–3048.

Kosslak, R.M. and Bohlool, B.B. 1984. Suppression of nodule development on one side of a split-root system of soybeans caused by prior inoculation of the other side. *Plant Physiol.* 75: 125–130.

Kosslak, R.M., Bookland, R., Barkei, J., Paaren, H.E. and Appelbaum, E.R. 1987. Induction of *Bradyrhizobium japonicum* common *nod* genes by isoflavones isolated from *Glycine max*. *Proc. Natl. Acad. Sci. USA.* 84: 7428–7432.

Kurkdjian, A.C. 1995. Role of the differentiation of root epidermal cells in Nod-factor (from *Rhizobium meliloti*) induced root hair depolarization of *Medicago sativa*. *Plant Physiol.* 107: 783–790.

Lakshmi Kumari, M., Singh, C.S. and Subba Rao, N.S. 1974. Root hair infection and nodulation in lucerne (*Medicago sativa* L.) as influenced by salinity and alkalinity. *Plant Soil.* 40: 261–269.

Larsen, P.O., Carnwell, K.L., Gee, S.L. and Bassham, J.A. 1981. Aminoacid synthesis in photosynthesizing spinach cells. *Plant Physiol.* 68: 1231–1236.

Lauchli, A. 1984. Salt exclusion in legumes from crops and pastures under saline conditions. In: *Salinity Tolerance in Plants: Strategies for Crop Improvements* (R.C. Staples, G.H. Toenniessen, eds.). John Wiley & Sons, New York, pp. 171–187.

Law, I.J. and Van Tonder, H.J. 1992. Localisation of mannose- and galactose-binding lectins in an effective peanut nodule. *Protoplasma* 167: 10–18.

Lerouge, P., Roche, P., Faucher, C., Maillet, F., Truchet, G., Prome, J.C. and Denarie, J. 1990. Symbiotic host specificity of Rhizobium meliloti is determined by a sulphated and acylated glucosamine oligosaccharide signal. Nature 344: 781–784.

Letchworth, M.B. and Blum, U. 1977. Effect of acute ozone exposure on growth, nodulation and nitrogen content of ladino clover. Environ. Pollut. 14: 303–312.

Libbenga, K.R. and Harkes, P.A.A. 1973. Initial proliferation of cortical cells in the formation of root nodules in Pisum sativum L. Planta 114: 293–299.

Libbenga, K.R., Van Iren, F., Bogers, R.J. and Schraag-Lamers, M.F. 1973. The role of hormones and gradients in the initiation of cortex proliferation and nodule formation in Pisum sativum L. Planta 114: 29–39.

Libbenga, K.R. and Bogers R.J. 1974. Root nodule morphogenesis. In Biology of Nitrogen Fixation (A Quispel ed.). North Holland Publication Co. Amsterdam pp. 431–472.

Ligero, F., Lluch, C. and Olivares, J. 1987. Evolution of ethylene from roots and nodulation rate of alfalfa (Medicago sativa L.) plants innoculated with Rhizobium meliloti as affected by the presence of nitrate. J. Plant. Physiol. 129: 461–467.

Ligero, F., Caba, J.M., Carmen, L. and Olivares, J. 1991. Nitrate inhibition of nodulation can be overcome by ethylene inhibitor aminoethoxy vinyl-glycine. Plant. Physiol. 97: 1221–1225.

Long, S.R. 1989a. Rhizobium-legume nodulation: Life together in the underground. Cell 56: 203–214.

Lorenzo, C. de, Jannetta, P.P.M., Fernandez-Pascaul, M., James, E.K., Lucas, M.M., Sprent, J.I., Witty, J.F., Minchin, F.R. and de-felip, F.R. 1993. Oxygen diffusion in lupin nodules. II. Mechanism of diffusion barrier operation. J. Exp. Bot. 44: 1469–1474.

Mackay, J., Simon, L. and Lalonde, M. 1987. Effect of substrate nitrogen on the performance of in vitro propagated Alnus glutinosa clone inoculated with Sp$^+$ and Sp$^-$ Frankia strains. Plant Soil. 103: 21–31.

Mahon, J.D. 1977a. Roof and nodule respiration in relation to acetylene reduction in intact nodulated peas. Plant Physiol. 60: 812–816.

Mahon, J.D. 1977b. Respiration and the energy requirement for nitrogen fixation in nodulated pea roots. Plant Physiol. 60: 817–821.

Malik, N.S.A., Calvert, S.E. and Bauer, W.D. 1987. Nitrate induced regulation of nodule formation in soybean. Plant Physiol. 84: 266–271.

Manhart, J.R. and Wong, P.P. 1980. Nitrate effect on nitrogen fixation (acetylene reduction) activities of legume root nodules induced by rhizobia with varied nitrate reductase activities. Plant Physiol. 65: 502–505.

Marcker, K.A., Bojsen, K., Jensen, E.Q. and Paludan, K. 1984. The soybean leghemoglobin genes. In: Advances in Nitrogen Fixation Research (C. Veeger and W.E. Newton, eds.). Martinus Nijhoff/Dr W. Junk Publ., The Hauge Wageningen, pp. 573–589.

Marinda-Ham, H.L. and Loyola-Vargas, V.M. 1994. Gutamate dehydrogenase and glutamine synthetase activities under water and salt stress. Phyton. 56: 7–15.

Marques, I.A., Obserholzer, M.J. and Erismann, K.H. 1983. Effects of different inorganic nitrogen sources on photosynthetic carbon metabolism in primary leaves of non-nodulated Phaseolus vulgaris. Plant Physiol. 71: 555–561.

McClure, P.R., Coker, G.T. and Schubert, K.R. 1983. Carbon-dioxide fixation in roots and nodules of Alnus glutinosa. I. Role of phosphoenol pyruvate carboxylase and carbamyl phosphate synthetase in dark CO_2 fixation, citrulline synthesis and N_2 fixation. Plant Physiol. 71: 652–657.

Miao, G.H., Hirel, B., Marsolier, M.C., Ridge, R.W. and Verma, D.P.S. 1991. Ammonia-regulated expression of a soybean gene encoding cytosolic glutamine synthetase in transgenic Lotus corniculatus. Plant Cell 3: 11–22.

Miller, R.W., Smith, B.E. and Eady, R.R. 1993. Energy transduction by nitrogenase-binding of MgADP to the MoFe protein is dependent on the oxidation state of the iron-sulphur 'P' cluster. Biochem. J. 291: 709–711.

Mishra, S.N., Bhutani, S. and Singh, D.B. 1994. Influence of nitrate supply on cadmium toxicity in *Brassica juncea* during early seedling growth. *Indian J. Plant Physiol.* 37: 12–16.

Moffat, A.M. 1990. Nitrogen-fixing bacteria find new partners. *Science* 250: 910–912.

Morrison, N. and Verma, D.P.S. 1987. A block in the endocytosis of *Rhizobium* allows cellular differentiation in nodules but affects the expression of some peribacterioid membrane nodulins. *Plant Mol. Biol.* 9: 195–196.

Mukhopadhyay, B., Ray, C. and Mukherji, S. 1990. Study on growth, nodulation, sugar, nitrogen and aminoacid contents in mungbean following application of penicillin as foliar spray. *Indian J. Plant Physiol.* 33: 197–203.

Munns, D.N. 1977. Mineral nutrition and legume symbiosis. In: *A Treatise on Dinitrogen Fixation*. Section IV: Agronomy and Ecology (R.W.F. Hardy and A.H. Gibson, eds.). John Wiley and Sons, New York, pp. 35–310.

Nandwal, A.S. and Bharti, S. 1982. The effect of kinetin and IAA on growth yield and nitrogen fixing efficiency of nodules in pea. *Indian J. Plant Physiol.* 25: 358–363.

Nap, J.P. and Bisseling, T. 1990. Developmental biology of plant prokaryote symbiosis. The legume root nodule. *Science* 250: 948–954.

Nelson, L. and Hynes, R. 1995. Traditional and novel legume inoculants. PBI Bulletin (NRC-CNRS) Nov., pp. 7–8.

Newcomb, W., Sippell, D. and Peterson, R.L. 1979. The early morphogenesis of *Glycine max* and *Pisum sativum* root nodule. *Can. J. Bot.* 57: 2603–2616.

Nirunsuksiri, W. and Sengupta-Gopalan, C. 1990. Characterization of a novel nodulin gene in soybean that shares sequence similarity to the gene for nodulin 24. *Plant Mol. Biol.* 15: 835–849.

Norel, F. and Elmerich, C. 1987. Nucleotide sequence and functional analysis of the two nif H copies of *Rhizobium* ORS 571. *J. Gen. Microbiol.* 133: 1563–1576.

Nutman, P.S. (ed.). 1976. *Symbiotic Nitrogen Fixation in Plants*. Cambridge Univ. Press, Cambridge.

Nutman, P.S. 1977. Study frameworks for symbiotic nitrogen fixation. In: *Recent Development in Nitrogen Fixation* (W. Newton, J.R. Postage and C. Radiquez-Barrueco, eds.). Acad. Press, London, pp. 443–447.

Pagan, J.D., Scowcroft, W.R., Dudman, W.F. and Gibson, A.H. 1977. Nitrogen fixation in nitrate reductase-deficient mutants of cultured rhizobia. *J. Bacteriol.* 129: 710–723.

Pankhurst, C.E. and Sprent, J.I. 1975. Effect of water stress on the respiratory and nitrogen-fixing activity of soybean root nodules. *J. Exp. Bot.* 26: 207–304.

Panwar, J.D.S., Chandra, R. and Sirohi, G.S. 1988. Effect of temperature on growth, nodulation and nitrogen fixation in summer mung (*Vigna radiata* L. Wilczek). *Ann. Plant Physiol.* 2: 121–129.

Peters, N.K., Trost, J.W. and Long, S.R. 1986. A plant flavone, luteolin, induces expression of *Rhizobium meliloti* nodulation genes. *Science* 233: 977–980.

Pfeiffer, N.E., Malik, N.S.A., Wagnor, F.W. 1983. Reversible dark induced senescence of soybean root nodule. *Plant Physiol.* 71: 393–399.

Phillips, D.A. 1971. Abscisic acid inhibition of root nodule initiation in *Pisum sativum*. *Planta* 100: 181–190.

Phillips, D.A. and Torrey, J.G. 1972. Studies on cytokinin production of *Rhizobium*. *Plant Physiol.* 49: 11–15.

Pierce, M. and Bauer, W.D. 1983. A rapid regulatory response governing nodulation in soybean. *Plant Physiol.* 73: 286–290.

Prabha, C. and Bharti, S. 1979. Effect of ascorbic acid treatments on nodulation, leghemoglobin and nitrogen content of cowpea nodules under simulated drought conditions. In: *Recent Researches in Plant Science* (S.S. Bir, ed.). Kalyani Publish., New Delhi, India, pp. 498–504.

Prakash, V. 1966. Effect of 2-chloroethyl L. trimethyl ammonium chloride and gibberellic acid on nodulation in *Trifolium alexandrinum*. *Indian J. Exp. Biol.* 4: 251.

Puhler, A., Agmilar, M.O., Hynes, M., Muller, P., Klipp, W., Priefer, U., Simon, R. and Weber, G. 1984. Advances in the genetics of free-living and symbiotic nitrogen fixing bacteria. In: *Advances in Nitrogen Fixation Research* (C. Veeger and W.E. Newton, eds.). Martinus Nijhoff/Dr. W. Junk Publ., The Hague-Wagenin, pp. 609–619.

Purohit, I., Prasad, P. and Nautiyal, A.R. 1996. Effect of water stress on nodulation and nitrogen fixation/assimilation in seedlings of two nitrogen fixing tree species. *Physiol. Mol. Biol. Plants* 2: 153–158.

Quinto, C., dela Vega, H., Flores, M., Leemans, J., Cevallos, M.A., Pardo, M.A., Azpiroz, R., de Laurdes, G.M., Calva, E. and Palacois, R. 1985. Nitrogenase reductase: a functional multigene family in *Rhizobium phaseoli*. *Proc. Natl. Acad. Sci. USA* 82: 1170–1174.

Quispel, A. (ed.). 1974. *The Biology of Nitrogen Fixation*. North-Holland Pub. Co., Amsterdam.

Rae, A.L., Perottos, S., Knox, J.P., Kannenberg, E.L. and Brewin, N.J. 1991. Expression of extracellular glycoproteins in the uninfected cells of developing pea nodule tissue. *Mol. Plant-Microbe Interact.* 4: 563–570.

Rawat, A.K., Khare, A.K. and Viashya, U.K. 1991. Dual effect of *Rhizobium* and mycorrhiza on inoculation and yield of chickpea. *Legume Res.* 14: 150–232.

Richardson, A.E., Simpson, R.J., Djordjevic, M.A. and Rolfe, B.G. 1988. Expression of nodulation genes in *Rhizobium leguminosarum* bv. trifolii is affected by low pH, Ca and Al. *Appl. Environ. Microbiol.* 54: 2541–2548.

Rigaud, J. and Puppo, A. 1977. Effect of nitrite upon leghemoglobin and interaction with nitrogen fixation. *Biochim. Biophys. Acta* 497: 702–706.

Roberts, D., Zimmerman, R.W., Stringer, J.W. and Carpenter, S.B. 1983. The effects of combined nitrogen on growth, nodulation and nitrogen fixation of black locust seedlings. *Can. J. For. Res.* 13: 1251–1254.

Robertson, J.G. and Farnden, K.J.F. 1980. Ultrastructure and metabolism of the developing legume root nodule. In: *The Biochemistry of Plants*, vol. 5 (B.J. Miflin, ed.). Academic Press, Inc. New York, pp. 65–113.

Roche, P., Debelle, F., Maillet, F., Lerouge, P., Faucher, C., Truchet, G., Denarie, J. and Prome, J.C. 1991. Molecular basis of symbiotic host specificity in *Rhizobium meliloti* nod N and nod PQ genes encode the sulfation of lipooligosaccharide signals. *Cell* 67: 1–20.

Röhrig, H., Schmidt, J., Walden, R., Czaja, I., Miklasevics, E., Wieneke, U., Schell, J. and John, M. 1995. Growth of tobacco protoplasts stimulated by synthetic lipochitooligosaccharides. *Science* 269: 841–843.

Rolfe, B.G. and Gresshoff, P.M. 1988. Genetic analysis of legume nodule initiation. *Ann. Rev. Plant Physiol. Plant Mol. Biol.* 39: 297–319.

Ronson, C.W., Bosworth, A., Genova, M., Gudbrandsen, S., Hankinson, T., Kwaitkowski, R., Ratclitte, H., Roble, C., Sweeney, P., Szeto, W., Williams, M., and Zablotowicz, R. 1990. Field release of genetically engineered *Rhizobium meliloti* and *Bradyrhizobium japonicum* strains. In: *Nitrogen Fixation: Achievements and Objectives* (P.M. Gresshoff, L.E. Roth, G. Stacey and W.E. Newton eds.). Chapman and Hall, New York, pp. 397–403.

Roponen, I. 1970. The effect of darkness on leghemoglobin content and amino acid levels in root nodules of pea plants. *Physiol. Plant* 23: 452–460.

Ryle, G.J.A., Powell, C.A. and Gordon, J.A. 1979. The respiratory costs of nitrogen fixation in soybean, cowpea and white clover. 1. Nitrogen fixation and the respiration of the nodulated root. *J. Exp. Bot.* 30: 135–144.

Sa, T.M. and Israel, D.W. 1991. Energy status and functioning of phosphorus-deficient soybean nodules. *Plant Physiol.* 97: 928–935.

Salzwedel, J.L. and Dazzo, F.B. 1993. p-Sym nod gene influence on elicitation of peroxidase activity from white clover and pea roots by rhizobia and their cell free supernatants. *Mol. Plant-Microbe Interact.* 6: 127–134.

Sanches-Roldan, C., Antonio, H. Victoriano, V. and Bukovac, M.J. 1990. Indole-3 acetic acid content and glutamine synthetase activity in the pericarp and peroxidase activity and isoenzymes in the meso- and exocarp of growing peach fruits. *J. Plant Growth Regulat.* 9: 171–174.

Sanjuan, J. and Olivares, J. 1991. Multicopy plasmids carrying the *Klebsiella pneumoniae* nif A gene enhance *Rhizobium meliloti* nodulation competitiveness on alfalfa. *Mol. Plant-Microbe Interact* 4: 365–369.

Scheres, B., Van de Wiel, C., Zalensky, A., Horvath, B., Spaink, H., Van Eck, H., Zwartkruis, F., Wolters, A.M., Gloudemans, T., Van Kammen, A. and Bisseling, T. 1990a. The ENOD 12 gene product is involved in the infection process during pea-*Rhizobium* interaction. *Cell* 60: 281–294.

Scheres, B., Van Engelen, F., Vander Knaap, E., van de Weil, C., Van Kammen, A. and Bisseling, T. 1990 b. Sequential induction of nodulin gene expression in the developing pea nodule. *Plant Cell* 2: 687–700.

Schmidt, J., Wingender, R., John, M., Wieneke, U. and Schell, J. 1988. *Rhizobium meliloti* nod A and nod B genes are involved in generating compounds which stimulate mitosis in plant cells. *Proc. Natl. Acad. Sci. USA.* 85: 8578–8582.

Semino, C.E. and Robbins, P.W. 1995. Synthesis of 'Nod' like chitin digosaccharides by the Xenopus developmental protein DG-42. *Proc. Natl. Acad. Sci. USA.* 92: 3498–3501.

Sengupta-Gopalan, C. and Pitas, J.W. 1986. Expression of nodule specific glutamine synthetase gene during nodule development in soybeans. *Plant Mol. Biol.* 7: 189–199.

Shah, V.K. and Brill, W.J. 1977. Isolation of an iron-molybdenum cofactor from nitrogenase. *Proc. Natl. Acad. Sci. USA.* 74: 3249–3253.

Shah, V.K., Madden, M.S. and Ludden, P.W. 1990. In vitro synthesis of the iron-molybdenum cofactor and its analogs: requirement of a non-nif gene product for the synthesis, and altered properties of dinitrogenase. In: *Nitrogen Fixation: Achievements and Objectives* (P.M. Gresshoff, L.E. Roth, G. Stacy and W.E. Newton, eds.). Chapman and Hall, New York, pp. 87–94.

Shah, V.K., Vgalde, R.H., Imperial, J. and Brill, W.J. 1984. Mo in nitrogenase. *Ann. Rev. Biochem.* 53: 231–257.

Shearman, C.A., Rossen, L., Johnston, A.W.B. and Downie, J.A. 1986. The *Rhizobium* gene nod F encodes a protein similar to acyl carrier protein and is regulated by nod D plus a factor in pea root exudate. *EMBO J.* 5: 647–652.

Sheng, C. and Harper, J.E. 1997. Shoot versus root signal involvement in nodulation and vegetative growth in wild-type and hypernodulating soybean genotypes. *Plant Physiol.* 113: 825–831.

Sheoran, I.S., Kaur, A. and Singh, R. 1988. Nitrogen fixation and carbon metabolism in nodules of pigeon pea (*Cajanus cajan*) under drought stress. *J. Plant Physiol.* 132: 480–483.

Sindhu, S.S. and Dadarwal, K.R. 1992. Symbiotic effectivity of cowpea miscellany *Rhizobium* mutants having increased hydrogenase activity. *Indian J. Microbiol.* 32: 411–416.

Sindhu, S.S. and Dadarwal, K.R. 1995. Molecular biology of nodule development and nitrogen fixation in *Rhizobium* legume symbiosis. In: *Nitrogen Nutrition in Higher Plants* (H.S. Srivastava and R.P. Singh eds.). Associated Publishing Co., New Delhi, India, pp. 57–129.

Singh, D.B. 1996. Amelioration of heavy metals and salinity stress by nitrate/ammonium and polyamines in *Brassica juncea*. Ph.D. thesis, M.D. Univ. Rohtak.

Singh, O.S., Parashar, V. and Suman Bala. 1984. Response of lentil of single and combined inoculation of *Rhizobium* and *Glomus*. In: *Symbiotic Nitrogen Fixation* (B.S. Ghai, ed.). U.S.G. Publishers and Distributors, India.

Singh, U., Nandwal, A.S., Bharti, S. and Kundu, B.S. 1994. Induced nodule senescence and nitrogen metabolism at various growth stages in chickpea (*Cicer arietinum* L.). *Indian J. Plant Physiol.* 37: 152–156.

Singleton, P.W. 1983. A split root growth system for evaluating the effect of salinity on the components of soybean — *Rhizobium japonicum* symbiosis. *Crop Science* 23: 259–262.

Sinha, S.K., Srivastava, H.S. and Mishra, S.N. 1988. Effect of lead on nitrate reductase activity and nitrate assimilation in pea leaves. *Acta. Soc. Bot. Poll.* 57: 457–463.

Smit, G., De Koster, C.C., Schripsema, J., Spaink, H.P., Van Brissel, A.A. and Kijne, J.W. 1995. Uridine, a cell division factor in pea roots. *Plant Mol. Biol.* 29: 869–873.

Smith, B.E. 1990. Recent studies in the biochemistry and chemistry of nitrogenase. In: *Nitrogen Fixation: Achievements and Objectives* (P.M. Gresshoff, L.E. Roth, G. Stacey and W.E. Newton, eds.). Chapman and Hall, New York, pp. 3–13.

Smith, G., De Koster, C.C. Schripsema, J., Spaink, H.P., Van Brissel, A.A. and Kinje, J.W. 1995. Uridine, a cell division factor in pea roots. *Plant Mol. Biol.* 29: 869–873.

Spaink, H.P., Sheeley, D.M., Van Brissel, A.A., Glushka, J., York, W.S., Tak, T., Gieger, O., Kennedy, E.P., Reinhold, U.N. and Lugtenberg, B.J.J. 1991a. A novel highly unsaturated fatty acid moiety of lipo-oligosaccharide signals determines host specificity of *Rhizobium*. *Nature* 354: 125–130.

Spaink, H.P., Geiger, O., Sheeley, D.A., Van Bruissel, A.A., York, W.S., Reinhold, U.N., Lugtenberg, B.J.J. and Kennedy E.P. 1991b. The biochemical function of the *Rhizobium leguminosarum* proteins involved in the production of host specific signal molecules. In: *Advances in Genetics of Plant-Microbe Interactions*, vol. 1 (H. Hennecke and D.P.S. Verma, eds.). Kluwer Academic Publishers, Dordrecht, pp. 142–149.

Sprent, J.I. 1972. The effect of water stress on nitrogen fixing root nodules. II. Effects of osmotically applied stress. *New Phytol.* 71: 451–460.

Sprent, J.I. 1976. Nitrogen fixation by legumes subjected to water and light stresses. In: *Symbiotic Nitrogen Fixation* vol. 7 (P.S. Nutman, ed.). Cambridge University Press, Cambridge, pp. 405–420.

Sprent, J.I. 1981. Nitrogen fixation. In: *The Physiology and Biochemistry of Drought Resistance in Plants* (L.G. Paleg and D. Aspinall, eds.). Acad. Press, New York, pp. 131–154.

Sprent, J.I. 1984. Effects of drought and salinity on heterotrophic nitrogen-fixing bacteria and on infection of legumes by rhizobia. In: *Advances in Nitrogen Fixation Research* (C. Veeger and W.E. Newton, eds.). Martinus Nijhoff/Dr W. Junk Publ., The Hague-Wagenin, pp. 295–302.

Srinivasan, P.S. and Gopalkrishna, S. 1977. Effect of planofix on NAA formulation on groundnut var. TM4-7. *Curr. Sci.* 32: 119–120.

Srivastava, H.S. and Ormrod, D.P. 1986. Effects of nitrogen dioxide and nitrate nutrition on nodulation, nitrogenase activity, growth and nitrogen content of bean plants. *Plant Physiol.* 81: 737–771.

Stacey, G. 1990. Workshop summary: Compilation of the nod, fix and nif genes of rhizobia and information concerning their function. In: *Nitrogen Fixation: Achievements and Objectives* (P.M. Gresshoff, L.E. Roth, G. Stacey and W.E. Newton, eds.). Chapman and Hall, New York, pp. 239–244.

Staehelin, C., Granado, J., Miller, J., Wiemken, A., Mellor, R.B., Felix, G., Regenass, M., Broughton, W.J. and Boller, T. 1994. Perception of *Rhizobium* nodulation factors by tomato cells and inactivation by root chitinases. *Proc. Natl. Acad. Sci. USA.* 91: 2196–2200.

Stephens, B.D. and Neyra, C.A. 1983. Nitrate and nitrite reduction in relation to nitrogenase activity in soybean nodules and *Rhizobium japonicum* bacterioids. *Plant Physiol.* 71: 731–735.

Stougaard, J., Marcker, K.A., Otten, L. and Schell, J. 1986. Nodule specific expression of a chimeric soybean leghemoglobin gene in transgenic *Lotus corniculatus*. *Nature* 321: 663–674.

Streeter, J.G. 1982. Synthesis and accumulation of nitrite in soybean nodules supplied with nitrate. *Plant Physiol.* 69: 1429–1434.

Streeter, J.W. 1985. Nitrate inhibition of legume nodule growth and activity. I. Long term studies with continuous supply of nitrate. *Plant Physiol.* 77: 321-324.

Streeter, J.G. 1988. Inhibition of legume nodule formation and nitrogen fixation by nitrate. *CRC Critical Rev. Plant Sci.* 7: 1-23.

Streeter, J.G. 1991. Transport and metabolism of carbon and nitrogen in legume nodules. *Adv. Bot. Res.* 18: 129-187.

Streeter, J.G. 1993. Translocation—a key factor limiting the efficiency of nitrogen fixation in legume nodules. *Physiol. Plant.* 87: 616-623.

Subba Rao, N.S. (ed.) 1979. *Recent Advances in Biology Nitrogen Fixation.* Oxford and IBH Pub. Co., New Delhi.

Subba Rao, N.S., Lakshmi Kumari, M., Singh, C.S. and Magu, S.P. 1987. Nodulation of lucerne (*Medicago sativa*) under influence of sodium chloride. *Indian J. Agric. Sci.* 42: 384-386.

Sutton, J.M., Lea, A.J.A. and Downie, L.A. 1994. The nodulation signaling protein nod O from *Rhizobium leguminosarum* biovar *viciae* forms ion channels in membranes. *Proc. Natl. Acad. Sci. USA.* 91: 9990-9994.

Swaraj, K. 1988. Effect of environmental stress on symbiotic nitrogen fixation. In: *Advances in Frontier Areas of Plant Biochemistry* (R. Singh and S.K. Sawhney, eds.). Prentice Hall of India Pvt. Ltd., New Delhi, pp. 354-376.

Swaraj, K. and Garg, O.P. 1970a. The effect of gibberellic acid (GA_3) when applied to the rooting medium on nodulation and nitrogen fixation in gram (*Cicer arietinum* L.). *Physiol. Plant* 23: 747-754.

Swaraj, K. and Garg, O.P. 1970b. The effect of ascorbic acid when applied to the rooting medium on nodulation and nitrogen fixation in gram (*Cicer arietinum* L.). *Physiol. Plant* 23: 889-897.

Swaraj, K., Kuhad, M.S. and Garg, O.P. 1986. Effect of dark treatments on symbiotic nitrogen fixation in *Cicer arietinum* L. (Chickpea). *Environ. Exp. Bot.* 26: 31-38.

Swaraj, K., Shishenko, S.V., Kazlova, G.I., Andreeva, I.N. and Zhiznevskaya, Gya. 1984. Effect of water deficit on symbiotic nitrogen fixation in soybean. *Fiziol. Rast.* 31: 833-840.

Sweet, G., Ganor, C., Voeglele, R., Wittekindt, N.A. Beuerie, J., Trunlger, V., Lin, E.C.C. and Winfried, B. 1990. Glycerol facilitator of *Escherichia coli*, cloning of glp F and identification of glp F product. *J. Bacteriol.* 172: 424-430.

Syono, K., Newcomb, W. and Torrey, J.G. 1976. Cytokinin production in relation to the development of pea root nodules. *Can. J. Bot.* 54: 2155-2164.

Takats, S.T. 1990. Early autoregulation of symbiotic root nodulation in soybeans. *Plant Physiol.* 94: 865-869.

Thiagrajan, T.R., Ames, R.N. and Ahmad, M.H. 1992. Response of cowpea (*Vigna unguiculata*) to inoculation with co-selected vesicular arbuscular mycorrhizal fungi and *Rhizobium* strains in field trials. *Can. J. Bot.* 38: 573-576.

Thorneley, R.N.F. and Lowe, D.J. 1984. The mechanism of *Klebsiella pneumoniae* nitrogenase action: Stimulation of the dependence of hydrogen evolution rate on component protein concentration and ratio and sodium dithionite concentration. *Biochem. J.* 224: 903-910.

Thummler, F. and Verma, D.P.S. 1987. Nodulin-100 of soybean in the subunit of sucrose synthase regulated by the availability of free heme in nodules. *J. Biol. Chem.* 262: 14730-14736.

Thurbur, G.A., Douglas, J.R. and Galston, A.W. 1958. Inhibitory effect of gibberellins on nodulation in dwarfbean (*Phaseolus vulgaris*). *Nature* 181: 1082-1083.

Tingey, D.T. and Blum, U. 1973. Effects of ozone on soybean nodules. *J. Environ. Qual.* 2: 341-342.

Tohan, Y.T. and Kennedy, I.R. 1989. Possible N_2 fixing root nodules induced in non-legumes. *Agricul. Sci.* 2: 57-59.

Trinchant, J.C. and Rigaud, J. 1980. Nitrite inhibition of nitrogenase from soybean bacterioids. *Arch. Microbiol.* 124: 49–54.
Trinchant, J.C., Guerin, V. and Regand, J. 1994. Acetylene reduction by symbiosomes and free bacterioids from broad bean (*Vicia faba*) nodules. *Plant Physiol.* 105: 555–561.
Triplett, E.W. 1988. Isolation of genes involved in nodule competitiveness from *Rhizobium leguminosarum* bv. trifolic T24. *Proc. Natl. Acad. Sci. USA.* 85: 3810–3814.
Triplett, E.W. and Barta, T.M. 1987. Trifolitoxin production and nodulation are necessary for the expression of superior nodulation competitiveness by *Rhizobium leguminosarum* bv. trifolii strain T24 on clover. *Plant Physiol.* 85: 335–342.
Truchet, G., Camut, S., de-Billy, F., Odorico, R. and Vesse, J. 1989. The *Rhizobium*-legume symbiosis: two methods to discriminate between nodules and other root-derived structures. *Protoplasma* 149: 82–88.
Tu, C.M. and Hietkamp, G. 1977. Effect of moisture on acetylene reduction (symbiotic nitrogen fixation) by *Rhizobium japonicum* and soybean root nodules in silica sand. *Commun. Soil. Sci. Plant Anal.* 8: 81–96.
Vanden Bosch, K.A., Rodgess, L.R., Scherrier, D.J. and Kishinevsky, B.D. 1994. A peanut nodule lectin in infected cells and in vacuoles and the extracellular matrix of nodule parenchyma. *Plant Physiol.* 104: 327–337.
Vanden Bosch, K.A., Bradley, D., Knox, J.P., Perotto, S., Butcher, G.W. and Brewin, N.J. 1989. Common components of the infection thread matrix and the intercellular space identified by immunocytochemical analysis of pea nodules and uninfected roots. *EMBO J.* 8: 335–342.
Van de Weil, C., Norris, J.H., Bochenek, B., Dickstein, R., Bisseling, T. and Hirsch, A.M. 1990a. Nodulin gene expression and ENOD 2 localization in effective, nitrogen-fixing and ineffective bacteria-free nodules of alfalfa. *Plant cell* 2: 1009–1017.
Van de Wiel, C., Scheres, B., Franssen, H., Van Lierop, M.J., van Lammeren, A., van Kammen, A. and Bisseling, T. 1990b. The early nodulin transcript ENOD 2 is located in the nodule parenchyma (inner cortex) of pea and soybean root nodules. *EMBO J.* 9: 1–7.
Vasconcelos, L., Miller, L. and Neycra, C.A. 1980. Free living and symbiotic characteristics of chlorate resistant mutants of *Rhizobium japonicum*. *Can. J. Microbiol.* 26: 338–342.
Vasse, J., de-Billy, F., Camut, S. and Truchet, G. 1990. Correlation between ultra-structural differentiation of bacterioids and nitrogen fixation in alfalfa nodules. *J. Bacteriol.* 172: 4295–4306.
Verma, D.P.S. 1992. Signals in root nodule organogenesis and endocytosis of *Rhizobium*. *Plant Cell* 4: 373–382.
Verma, D.P.S. and Long, S. 1983. The molecular biology of Rhizobium-legume symbiosis. In: *International Review of Cytology, Supplement 14: Intracellular symbiosis* (K. Jeon, ed.). Academic Press Inc., New York, pp. 211–245.
Verma, D.P.S. and Delauney, A.J. 1988. Root nodule symbiosis: Nodulins and nodulin genes. In: *Plant Gene Research: Temporal and Spatial Regulation of Plant Genes* (D.P.S. Verma and R. Goldberg, eds.). Springer-Verlag, New York, pp. 169–199.
Verma, D.P.S., Kazazian, V., Zogbi, V. and Bal, A.K. 1978. Isolation and characterization of the membrane envelop enclosing the bacterioids in soybean root nodules. *J. Cell Biol.* 78: 919–936.
Verma, D.P.S., Fortin, M.G., Stanley, J., Mauro, V., Purohit, S. and Morrison, N. 1986. Nodulins and nodulin genes of *Glycine max*. *Plant Mol. Biol.* 7: 51–61.
Vogel, C.S. and Dawson, J.W. 1991. Nitrate reductase activity, nitrogenase activity and photosynthesis of black alder exposed to chilling temperature. *Physiol. Plant.* 82: 551–558.
Wahab, A.M.A. and Zahran, H.H. 1979. The effect of water stress on N_2 (C_2H_2) fixation and growth of four legumes. *Agricultura (Hoverlee)* 28: 383–400.

Wang, S.P. and Stacey, G. 1991. Studies of the *Bradyrhizobium japonicum* nod Di promoter: a repeated structure for the nod box. *J. Bacteriol.* 173: 3356–3365.

Webb, J. and Sheehy, J.E. 1991. Legume nodule morphology with regard to oxygen diffusion and nitrogen fixation. *Annal. Bot.* 67: 77–83.

Werner, D., Mellor, R.B., Hahn, M.G. and Grisebach, H. 1985. Soybean root response to symbiotic infection: Glyceollin I accumulation in an ineffective type of soybean nodules with an early loss of the peribacterial membrane. *Z. Naturforsch.* 40c: 179–181.

Wilson, R.C., Long, F., Maruoka, E.M. and Cooper, J.B. 1994. A new proline-rich early nodulin from *Medicago truncatula* is highly expressed in nodule meristematic cells. *Plant Cell* 6: 1265–1275.

Witty, J.F. and Minchin, F.R. 1990. Oxygen diffusion in the legume root nodules. In: *Nitrogen Fixation: Achievement and Objectives.* (P.M. Gresshoff, L.E. Roth, G. Stacey, and W.E. Newton eds.). Chapman and Hall, New York, pp. 285–292.

Wong, P.P. and Evans, H.J. 1971. Poly-β-hydroxy-butyrate utilization by soybean (*Glycine max.* Merr). Nodules and assessment of its role in maintenance of nitrogenase activity. *Plant Physiol.* 47: 750–755.

Worrall, V.S. and Roughley, R.J. 1976. The effect of moisture stress on infection of *Trifolium subterreum* L. by Rhizobium trifolli. *J. Exp. Bot.* 27: 1233–1241.

Yang, W.C., Katinakis, P., Hendriks, P., Smolders, A., DeVries, F. Spee, J., Van Kammen, A., Bisseling, T. and Franssen, H. 1993. Characterization of Gm ENOD 40, a gene showing novel patterns of cell specific expression during soybean nodule development. *Plant J.* 3: 573–585.

Yates, M.G. 1980. Biochemistry of nitrogen fixation. In: *The Biochemistry of plants,* vol. 5 (B.J. Miflin, ed.). Academic Press Inc. New York, pp. 1–63.

Yoneyama, T.T., Totsuka, A., Hasimato, J. and Yazaki, J. 1979. Absorption of atmospheric NO_2 by plants and soils. III. Changes in concentration of inorganic nitrogen in the soil fumigated with NO_2. The effect of water conditions. *Soil Sci. Plant. Nutr.* 25: 337–347.

Young, J.P.W. and Johnston, A.W.B. 1989. The evolution of specificity in the legume-Rhizobium symbiosis. *Trend. Ecol. Evol.* 4: 341–349.

Yousef, A.N. and Sprent, J.I. 1983. Effect of NaCl on growth, nitrogen incorporation and chemical composition of inoculated and NH_4NO_3 fertilized *Vicia faba* L. plants. *J. Exp. Bot.* 34: 941–950.

Zeroni, M. and Hall, M.A. 1980. Molecular effects of hormone treatment on tissue. In: Hormonal regulation of development. In: *Molecular Aspects of Plant Hormones* (J. MacMillan, ed.). Springer-Verlag, New York, pp. 511–586.

Chapter 4

FOLIAR ABSORPTION AND USE OF AIRBORNE OXIDIZED NITROGEN BY TERRESTRIAL PLANTS

S. Nussbaum, M. Ammann and J. Fuhrer

I. Oxidized Nitrogen in the Troposphere
II. Methods to Measure Exchange of Gases and Particles with Vegetated Surfaces
III. Deposition, Absorption and Uptake of Atmospheric Oxidized Nitrogen
    A. Deposition
    B. Cuticular Adsoption
    C. Substomatal Cavity Adsorption
    D. Uptake
IV. Physiological Consequences of Foliar Absorption of Oxidized Nitrogen
    A. Nitrogen Metabolism
    B. Nitrogen Containing Compounds: Pool Sizes and Cycling
    C. Effects on Chloroplasts
    D. Other Effects on Plants
V. Damage or Benefit?
VI. Conclusions and Future Prospects

Literature Cited

Abbreviations: DR, dark respiration; fAA, free amino acids; GDH glutamate dehydrogenase; GOGAT; glutamine oxoglutarate aminotransferase; GS, glutamine synthetase; LAR, leaf area ratio; MDA, malondialdehyde; NAR, net assimilation rate; NiR, nitrite reductase; NR, nitrate reductase; PA, phosphatidic acid; RGR, relative growth rate.

I. OXIDIZED NITROGEN IN THE TROPOSPHERE

Reactive gaseous nitrogen compounds, often referred to as odd-nitrogen (NO_y), play a key role in the chemistry of the troposphere which is dominated by oxidizing and acidic conditions (Roberts, 1995). The abbreviation NO_y includes the simple oxides of nitrogen (NO_x, i.e., nitric oxide, NO, and nitrogen dioxide, NO_2) and all products from their reactions in the atmosphere. The most important are nitric acid (HNO_3), nitrous acid (HONO), nitrate radical (NO_3), dinitrogen pentoxide (N_2O_5), and a large group of organic species including peroxyacetyl nitrates (PAN, $CH_3C(O)OONO_2$) and further compounds of the form $RONO_2$, $ROONO_2$ and RNO_2.

Although nitrous oxide ('laughing gas', N_2O) is the most abundant form of oxidized nitrogen in the atmosphere, occurring in concentrations of 1–300 ppb (Hidy and Mueller, 1986), it is not involved in tropospheric chemistry and is therefore not considered a reactive nitrogen compound. Its lifetime is estimated to be up to 150 years (Watson et al., 1990). It is emitted predominantly by soil microbial activity (nitrification, denitrification; Williams, 1992; Robertson, 1993). These processes may be influenced by effects of agricultural practices, such as land use, fertilization and soil compaction. The only known industrial process to contribute to N_2O emission in appreciable amounts is nylon production (Thiemens and Trogler, 1991). N_2O accounts for 6 to 8% of the present greenhouse forcing rate (IPCC, 1990). The main tropospheric elimination process is transport to the stratosphere where it is photochemically destroyed. This initiates a complex set of gas-phase and heterogeneous reactions, which play a key role in the chemical destruction of stratospheric ozone (Shen et al., 1995). There are no known absorption or uptake processes for nitrous oxide by plants. For this reason, this form of oxidized nitrogen will not be treated further in this article.

Nitric oxide (NO) and, to a lesser extent nitrogen dioxide (NO_2), are primary reaction products of oxidation of atmospheric dinitrogen (N_2) or reduced nitrogen by oxygen. Besides natural sources, such as lightning, microbial activity in the soil, or atmospheric oxidation of ammonia, anthropogenic emission by fossil fuel combustion or biomass burning contributes substantially to the total input of NO and NO_2 into the atmosphere (Logan, 1983; Davidson, 1991).

In the presence of oxidants, such as ozone (O_3), NO is rapidly oxidized to NO_2. In an unpolluted atmosphere, NO, NO_2, and O_3 are assumed to be in a photostationary equilibrium preventing excess accumulation of any one of the constituents (Leighton, 1961):

$$NO + O_3 \rightarrow NO_2 + O_2$$
$$NO_2 + h\upsilon \rightarrow NO + O$$
$$O + O_2 + M \rightarrow O_3 + M$$

where $h\upsilon$ represents a photon of light of $\lambda \leq 430$ nm and M represents a molecule (e.g. N_2 or O_2) absorbing excess vibrational energy. Under typical midday tropospheric conditions the characteristic reaction times are in the range of minutes. In the polluted troposphere, the presence of volatile organic compounds (VOC), hydroxyl-radicals, and water result in a net ozone production not accounted for by this simple cycle, thereby initiating the complex NO_y chemistry (Roberts, 1995).

The concentration ranges of the most important products formed through gas phase or heterogeneous reactions (HNO_3, HONO and the most abundant of the organic nitrates, PAN) are given in Table 4.1. While for NO and NO_2 the ranges are well known, concentration figures for the other gases still rely heavily on modelling. Experimental confirmation is relatively scarce. In air polluted with NO_x and VOC, enhanced levels of all of them together with radiation-driven increased net ozone production produce a cocktail often referred to as photochemical smog.

Due to their reactive nature, typical average tropospheric residence times of NO, NO_2, HONO and HNO_3 are calculated to be less than 1 day (Seinfeld, 1986). Dry deposition or heterogeneous loss followed by wet deposition to different surfaces are the main tropospheric elimination pathways for these gaseous oxides of nitrogen. Logan (1983) calculated these fluxes as amounting to 11–22 and 12–42 Tg N a^{-1}, respectively. The lifetime of a PAN molecule prior to its redissociation to NO_2 and $CH_3C(O)OO$ is highly temperature dependent (30 min at 300 K, one month at 260 K) and consequently, at lower temperature in the middle or upper troposphere, it can act as a reservoir for NO_x as do N_2O_5 or NO_3, both being formed through oxidation by ozone during the night. Though chemically very important, such compounds constitute only minor traces in the atmosphere. Their behaviour towards plants is practically not known.

Table 4.1. Concentration ranges of the most important oxidized nitrogen species in clean and polluted air (Demerjian, 1986). Figures specific for Europe can be taken from Sandness and Styve (1992)

Species	Clean (nl l^{-1})	Polluted (nl l^{-1})
NO	0.01–0.05	60–740
NO_2	0.1–0.5	40–220
HNO_3	0.02–0.3	6–20
HONO	10^{-3}–10^{-1}	4–21
PAN	<1	10–65

As the main gaseous oxidized compounds, NO_x, are comparatively unreactive towards dry or aqueous surfaces, their oxidation to the highly reactive HNO_3 or particle-bound nitrate is very important from the point of view of deposition processes. Gaseous HNO_3 is eliminated very efficiently by dry deposition or by scavenging by particles, cloud droplets or snow. NO_2 or HNO_3 may react with salt particles to form nitrate. In densely populated areas, ammonium nitrate partitioned to the gas and particle phase is the main source of nitrate in rainwater. More generally, particulate nitrate originates from gaseous NO_y precursors, and therefore appears mainly in the accumulation mode, i.e. particles in the size range of 0.1 to 1 µm, and to a lesser extent in coarse dust particles.

II. METHODS TO MEASURE EXCHANGE OF GASES AND PARTICLES WITH VEGETATED SURFACES

Gases or particles may be deposited directly onto a surface by a process termed dry deposition. Alternatively, they may be scavenged by water droplets and subsequently be deposited indirectly to the ground by gravitation. This process is known as wet deposition. Cloud droplets may also be deposited by turbulent transfer and captured by interception. This form of deposition is sometimes referred to as occult deposition. The behaviour of the different oxidized nitrogen molecules in these deposition processes is extremely diverse and complete determination of their atmospheric deposition to leaves is difficult. Whole canopy approaches have to be combined with studies on single plants, branches, leaves, or even smaller structures.

Using the 'big leaf' approach (Baldocchi et al., 1987), the complex three-dimensional structure of a canopy is considered as a homogenous leaf and net fluxes from and to this surface are analyzed. Experimentally, continuous measurements over extended periods of time allow quantification of net fluxes for the selected area and characterization of the deposition or emission processes for each compound. There are two approaches to continuously determine exchange rates of gases or particles between the atmosphere and extended surfaces: surface collection methods (throughfall/stemflow) below the canopy, and micrometeorological methods above the canopy. It is important to note that they are continuous or quasi-continuous on different time scales.

In the surface collection methods the amount and composition of the deposition collected above the canopy is compared with the throughfall and stemflow collected below the canopy (Lindberg et al., 1986). The difference between the flux above and below is attributed to dry and occult deposition. In contrast to sulphate, for which these methods provide good estimates of the total inputs (Lindberg et al., 1992), this approach seems not very promising for oxidized nitrogen compounds as canopy uptake of gases is not addressed directly and the relevance of canopy exchange of ionic nitrogen species has not been sufficiently assessed (see Lovett and Lindberg, 1993 and critical review by Duyzer and Fowler, 1994). In addition, the complex chemical and meteorological environment of a canopy does not allow attribution of the throughfall nitrate to its atmospheric precursor (Erisman et al., 1994). However, valuable information about possible exchange processes taking place within the canopy, especially concerning wet and occult deposition, may be obtained (Lindberg et al., 1988) when these tools are combined with the others.

An alternative approach to measure dry deposition of gases is based on the assumption that under certain conditions (see Duyzer and Fowler, 1994) the vertical flux of trace gases in the boundary layer is constant with height and can be derived from measurement of the concentration and micrometeorological parameters. Two different methods are currently used for gaseous flux determinations: eddy correlation (Panofsky and Dutton, 1984; Stocker et al., 1993) and flux gradient (Bowen ratio; Tanner, 1960). The advantage over surface collection methods is that homogeneous areas of several hectares can be studied without interference with the actual exchange processes and that mechanistic models for underlying deposition and emission processes can be derived based on continuous observation during several days or even weeks. Due to the gas phase reactivity of nitrogen compounds the surface flux determined by these methods is different from the flux measured above the surface. Careful modelling taking into account these processes is necessary to derive the 'real' fluxes (Duyzer et al., 1995). In some cases these modelling corrections of the measured fluxes can result in a change in the direction of the flux. The equipment to verify the calculated fluxes is not yet available; therefore, vertical fluxes derived from deposition measurements using micrometeorological methods may be biased by chemical reactions and should be interpreted with care, especially when dealing with complex canopies, such as forests. Routine micrometeorological methods for measuring the dry deposition of particles are still under development (Erisman et al., 1994; Peters and Bruckner-Schatt, 1995; Peters and Eiden, 1992). A fast chemical and physical characterization of particles is difficult to achieve. The same holds for methods to measure occult deposition, which have received increased attention recently (Fowler et al., 1991; Lovett and Kinsman, 1990). Different techniques have been applied to separate cloud droplet deposition from dry or wet deposition. As dew contains very low amounts of pollutants, it mainly influences dry deposition processes via increasing the surface wetness.

When estimating foliar deposition from micrometeorological exchange measurements, knowledge of the magnitude of the resistances for each compound is crucial. Wesely (1989) presents an extensive overview over a parametrization scheme which takes into account different land-use types in different seasons and gives parametrization approaches for different surface resistance components. The resolution of this model is adequate to describe deposition to a canopy as a whole but is not capable of giving sufficiently precise figures for a specific branch or leaf. Also, it should be kept in mind that the 'big leaf' approach yields net fluxes only and does not take into account internal cycling within the canopy. For example, NO_x emissions from soils (Williams, 1992) are not directly accessible. Thus, the total uptake of these compounds by the leaves may be underestimated when deriving it from micrometeorologically determined deposition fluxes alone. Given their phytotoxic potential (Wellburn, 1990), this could result in a misinterpretation of the potential risk of NO_x deposition for plants. Corrections based on the parametrization of NO_x emission by soils and subsequent chemical reactions below and within the canopy are necessary, but experimental data is scarce.

Mechanistic studies using enclosure methods under controlled conditions, in combination with micrometeorological deposition measurements, have provided a general understanding of the main foliar deposition pathways for each compound. Commonly, plants or plant parts are enclosed in containers and are exposed to known concentrations of the gas of interest in a stream of air of known mass flow. Ventilators usually provide a homogeneous turbulent mixture of the air. After subtraction of the amount absorbed by other surfaces, concentration differences between inflow and outflow are attributed to deposition to the enclosed plant parts. Alternatively, isotopic nitrogen tracers (^{13}N, ^{15}N) can be used to monitor deposition (for references see section IV). They provide a direct quantification of the influx of airborne nitrogen compounds into the plant, and the distribution of the nitrogen moiety of the applied compound within the plant and its assimilation can be followed. Widespread use of tracers has been restrained by the cost of the compounds and/or the analytic equipment needed. In the case of ^{13}N the short half-life (9.96 min) requires proximity of a production facility for this isotope. Enclosure methods enable a study of the parameters influencing deposition processes in detail by controlled manipulation. They yield valuable information for parametrization of resistances in deposition models, but extrapolation to ambient conditions may be difficult. Skärby et al. (1981), Johansson (1987), Dasch (1989), Vose and

Swank (1990) and Rondon et al. (1993), successfully applied enclosures of branches in the open field, but prolonged use of any enclosure system results in significant deviations in micrometeorological parameters from ambient conditions. In turn, this alters the absorption characteristics of a plant. Hence, enclosure methods are not useful for monitoring purposes. A more promising alternative seems to be exposure in open top chambers placed in the field and 'enclosing' groups of plants or trees in their natural habitat. Most of the mentioned disadvantages of each method are inherent and in many cases there are no accepted ways to translate results from one method to the other. Hence, in many cases listing results together with the applied method is the best we can do. In many of the studies addressing foliar deposition processes of NO_y or physiological responses of plants to NO_y, unrealistically high concentrations of the gases were used. Extrapolating results of such experiments to lower concentrations is prone to errors due to the non-linearity of the responses of plants to environmental changes.

III. DEPOSITION, ABSORPTION AND UPTAKE OF AIRBORNE OXIDIZED NITROGEN

The following terminology is used in this text: The term 'exchange' is used as the generic term for deposition, absorption and uptake for one direction of flux, and for emission, desorption and excretion for the reverse processes. 'Deposition' is the transfer of a molecule from the air to any surface capable of interacting with this molecule. 'Absorption' is the capturing of a molecule by a specific surface. For plants, we extend this process to all reactions which take place between the air and the lipid bilayer of the plasmalemma. 'Uptake' is the active or passive transport of a molecule through the lipid bilayer of the plasmalemma into the cytosol. 'Emission', 'desorption' and 'excretion' are used for the respective reverse processes.

A. Deposition

The phase of the compound (gas, aerosol or liquid) and the morphology of the receptor (i.e., plant leaf characteristics) determine the deposition pathway, the rate of deposition, and the fate on or in the plant leaf (Fig. 4.1). The gases generally reach all plant-atmosphere interfaces via the dry deposition pathway, i.e., turbulent transport to the viscous sublayer (leaf boundary layer), molecular diffusion through the viscous sublayer to the cuticle or further through the stomata into the apoplastic space within the substomatal cavity. Particles have a

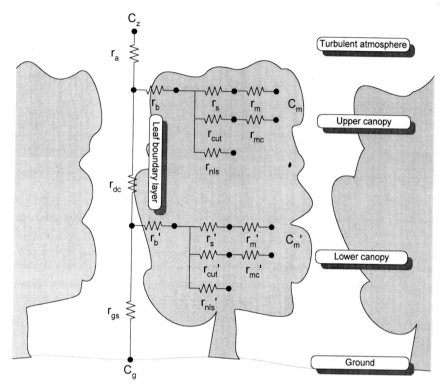

Fig. 4.1. Schematic diagram of pathway resistances for the dry deposition of a gas to a vegetated surface. C_z is the concentration of the gas at height z, C_g concentration at the ground. C_m and C_m' are the concentrations in the mesophyll in the upper and lower canopy, respectively. r_a is the resistance to transport through the turbulent layer of air between z and the vegetation (aerodynamic resistance), r_b and r_b' the resistances to transport through the quasi-laminar layer surrounding vegetation elements (leaf boundary layer resistance), r_{dc} resistance for a gas transfer affected by buoyant convection within the canopy, r gs resistance of the soil and any surface covering the soil (leaf litter etc.), r_s and r_s' stomatal resistances to transport to the substomatal cavity (see Fig. 4.2), r_c and r_c' resistances to absorption by the cuticle, r_m and r_m' mesophyll resistances in the substomatal cavity, r_{mc} and r_{mc}' mesophyll resistances below the cuticle and r_{nls} and r_{nls}' resistances for absorption to exposed surfaces other than leaves, such as the bark. The mesophyll resistances are a combination of resistances to absorption in the extracellular fluid and chemical or enzymatic reactions therein, diffusion through the cell wall, transport through the lipid bilayer of the plasmalemma, and chemical and enzymatic reactions within the cytosol, including exchange with cell organelles. Due to the complex structure of canopies, resistance models for vegetated surfaces become more and more detailed (see Kramm et al., 1995).

considerably smaller diffusion coefficient than gases. Their dry deposition flux is dominated by turbulent transport and impaction on the cuticle. Similarly, cloud droplet interception occurs via occult deposition. Ions dissolved in rainwater reach the vegetation by wet deposition.

Thus for particulate, rain and cloud droplet nitrate, diffusive transfer to the interface is not important.

Models of dry deposition of gases describe the transfer between atmosphere and surface in analogy to an electrical resistance scheme, wherein the concentration gradient between the free air at a reference height and the surface acts as the driving force. There are three resistances to be considered: aerodynamic resistance (r_a), boundary layer resistance (r_b) and surface resistance ($r_c = r_{tot} - [r_a + r_b]$). r_a, representing the resistance to transport through the turbulent layer of air between the reference level and the vegetation, is equal for all gases; r_b, the resistance to transport through the viscous sublayer surrounding vegetation elements depends on the diffusion coefficient. Often, r_b can be neglected, when compared with r_a and/or r_c. Both r_a and r_b are related to wind speed, surface roughness and atmospheric stability. The surface resistance r_c has to incorporate a set of different processes (discussed below) represented by a set of parallel and serial resistances (Fig. 4.2) which exhibit different characteristics towards each compound. They change as a function of position within the canopy, micrometeorological conditions at this position and plant status. In this context, gas deposition velocity (v_d) describing deposition to whole canopies is formulated as follows:

$$v_d \equiv -F_c/(C_z - C_s)$$

where F_c is the flux density (n mol m^{-2} s^{-1}), C_z is the concentration at height z and C_s is the concentration on the surface (n mol m^{-3}) which can be set to zero when no emission occurs. The SI unit for v_d is m s^{-1}. The review by Hanson and Lindberg (1991) summarizes dry deposition data and reports values of v_d for most reactive nitrogen compounds. These vary considerably depending on the method used and the type of vegetation. At the leaf level, conductance (K_l, m s^{-1}) is used in analogy to v_d and the corresponding equation reads:

$$K_1 \equiv -F_1/(C_a - C_i)$$

where F_1 is the flux towards the leaf (n mol m^{-2} s^{-1}), C_a is the concentration of the gas in the air surrounding the leaf and C_i is the concentration of the same gas on or in the leaf (n mol m^{-3}), which again is set to zero in the absence of detectable emission. The leaf level conductances are typically 4–20 times lower than deposition velocities, which is mainly a result of the leaf area associated with the canopy (Hanson and Lindberg, 1991). A crude conversion of K_1 to the canopy level deposition velocity can therefore be obtained by multiplication with the leaf area index (LAI, m^2 m^{-2}),

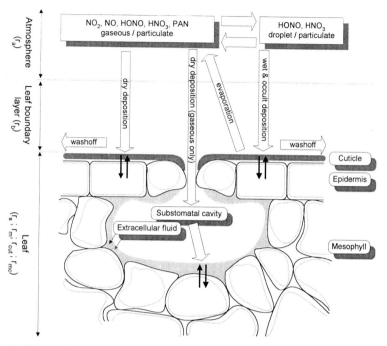

Fig. 4.2. Schematic leaf cross-section with an opened stomate. Open arrows show important pollutant fluxes outside the plant (deposition, emission). Black arrows symbolize absorption and uptake processes. Width of the arrows is not representative of flux magnitudes. r_a, r_b, r_s, r_m, r_{cut} and r_{mc} are pathway resistances (for notation see Fig. 4.1).

$$v_d = K_1 * \text{LAI}.$$

More sophisticated approaches are given by Baldocchi et al. (1987), Baldocchi (1988) and Kramm et al., (1995). The most advanced of these include a detailed description of the viscous sublayer as function of surface structure, canopy position and micrometeorological conditions. K_1 and v_d are inversely correlated with the sum of the resistances to deposition (r_a and r_b), absorption and uptake (r_c). The simplest formula for this relationship reads,

$$|v_d| = (r_a + r_b + r_c)^{-1}.$$

If the processes following dry deposition are very fast, the aerodynamic resistance, r_a, is limiting, so that r_a and v_d may be derived via micrometeorological methods from flux measurements directly. A zero surface resistance is generally assumed for HNO_3 (Walcek, 1986; Dollard et al., 1987; Harrison et al., 1989) although this assumption has only been tested for a limited number of vegetation types (Huebert and Robert,

1985; Lindberg and Lovett, 1985; Dasch and Cadle, 1986; Johansson and Granat, 1986; Meixner et al., 1988). Meyers et al. (1989) reported a mean value for v_d of 4 cm s^{-1} over a fully developed canopy of a deciduous forest with an estimated LAI of 7, ranging from 2.2 to 6 cm s^{-1}. Harrison et al. (1989) pointed out that these estimates are not hampered by the heterogeneous equilibrium of HNO$_3$ with NH$_3$ and solid NH$_4$NO$_3$. Their values of v_d ranged between 0.4 and 7.7 cm s^{-1}. On an annual basis v_d reaches maximum values of 2 cm s^{-1} over low vegetation and 4 cm s^{-1} over forests (Duyzer and Fowler, 1994). For agricultural crops with a large LAI, v_d may even reach 13.5 cm s^{-1} (Dollard et al., 1987). These authors also pointed out that in the case of a negligible surface resistance, r_c, r_b may not be neglected. In non-homogeneous terrain under the influence of canopy edges, even the aerodynamic resistance is reduced considerably (McMillen, 1990). Therefore, large uncertainties are associated with HNO$_3$ deposition estimates.

The very few studies dealing with HNO$_3$ deposition at the leaf level reveal interesting features about the processes involved. Cadle et al. (1991), working with a reliable HNO$_3$ source and analysis system in fumigation chambers, concluded from measurements with white pine, Norway spruce and red spruce seedlings that the leaf level conductance of 3.1 cm s^{-1} for HNO$_3$ amounted to about 80% of the conductance due to the aerodynamic and boundary layer resistance, which means that an HNO$_3$ molecule has a chance of 20% of not interacting with a clean leaf surface. This chance increased up to 95% during continued exposure leading to a considerable decrease in conductance down to 0.01 cm s^{-1}, which was then assigned to a transcuticular resistance. In addition, these authors were able to resolve a stomatal component, which made up about 7% of the total conductance during the day. In contrast, the dry deposition flux of NO$_2$ is limited by diffusion through the stomata to the substomatal cavity (Duyzer and Fowler, 1994). There, the absorption process (see below) is fast enough so that the surface concentration is near zero. Therefore, the resistance model should include the sum of the aerodynamic and the boundary layer resistance, $r_a + r_b$, and the stomatal resistance, r_s only. Since the stomatal resistance for NO$_2$ is linearly correlated with the resistance for water vapour (i.e., the same diffusional barrier for the opposite flux), its behaviour as a function of environmental parameters, water status, canopy structure, vegetation type and species is comparatively well known. A set of species-specific constants for the parametrization can be incorporated in advanced canopy stomatal resistance models (Baldocchi et al., 1987; Wesely, 1989; Kramm et al., 1995) and is provided by Turner and Begg (1973), Grace et al. (1975) and Jarvis (1976). Still, considerable discrepancies remain between measured fluxes and models, and a broad range of values of v_d

for NO_2 even for similar canopies have been reported. With micrometeorological methods, NO_2 may be the best studied member of the group of oxidized nitrogen compounds, and its deposition fluxes on the canopy level have been extensively studied throughout Europe and the United States. Hanson and Lindberg (1991) provided an extensive overview over v_d values found in field studies ranging up to 1 cm s^{-1}. Stomatal control of the NO_2 deposition flux has been verified by numerous studies. Maximum values of the pronounced diurnal cycles of 1.25 cm s^{-1} were observed by Hargreaves et al. (1992) over pasture. Eugster and Hesterberg (1996) reported daily average values between 0.11 and 0.24 cm s^{-1} over a litter meadow. Maximum v_d was 0.83 cm s^{-1} and 0.39 cm s^{-1} under moist and dry conditions, respectively. They tested different parametrizations and found significant differences in performance, especially when the water status seemed to play an important role. Wesely et al. (1982) determined v_d values of 0.77 cm s^{-1} during the day and 0.05 cm s^{-1} during the night above a soybean field. A similar range was observed for a wheat field by Hummelshoj et al. (1994). The complications inferred by chemical reactions of NO_2 within the canopy are addressed by Duyzer et al. (1995). The photochemical reactions of NO and NO_2 can result in bi-directional fluxes of NO_2. Therefore, a correction procedure minimizing the difference between derived and true fluxes to 5% during the day and 20% during the night was included. Choularton et al. (1994) reported corrected values of 0.4 to 0.5 cm s^{-1} measured in early morning and <0.1 cm s^{-1} during nocturnal periods over a Sitka spruce forest. With respect to finite fluxes in the night, Kerstiens et al. (1992) pointed out that the observed nocturnal fluxes above canopies may not only be the consequence of incomplete closure of stomata, but also of a non-negligible cuticular resistance (Kisser-Priesack et al., 1987; Kisser-Priesack and Gebauer, 1991).

At the leaf level, conductances and their stomatal control were the topic of studies using fumigation techniques in climate-controlled chambers and enclosure cuvettes in field or laboratory experiments. The conductance to NO_2 can be derived from the NO_2 removal in a flow-through system, or from the amount of a tracer, usually ^{15}N, absorbed by the leaves, provided the absorption processes are known (see below). Thoene et al. (1991) found a close linear relation between NO_2 uptake rate and the water transpiration rate with young spruce trees (*Picea abies* (L.) Karst.) for a broad range of concentrations. They mentioned the possibility for a compensation point corresponding to a non-zero surface concentration, C_i, within the substomatal cavity, which may have been the result of analytical problems at very low concentrations and of the concurrent NO fumigation applied. In a modelling approach based on a large number of data a compensation point for NO_2 deposition of

3.2 nl l^{-1} was found in *Picea abies* (Slovik et al. 1996). A non-zero compensation point would decrease the conductance considerably at low concentrations and the derivation of deposition velocities would be biased by an internal resistance. Rondon et al. (1993) addressed this question carefully using a dynamic chamber system in the field with Scots pine and spruce trees in Sweden. According to their results, an internal resistance would always be well below 0.2 nl l^{-1} NO$_2$, which was confirmed in the laboratory (Rondon and Granat, 1994). Still, this would have important implications for the modelling of NO$_2$ dry deposition in remote areas, as pointed out by Johansson (1987), who argued that with an LAI of 4, one would arrive at a canopy deposition velocity below 0.1 cm s^{-1}. This value is considerably lower than those derived from flux measurements over canopies cited above. Above 1 nl l^{-1} NO$_2$, Rondon et al. (1993) confirmed the linearity between NO$_2$ deposition rate and transpiration rate, although the absolute flux through the stomata seemed to be slightly lower than the flux calculated from the transpiration rate. This may be linked to the absorption processes, as discussed below. The same observation was made by Bruckner et al. (1993) with spruce trees. Stomatal control of NO$_2$ deposition was also obtained for sunflower (*Helianthus annuus* L.) and tobacco plants (*Nicotiana tabacum* L.) by Neubert et al. (1993) at 50 nl l^{-1}, and by Segschneider et al. (1995) for *Helianthus annuus* between 5 and 73 nl l^{-1}. With *Triticum aestivum*, Weber and Rennenberg (1996) were able to detect emission below 2 nl l^{-1}. With a stomatal conductance of 0.2 cm s^{-1} the observed emission flux was 3.7 ng N m^{-2} s^{-1}. Okano et al. (1988, 1989) also observed stomatal control of NO$_2$ deposition for sunflower (*Helianthus annuus*), radish (*Raphanus sativus*), tomato (*Lycopersicon esculentum*), tobacco (*Nicotiana tabacum*), cucumber (*Cucumis sativus*), kidney bean (*Phaseolus vulgaris*), maize (*Zea mays*) and sorghum (*Sorghum vulgare*), and the broad-leaved tree species *Populus, Cinnamomum* and *Viburnum*, but at rather high concentrations, between 300 and 500 nl l^{-1}.

For NO, the situation seems more complex and far less well investigated than for NO$_2$. The assumption of a zero surface concentration, C_s is not fulfilled, and surface resistance cannot be described by stomatal conductance alone. The observed deposition fluxes are considerably smaller than for NO$_2$ at comparable concentrations. In addition, soil NO emissions are strong enough to compete with deposition in most ecosystems so that net NO fluxes measured with micrometeorological methods cannot be converted to deposition velocities without large uncertainties (Johansson, 1987, Duyzer and Fowler, 1994). Most knowledge about deposition to the leaves must be derived from leaf-level measurements to be incorporated into NO dry deposition models. Skärby et al. (1981) and Saxe (1986b) found a dependency on stomatal conductance, but the flux to leaves seemed to be limited by an internal resistance. Neubert et al. (1993) studied the relation of NO deposition to stomatal conductance in more detail for sunflower (*Helianthus annuus* L.) and tobacco (*Nicotiana tabacum* L.) and assigned the observed limitation to an internal resistance for concentrations above 20 nl l^{-1}. The deposition velocities obtained for sunflower (0.15 cm s^{-1} in light and 0.09 cm s^{-1} in darkness) and tobacco (0.17 cm s^{-1} and 0.09 cm s^{-1}) were in the range of values reported earlier (e.g. Hanson and Lindberg, 1991).

Very little is known about dry deposition of PAN. Older work by Garland and Penkett (1976) revealed deposition velocities of about 0.25 cm s^{-1} over grassland, and the surface resistances were much larger than the aerodynamic resistance. Okano et al. (1990) presented reliable measurements of the uptake rate in herbaceous species using a novel, pure PAN source in a fumigation system at concentrations of 100 nl l^{-1}. They derived deposition velocities between 0.15 and 2.0 cm s^{-1} and found a strong correlation between uptake rate and stomatal conductance, and only minor adsorption to the surface.

Due to low diffusion coefficients, diffusive transfer of particles through the viscous sublayer is not an efficient process. Collection by the receptor surface occurs via inertial impaction, a process which strongly depends on the thickness of the viscous sublayer, the mass and the size of the particles. Recent field studies to quantify particle dry deposition clearly showed the influence of edge effects, canopy turbulence and structural characteristics of the canopy (e.g. Draaijers et al., 1992). Modelling (Kramm et al., 1995) cannot rely on zero surface concentrations and the resistance scheme in the form described above should be applied with care. Beswick et al. (1994) described the problems associated with the experimental approach to measure deposition fluxes. Their estimates of the deposition velocity for particles in the size range between 0.1 and 3.0 µm range from 0.5 to 3 cm s^{-1} over a forest canopy. Wyers et al. (1994) used the ^{214}Pb isotope as a tracer and arrived at 0.75 to 1.5 cm s^{-1} over the same forest for particles between 0.5 and 1.5 µm in diameter. Depending on the main processes governing nitrate formation in the tropospheric aerosol, nitrate may be associated with different modes of the size distribution. Nitrate on coarse particles is deposited much more efficiently (v_d > 10 cm s^{-1} — Sievering et al., 1993) than nitrate on fine particles (v_d < 0.05 cm s^{-1} — Lindberg et al., 1986). Over a forest canopy, particulate nitrate dry deposition velocities between 2 and 8 cm s^{-1} may be realistic (Sievering et al., 1993). Deposition of coarse particulate nitrate can make up a substantial fraction of the total nitrogen dry deposition, especially at remote sites (Lovett and Lindberg, 1993). Deposition of submicrometer aerosol particles appears to be inefficient but at high elevations such particles grow to haze droplets of 10 to 20 µm (diameter), which are efficiently captured by forest canopies. In general, particle deposition is rarely measured and the corresponding lack of knowledge may cause a significant underestimation of nitrogen dry deposition in some areas. To our knowledge, no leaf-level experiments have been performed to date to obtain more local information about the particle deposition processes.

B. Cuticular Adsorption

The net flux associated with the deposition process depends on the concentration gradient, which is strongly influenced by the absorption process on the cuticle or within the substomatal cavity. HNO_3 adheres readily on the cuticle, presumably as nitrate, upon hydrolysis which virtually results in a zero gas concentration above the surface. A small fraction is able to diffuse through the stomata (Cadle et al., 1991). With continued exposure, HNO_3 adsorption seemed to saturate, and the authors were able to quantify the maximum HNO_3 surface loading ranging from 0.7 to 1.3 µg cm^{-2}. Of the total amount of HNO_3 deposited, the fraction immediately bound (i.e., no longer available for washoff or revolatilization) to the cuticle was about 45% for hardwood foliage and only 3% for pine needles (Hanson and Garten, 1992), a strong indication that cuticular characteristics may have a strong influence on foliar exchange of deposited HNO_3. The affinity of NO and NO_2 to sorption on the epicuticular waxes is assumed to be small (Kisser-Priesack and Gebauer, 1991). The heterogeneous conversion of NO_2 to HONO and HNO_3 may become important at night (Sakamaki et al., 1983; Febo and Perrino, 1991), but may be of greater significance for atmospheric chemistry than for the plants. It may be assumed that particles reaching any foliar surface will adhere to it (Seinfeld, 1986). Only for dust particles may bounce-off become important. The amount of nitrate depositing on leaves from wet deposition cannot be derived from throughfall measurements due to the strong bias by washoff of dry deposited nitrate (Lovett and Lindberg, 1993). According to a laboratory study by Garten and Hanson (1990) using simulated rain, 13% and 27% of the nitrate applied was retained by the leaves of red maple (*Acer rubrum*) and white oak (*Quercus alba*) respectively. Similarly, Evans et al. (1986) found a retention between 2.4 and 24.8% on leaves of *Phaseolus vulgaris*. The tendency towards increased retention with decreasing pH was not significant. However, Momen and Helms (1996) were not able to detect any retention by needles of *Pinus ponderosa* (Dougl.). The smaller retention of rain nitrate compared to that of gaseous HNO_3 by comparable hardwood foliage is certainly due to the hydrophobic behaviour of epicuticular waxes. Again, as in the case of gaseous HNO_3, different morphologies may be responsible for the differences between conifer needles and other foliage.

Three absorption pathways are open to nitrogen compounds adhering to the cuticle from any deposition pathway:

1) Gases may be revolatilized and thereby enter the dry deposition or emission pathway; even nitrate may evaporate as ammonium nitrate, which may become important during dry periods (Kramm et al., 1995). Dasch (1989) argued, based on laboratory studies with elm (*Ulmus americana* L.) and oak (*Quercus palustris*, Muenchh.), that revolatilization of nitrate is generally not important except for detached leaves or during dormant periods.

2) Nitrate may be washed off and reach the soil via throughfall and stemflow, which seems to be the fate of 10 to 90% of the total dry deposited nitrate in forest canopies in temperate climates (Lovett and Lindberg, 1993). The net throughfall flux correlates linearly with nitrate dry deposition, especially for coniferous canopies. According to the results presented by Hanson and Garten (1992) with HNO_3 in laboratory fumigations, post-exposure uptake (i.e., transfer through the cuticle) was 1 to 2 orders of magnitude lower than total deposition during exposure, so that on hardwood foliage more than 50% and for conifers more than 90% of the deposited HNO_3 (-nitrate) should be available for washoff. Dasch (1989) and Cadle et al. (1991) arrived at similar conclusions. The differences between tree species also points to the influence of the cuticular composition on the fate of nitrate on leaves.

3) The leaf cuticle, the plant's major barrier against water loss, exhibited non-zero permeation coefficients for a variety of volatile and non-volatile organic and inorganic compounds in a number of studies (Guenthardt, 1984; Lendzian and Kerstiens, 1991; Kerstiens et al., 1992; Percy et al., 1993; Reynhardt and Riederer, 1994). Neubert et al. (1993) quantified the cuticular conductance for NO and NO_2 for sunflower (about 0.004 cm s^{-1}) and tobacco (about 0.002 cm s^{-1}) to be of the same order of magnitude as for water. Cadle et al. (1991) derived a cuticular conductance for HNO_3 (presumably as nitrate) of 0.01 to 0.02 cm s^{-1} for white pine, red spruce and Norway spruce. Still, the latter authors concluded that in rural areas most of the nitric acid deposited remains on the surface and thus is available for washoff. Dasch (1989) arrived at similar conclusions for elm (*Ulmus americana* L.) and oak (*Quercus palustris*, Muenchh.). Momen and Helms (1996) found no nitrate transfer into needles of *Pinus ponderosa* (Dougl.), in contrast to other studies in which the transfer of rain nitrate to the leaves via the soil was not separated (Reich et al., 1988; Wright et al., 1990). The chemical structure of the cuticular layer exhibits a range of functional groups providing a target for oxidative attack. Some nitration of cuticular components by NO_2 and HNO_3 seems to occur, mostly in the inner, hydrolyzable part of the cuticle (Lendzian and Kahlert, 1988; Cadle et al., 1991; Kisser-

Priesack and Gebauer, 1991), the binding rate being dependent on NO_2 concentration. Epiphytic lichens associated with the cuticle may further increase the nitrate consumption of forest canopies (Lovett and Lindberg, 1993) from which at least part may be released again as organic nitrogen to the throughfall and thereby offset part of the effective foliar nitrate consumption derived from deposition measurements. Permeability of the cuticle seems to be of quantitative significance for NO, HNO_3 and nitrate but not for NO_2. The irreversible binding of NO_2 and HNO_3 may contribute to chemical alterations of the cuticle. On average, forest canopies absorb about 50% of the deposited nitrate (Lindberg et al., 1990), mostly via the described cuticular pathway.

C. Substomatal Cavity Absorption

No significant barrier, such as the cuticle, seems to be present in the substomatal cavity, so that the predominating absorption process is the transfer of gaseous NO, NO_2, HNO_3, PAN and HONO to the apoplastic fluid. This is mainly a solution process, eventually followed by reaction (e.g. dissociation or capture by organic molecules) and diffusion to the lipid bilayer of the plasmalemma. Often the solubility of the molecule is rate limiting at this stage. The solubility of NO in aqueous solutions is very small and the corresponding transfer resistance is larger than the stomatal resistance. In the case of NO_2, uptake rates are limited by diffusion through the stomata only, so that an efficient removal of gaseous NO_2 and conversion to a dissolved anionic form is required. The measured uptake rates could not be explained by aqueous disproportionation of NO_2 to nitrite and nitrate with the kinetic parameters given by Lee and Schwartz (1981). As an alternative mechanism, scavenging of NO_2 by an apoplastic antioxidant was proposed on the basis of high ascorbate concentrations found in the apoplastic fluid (Ramge et al., 1993), but to our knowledge this has not yet been verified experimentally. Furthermore, a discrepancy remained between the presence of nitrate postulated from the physiological response to NO_2 exposure and such a radical H-abstraction yielding nitrite only. Using radioactively labelled NO_2, Ammann et al. (1995a) recently showed that in spruce trees exposed to 80 nl l^{-1} NO_2 only nitrate was formed in detectable amounts. They argued that in an acidic solution, as is the case for the apoplast of conifers with pH between 5.0 and 5.5 (Pfanz and Dietz, 1987), nitrite may be revolatilized as HONO in analogy to the corresponding reaction of NO_2 with cloud droplets studied by Bambauer et al. (1994). In contrast to Lee and Schwartz (1981), Bambauer et al. (1994) found a pseudofirst-order kinetics in NO_2, which corresponds to the linear relation found between NO_2 uptake rates and NO_2 concentration down to very low levels. Therefore, the amount of nitrite formed depends on the pH of the apoplastic solution, which for most plants is higher than for conifers (White et al., 1981, Hartung et al., 1988); thus nitrite may often be formed at the same rate as nitrate. Various authors have reported the presence of nitrite in plant extracts (without indication of localization, however; see section IV). This nitrite seemed to be the result of nitrate reduction and/or inhibited or incomplete nitrite reduction within the symplast, rather than a nitrite-forming mechanism in the apoplast.

To our knowledge, no mechanistic study addressing the processes of NO absorption beyond the substomatal cavity has been undertaken. The second-order rate constant for the aqueous phase NO reaction forming nitrite with concomitant absorption of NO_2 is an order of magnitude lower than that of NO_2 alone (Seinfeld, 1986). Nevertheless, for most relevant situations (except in the total absence of NO_2), the nitrate ion will be the preferred state of the dissolved gases at equilibrium.

The dissociation of HNO_3 and HONO to nitrate and nitrite respectively, again depends on the pH of the apoplastic solution, and it may well be dissociation rather than solubility, which limits the formation of nitrate and nitrite, and thereby promotes the transport of the undissociated acids in either direction.

Although PAN is not very soluble in water, correlation of the deposition flux with stomatal conductance indicates that the dissolution and removal process in the apoplast is fast enough to cope with the stomatal influx.

D. Uptake

As the most relevant absorption processes of oxidized nitrogen compounds finally yield HNO_3, HNO_2, nitrite and nitrate in the extracellular fluid, the biological fate of deposited oxidized nitrogen depends on transport through the lipid bilayer of the plasmalemma. Gutknecht and Walter (1981) investigated non-ionic diffusion of molecular HNO_3 and found permeabilities ranging from 0.0001 to 0.001 cm s^{-1} whereas the permeability for the nitrate ion is five to seven orders of magnitude lower. It may therefore be assumed that uptake of oxidized nitrogen absorbed by leaves happens by an active transport. However, no report of such a transport system in leaves is available.

IV. PHYSIOLOGICAL CONSEQUENCES OF FOLIAR ABSORPTION OF OXIDIZED NITROGEN

As outlined above, the compounds entering the symplast from foliar absorption of airborne oxidized nitrogen are assumed to be predominantly nitrate and nitrite or the undissociated corresponding acids. Once inside the symplast, both ions are substrates or products of the enzymatic assimilatory nitrate reduction to ammonia taking place in plant cells. Subsequent incorporation into amino acids and other organic compounds removes ammonia, a toxic agent for plants. Andrews (1986) states that this pathway is found in leaves of all terrestrial higher plants, but the distribution of nitrate reduction between root and shoot varies considerably between species (Smirnoff et al., 1984, Andrews, 1986). Thus, airborne nitrogen absorbed by the leaves may contribute to the nutrition of terrestrial plants, as demonstrated by incorporation studies using labelled nitrogen oxides (Durmishidze and Nutsubidze, 1976; Rogers et al., 1979; Yoneyama and Sasakawa, 1979; Kaji et al., 1980; Rowland et al., 1987; Latus et al., 1990; Nussbaum et al., 1993; Ammann et al., 1995a; Segschneider et al., 1995; Weber et al., 1995). Faller (1972) reported sustained growth of N-deficient *Helianthus annuus* L. for 3 weeks, NO_2 (800–3,100 nl l^{-1}) being the only nitrogen source. On the other hand, Spierings (1971) registered growth and yield reductions of *Lycopersicon esculentum* Mill. exposed to 250 nl l^{-1} NO_2 for 4 months, illustrating that dry deposition of this gas can have negative effects as well. This ambivalence has been the subject of many studies since and discussion about the relative importance of the two aspects is still ongoing (see Wellburn, 1990; Srivastava et al., 1995). We shall give an overview of the literature with a special emphasis on 'real world' conditions.

A. Nitrogen Metabolism

1. *Nitrate Reductase and Nitrate Content in Shoots and Roots*: Reduction of nitrate to nitrite is catalyzed by NAD(P)H-dependent nitrate reductase (NR; E.C. 1.6.6.1 and 1.6.6.2) localized in the cytosol. This enzyme is nitrate-inducible and inhibited by NH_4^+ and some end-products of amino acid biosynthesis, such as glycine glutamine and asparagine (Radin, 1975; Beevers and Hageman, 1980; Rajasekhar and Oelmüller, 1987; Vincentz et al., 1993). Nitrate reduction is thought to be the rate limiting step of assimilatory nitrate reduction. Changes in NR activity may be an indication of changes in the metabolic pool of nitrate. A number of studies on the effects of airborne oxidised nitrogen compounds have reported activity changes in both shoot and root (Table 4.2). An increase in NRA is generally accompanied by an increased nitrate content of the same tissue. Part of this accumulation may result from an increase in transitory storage pools. Srivastava et al. (1994b) reported that NR activity in *Hordeum vulgare* L. (cv Zephyr and NO_2 tolerant mutants) leaves was still significantly enhanced 3 d after termination of fumigation with

Table 4.2. Effect of NO_x on nitrate reductase activity (NRA) and nitrite reductase activity (NiRA) in leaves (arrow before slash) and roots (arrow after slash); arrows are missing when no measurements are reported. ↑: significant increase compared to clean air control; →: no effect compared to control; ↓: significant decrease compared to control. Arrows in parentheses are substantial but not significant changes compared to the control. Nitrate and nitrite concentrations are given when measured simultaneously. Root N-supply form: s = solid substrate (soil, perlite etc.) with form of applied fertilizer; h = hydroponic culture systems

Species	Root N-supply form	mM	Pollutant	Conc. nl l⁻¹	Duration	NO_3^-	NRA	NO_2^-	NiRA	Reference
Clenidium molluscum	s		NO	35	24 h		↓/			Morgan et al., 1992
Homalothecium sericeum							↓/			
Pleurozium schreberi							↓/			
Hylocomium splendens							↓/			
Clenidium molluscum			NO_2				↑/			
Homalothecium sericeum							↑/			
Pleurozium schreberi							↑/			
Hylocomium splendens							↑/			
Hordeum vulgare cv Patty	h/NO_3^-	0.01	NO_2	300	9d	↑/→	↑/↓		↑/→	Rowland et al., 1987
	h/NH_4^+	0.1				↑/→	↑/↑		↑/→	
Hordeum vulgare cv Steptoe	h/NO_3^-	5.0	NO_2	300	9d		↓/→			Rowland-Bamford et al., 1989
	h/NH_4^+						↑/→			
	h/NO_3^-						↓/↓			
nar1a (NR deficient)	h/NH_4^+						↓/↓			
	h/NO_3^-						↓/↓			
nar2a (NR deficient)	h/NH_4^+						→/↓			
Hordeum vulgare cvs Maris	—	0	NO_2	500	3d		↑/↑			Srivastava et al. 1994b

Species			Exposure						Reference
Mink/Zephyr and 2 NO$_2$ resistant mutants		h/NO$_3^-$ 5.0				↑/↑			
Lolium perenne			NO$_2$	110 140 d/15 h d^{-1}				↑/ ↑/ ↑/ ↑/	Wellburn et al., 1981
Dactylis glomerata									
Phleum pratense									
Poa pratensis									
Lycopersicon esculentum s									Murray and Wellburn, 1985
cv Ailsa Craig			NO$_2$	1,500	18 h	↑/→	↑/→	↑/→	
			NO			→/→	→/→	↑/	
cv Eurocross BB			NO$_2$			↑/→	↑/→	→/↑	
			NO			→/→	→/	→/	
Capsicum annuum									
cv Bell Boy			NO$_2$			→/(↓)	→/↑		↓/→
			NO			→/→	→/		→/
			NO$_2$			→/	→/		↑/
cv Rumba									
Lycopersicon esculentum									Wellburn et al., 1980
cv. Ailsa Craig s			NO	400	28 d	↓/			↑/
Phaseolus vulgaris s			NO$_2$	4,000	6 h (light)			↑/ ↑/ ↑/	↑/ ↑/ ↑/ ↑/ ↑/	Yoneyama et al., 1979
Helianthus annuus									
Zea mays									
Phaseolus vulgaris					6 h (dark)			↑ ↑ ↑	↑/ ↑/ ↑/
Helianthus annuus									
Zea mays									
Phaseolus vulgaris s/NO$_3^-$/NH$_4^+$			NO$_2$	39	34 d (1988)	→/	→/		→/	Bender et al., 1991
				31	34 d (1989)	→/	↑/		↑/
				39	48 d (1988)	→/	→/		↑/
				31	48 d (1989)	→/	→/		↓/

(contd.)

Table 4.2 (contd.)

Species	Root N-supply form	mM	Pollutant	Conc. nl l⁻¹	Duration	NO_3^-	NRA	NO_2^-	NiRA	Reference
Picea abies	s		NO_2	12	3 d		↑/			von Ballmoos et al., 1993
Picea abies	s		NO_2	78* 125*	72 h		↑/ (↑)/		(↑)/ →/	Thoene et al., 1991
Picea abies	s/NO_3^-		NO	1,400	6 d		↑/		→/	Nussbaum et al., 1991
Picea abies	s		NO_2	60	4 d		↑/			Nussbaum et al., 1993
Picea abies	h/NO_3^-	0.72	NO_2	500	77 d	↑/→	↑/↓		↑/↓	Tischner et al., 1988
Picea abies	s/NO_3^-/NH_4^+	2.0	NO_2	100	28 d 49 d	→/↑ ↑/↑	→/↑ ↑/↑			Muller et al., 1996
Picea rubens	s		NO_2 HNO_3	75	11 d 3d		↑/ ↑/	→/ →/		Norby et al., 1989
Pinus sylvestris	s		NO/NO_2	8-12/ 70-80	14 d	↑/	↑/			Wingsle et al., 1987
Pisum sativum	s/NH_4^+		NO_2	4,000/ 8,400	1 h	↑/	↑/	↑/		Zeevaart, 1974 Zeevaart, 1976
Spinacia oleracea	s		NO_2	8,000	6 h (light) 6 h (dark)			→/ ↑/	→/ ↑/	Yoneyama and Sasakawa, 1979
Triticum aestivum	s/NO_3^-/NH_4^+		NO_2	60	2 d		↑/		↑/	Weber et al., 1995
Zea mays	h	0	NO_2	100 520	14 d 1–7 d		→/ ↑/			Kast et al., 1995

* Identical pollutant deposition at both concentrations.

500 nl l^{-1} NO$_2$ for 3 d. Transitory storage pools were also discussed by Rowland-Bamford et al. (1989). Using NR-deficient mutants, these authors found that NR activity is not a requirement for NO$_2$ absorption by barley. Weber et al. (1995) found only a transitory increase in foliar NR activity during the first 72 h of fumigating *Triticum aestivum* L. (cv Obelisk) with 60 nl l^{-1} ^{15}NO$_2$. The decrease in NR activity coincided with a decrease in the incorporation of ^{15}N-label into free amino acids. This could be another indication for the existence of storage pools for nitrate, since stomatal conductance did not change during the experiment, and NO$_2$ absorption is proportional to this parameter (Weber, and Rennenberg, 1996). Unfortunately nitrate content and its labelling were not measured. Murray and Wellburn (1985) showed that the effect of NO$_2$ on NR activity in leaves of *Lycopersicon esculentum* Mill. was cultivar specific. This parameter was not affected in either *Capsicum annuum* L. cultivars exposed in the same experiment. Exposure to the same concentration of NO produced no effect on NR activity and nitrate content in the shoots of the two species. Interestingly, exposing *L. esculentum* (cv Ailsa Craig) to NO for a longer time at lower concentrations depressed NR activity (Wellburn et al., 1980). Bender et al. (1991) found in long-term studies with *Phaseolus vulgaris* that the response of NR activity to low NO$_2$ concentrations also depended on the duration of the fumigation, indicating the importance of either the developmental stage of the plant or the applied dose. No effect of NO$_2$ on root NR activity or nitrate concentration was observed in any of these studies. Such effects were obvious in *Hordeum vulgare* L. fumigated with NO$_2$ (Rowland et al., 1987; Rowland-Bamford et al., 1989). In cv Patty they depended on nitrate concentration in the nutrient solution, and in cv Steptoe on the form of root nitrogen-supply. The reaction of foliar NR activity in cv Steptoe was the opposite on nitrate-containing nutrient solution compared to ammonium as the nitrogen-source. On nitrate-containing nutrient solution, a decrease as opposed to an increase in cv Patty was observed. Thus, *Hordeum vulgare* showed cultivar-specific responses when grown with nitrate as the sole nitrogen-source. However, it has to be pointed out that the nitrate concentrations used were not identical.

Interactions between atmospheric and pedospheric N-input on the reaction of foliar NR activity were also monitored by other researchers. Srivastava and Ormrod (1989) noted that the increase in foliar NR activity of *Phaseolus vulgaris* L. was the same with or without exposure to 500 nl l^{-1} NO$_2$, when 10 mM nitrate was added to nitrogen-starved plants. Srivastava and Ormrod (1984) thoroughly investigated the impact of nutrient-nitrogen concentrations on the reaction of foliar NR activity and foliar nitrate concentration of *Phaseolus vulgaris* to different

NO_2 concentrations. Their results are summarized in Fig. 4.3. The foliar nitrate reduction capacity appears to become saturated by increasing amounts of nitrate imported from the roots at increasing root nitrate supply. Therefore, at high nitrate supply the additional nitrate supplied by NO_2 cannot be efficiently removed and is accumulated. It is possible that with prolonged fumigation with high NO_2 concentrations this saturation eventually leads to a measurable increase in internal resistance to NO_2 absorption. Such resistances were reported by Srivastava and coworkers (1975a, b) for *Phaseolus vulgaris* L. (cv Pure Gold) at NO_2 concentrations between 3000 and 7000 nl l^{-1} and by Skärby et al. (1981) in *Pinus sylvestris* L. for much lower NO_2 concentrations (320 nl l^{-1}). The lower concentration for Scots pine may be an expression of differences between the two species in the distribution of nitrate reduction between shoot and root. Rowland et al. (1987) and Srivastava et al. (1994b), detected no difference in the reaction of foliar NR activity to different root nitrate supplies in barley. According to Srivastava and Ormrod (1984) this was not expected in bean with the nitrate concentrations used.

Cotyledons apparently exhibit different reaction patterns. NR activity in *Cucurbita maxima* Duch. cotyledons was reversibly reduced by 4000 nl l^{-1} NO_2 applied for 2 to 6 h, both *in vivo* (–50 %) and *in vitro* (–20 %), with no change in nitrate content (Takeuchi et al., 1985). In contrast, no change in activity occurred in the primary leaves, where nitrate content slightly decreased. NO_2 concentrations in the range between 500 and 4000 nl l^{-1} progressively reduced light-induced NR activity increase in squash cotyledons after 34 h of darkness (Hisamatsu et al., 1988). Apparently this was due to a suppressed *de novo* synthesis of the NR protein, as indicated by immunoblotting.

In annual crops with a relatively high proportion of foliar nitrate reduction, no generalization is possible for the reaction of NR to atmospheric oxides of nitrogen. Increases in NR activity were only reported for exposures to NO_2 whereas NO fumigations prompted no reaction (Murray and Wellburn, 1985), or reductions in leaf NR activity (Wellburn et al., 1980). The effects depended on species, cultivar, plant organ, developmental stage, photosynthetic photon flux density (Srivastava et al., 1990), N-nutrient form and concentration, and also on the concentration of the pollutant and the duration of exposure. Most of the reported effects on NR activity and nitrate concentrations were induced by NO_x concentrations above 200 nl l^{-1} applied over short periods of time with plants grown with low nitrogen supply. These are conditions which are not likely to occur at locations where annual crops are grown in the open field. Studies employing realistic conditions are scarce. Also,

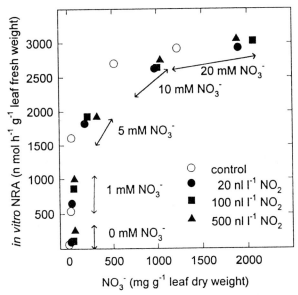

Fig. 4.3. Foliar *in vitro* nitrate reductase activity (NRA) of *Phaseolus vulgaris* seedlings grown with solutions containing different nitrate concentrations and exposed to different NO_2 concentrations for 5 d plotted as a function of foliar nitrate concentration. Different symbols show different pollutant concentrations (see legend). Arrows mark groups of plants grown at the same nitrate concentration, as indicated in the accompanying text. Results are from Srivastava and Ormrod (1984).

we are not aware of any reports about fumigation of annual crops with HNO_3, HONO or PAN addressing the question of induced changes in NR activity or nitrate pools.

Morgan et al. (1992) showed differential responses of NR activity in bryophytes exposed to 35 nl l⁻¹ NO or NO_2 for 21 d. Bryophytes differ from higher terrestrial plants in their lack of stomata, cuticles, and vacuoles for storing excess cytosolic nitrate, as well as in the absence of a long-distance transport system. NR activity was increased transiently in response to NO_2 and was permanently reduced by NO. NR inducibility by nitrate spraying at the end of the experiment was severely impaired by both treatments. This effect was reversed in clean air. The authors speculate that toxic nitrite accumulation and subsequent reduction in clean air was responsible for the observed effects.

Plants which reduce nitrate predominantly in the roots may show different characteristics in the reaction to oxidized atmospheric nitrogen compounds compared to plants which are capable of reducing at least some of the nitrate taken up by the roots in their leaves. Most conifers belong to the first category. In this case foliar NR is barely inducible by nitrate supply to the roots (Adams, and Attiwill, 1982; Smirnoff et al., 1984; Wingsle et al., 1987; L'Hirondelle et al., 1992; von Ballmoos, 1995). Consequently, in clean air leaf NR activity is invariably low and there is no need for a quick reaction to changing nitrate levels. Norby et al. (1989) and Brunold et al. (1997) registered no NR activity increase in needles of *Picea rubens* Sarg. and *Picea abies* (L.) Karst., respectively, within the first 4 h of fumigation with NO_2 < 80 nl l^{-1}. After that both authors reported significant NR increases. This time lag is sufficient for *de novo* synthesis of the enzyme (Galangau et al., 1988). Ammann et al. (1995a) presented evidence for a constitutive small level of NR in *Picea abies* needles. After only 5 min of fumigation of trees preconditioned in filtered air with 81 nl l^{-1} $^{13}NO_2$, the label was found in organic fractions. In contrast to conifers, the short NR activity induction times reported by Zeevaart (1974; 10 min with 4,000 nl l^{-1} NO_2) in *Pisum sativum* L. and by Srivastava and Ormrod (1989; 1 h with 500 nl l^{-1} NO_2) in *Phaseolus vulgaris* point to an activation/deactivation cycle, such as the one described by Kaiser et al. (1993). However, there is no induction study with annual crops using equally low NO_2 concentrations, so there is still the possibility that the observed difference was due to the pollutant concentrations applied.

In Norway spruce (*Picea abies*) needles a significant increase in NR activity was measured at an NO_2 concentration of 12 nl l^{-1} (von Ballmoos et al., 1993). In the range between 0 and 85 nl l^{-1} changes in NO_2 concentrations over 3-d intervals correlated linearly with NR activity in needles, and $^{15}NO_2$ absorption was linearly related to NR activity. This was confirmed in a field study along an NO_2 gradient generated by a highway, by measuring NR activity in 10 to 20-year-old spruce trees on days with comparable PAR (Ammann et al., 1995b). In a two-year measuring campaign, NR activity of needles of 25-year-old Norway spruce growing at 10 m distance from the same highway was consistently higher than at 1300 m. These findings suggest that NR activity in Norway spruce needles could be an indicator for NO_2 absorption under otherwise comparable conditions, provided that NR changes persisted over extended periods of time. This was shown by Egger et al. (1988) and von Ballmoos (1995) for periods of several days. Tischner et al. (1988) measured an increased NR activity in needles of Norway spruce seedlings after exposure to 500 nl l^{-1} NO_2 for 11 weeks. However, conflicting evidence is presented by Thoene et al. (1991) who fumigated single twigs of 8–9-year-old Norway spruce. NR activity rose to a maximum after 2 d, before returning to control levels after 4 d. Stomatal aperture and NO_2 absorption was not affected by the treatment. Qualitatively similar NR induction curves were measured by Wingsle et al. (1987) using NO_2 and Norby et al. (1989) using either NO_2 or HNO_3

although none of these studies reported return of NR activity to control levels. This again may be an indication of the existence of both metabolically active and inactive pools of nitrate and regulated fluxes between them due to a relatively high nutrient-nitrogen status of the plants. Norby et al. (1989) and Thoene et al. (1991) used commercial soil mixtures without indicating the nutrient contents, and Wingsle et al. (1987) mentioned regular application of fertilizer prior to the experiment. The nitrogen content of the soil used by von Ballmoos et al. (1993) was relatively low and about the same as reported by Ammann et al. (1995b) for a 30-year-old stand (0.3%). Muller et al. (1996) measured no reaction in either shoots or roots when exposing seedlings to 100 nl l^{-1} NO$_2$ for 4 weeks when cultivated on a perlite/clay mixture in contact with nutrient solution containing 2 mM nitrogen. Three weeks later NR activity in both shoots and roots was roughly double the control value. Likewise nitrate contents were no different after 4 weeks whereas a significant increase after NO$_2$ fumigation was found in both shoots and roots after 7 weeks.

Overall, the reaction to airborne oxidized nitrogen species of plants which predominantly reduce nitrate in the roots is not as well documented as for plants with concurrent leaf reduction under clean air conditions, but the concentrations the plants were exposed to were usually more realistic. Compared with annual crops, conifers tend to show a more pronounced relative increase in NR activity induced at lower concentrations of NO$_2$, but the reaction to fast changes in pollutant levels may be delayed. Such changes occur regularly along car traffic routes during the rush hour and may impose a comparatively higher stress on conifers growing nearby. As in annual crops, increased nitrate supply to the roots reduced NR induction in needles of Norway spruce (von Ballmoos, 1995). Nitrate concentrations of the plant tissues are seldom reported together with NR activity but the results presented show similar tendencies as in annual crops, in which measured NR activity in roots was influenced by atmospheric NO$_2$ exposure. Unlike for annual crops, there are reports of increased NR activity in conifers by NO (*Picea abies*, 1400 nl l^{-1}; Nussbaum et al., 1991) and HNO$_3$ vapour (*Picea rubens*, 75 nl l^{-1}; Norby et al., 1989). Norby et al. (1989) also applied nitrate-containing acidic mist (droplet size 50 µm) and found no significant increase in NR activity in the needles. L'Hirondelle et al. (1992) misted the same species with a pH 3.0, 10:1 nitrate sulphate (w/w) solution for one month (five 1 h events per night, droplet size ca 20 µm) and found an NR increase in both shoot and roots, which was more pronounced in plants at a low nitrogen-supply. Pearson and Soares (1995) likewise reported enhanced NR activity 24 h after exposing *Populus deltoides* for 1.5 h to nitrate-containing acid mist (droplet size < 5 µm) but the observed NR increase in *Picea sitchensis, Betula pendula, Quercus robur* and *Fagus sylvatica* was not significant. Most other studies with acid mist or acid rain application used combinations of N and S containing solutions in low component ratios, or they did not separate the soil from the atmosphere. The question whether nitrate and nitrite contained in mist droplets gain direct access to mesophyll cells is yet unresolved. Apparently the droplet size is of importance in eliciting NR responses. It has to be kept in mind that mist applications always include a certain amount of gaseous pollutant, which may reach the substomatal cavity; this fraction is probably enhanced with decreasing droplet size. No study of the effect of PAN or HONO on NR activity or nitrate contents in conifers is known to us.

2. *Nitrite Reductase and Nitrite Content in Shoots and Roots*: Nitrite reductase (NiR; E.C. 1.7.7.1) in non-green tissues is localized in plastids. In green tissues it is found in the chloroplasts (Emes and Fowler, 1979; Wallsgrove et al., 1979). It reduces nitrite to ammonium in a six-electrons consuming step. The electrons are supplied via ferredoxin, a primary electron acceptor of PSI, and NAD(P)H the actual reductant (Beevers and Hageman, 1980). It is thought to be induced by nitrate (Gupta and Beevers, 1983). Wellburn (1990) suggested that the increase in nitrite reductase (NiR) activity without concurrent NR increase observed in tomato (cv Ailsa Craig) exposed to 1500 nl l^{-1} NO (Murray and Wellburn, 1985; Table 4.2) may be an indication of inducibility by nitrite. Its specific activity in leaves is reported to be several times higher than that of NR (Wellburn et al., 1980; Murray and Wellburn, 1985; Weber et al., 1995; Rowland et al., 1987; Tischner et al., 1988; Nussbaum et al., 1991; Thoene et al., 1991). It is therefore thought that in undisturbed systems the removal of nitrite originating from nitrate reduction is efficient enough to prevent toxic accumulation of this compound. Since NO_2, NO and HONO absorption by leaves may result in enhanced nitrite levels in the cytoplasm, and substrate induction of NiR is still questionable, their uptake could impose some stress in plants.

Nitrite accumulations were reported by Zeevaart (1976), Yoneyama and Sasakawa (1979), Yoneyama et al. (1979), Kaji et al., (1980), Srivastava and Ormrod (1984), Takeuchi et al. (1985), Murray and Wellburn (1985), Wellburn (1985), Yu et al. (1988) and Shimazaki et al. (1992) fumigating annual crops with NO_2 in the concentration range between 1000 and 12,000 nl l^{-1} for a few hours (Table 4.2). Visible injury often coincided with a high nitrite content (Zeevaart, 1976; Yoneyama et al., 1979; Kaji et al., 1980). All but one of the species tested by Zeevaart (1976) responded with higher nitrite contents in darkness and increased damage compared to fumigations in light. Shimazaki et al. (1992) measured enhanced nitrite contents in darkness in *Spinacia oleracea* L. and *Phaseolus vulgaris* L. exposed to 8,000 nl l^{-1} NO$_2$, which returned to control levels within two hours after the onset of illumination. *Nicotiana glutinosa* behaved differently, showing more visible damage when nitrite contents in leaves were low during exposure in darkness (Zeevaart, 1976). The same reaction of nitrite content to illumination was reported for *Phaseolus vulgaris* L. (cv Edogawa), *Helianthus annuus* L. (cv Russian Mammoth) and *Zea mays* L. (cv Dento; Yoneyama et al., 1979). Besides differences between species in response to light treatment, there may also be differences within species, as suggested by studies with *Phaseolus vulgaris* (Yoneyama et al., 1979; Yu et al., 1988; Shimazaki et al., 1992). However, there are no investigations published that directly address cultivar specificity in this respect; therefore the possibility of differences being due to different cultivation/exposure protocols still exists.

Only one report of an increase in foliar nitrite content when the NO$_2$ concentration was below 1000 nl l^{-1} was found in the literature (Fuhrer and Erismann, 1980—Phaseolus vulgaris, 310 nl l^{-1} for 2 h). Some researchers have shown no increase or even a slight decrease at lower concentrations (Spierings, 1971—*Lycopersicon esculentum*, 250 nl l^{-1}; Yu et al., 1988—*Spinacea oleracea*, 1,000 nl l^{-1} in darkness; Norby et al., 1989—*Picea rubens*, 75 nl l^{-1}; Srivastava and Ormrod, 1989—*Phaseolus vulgaris*, 500 nl l^{-1}). Consequently, nitrite accumulation may occur only at exceedingly high concentrations, which is not representative for the conditions under which plants grow in the open. Apparently, they are capable of controlling nitrite concentrations at moderate pollutant levels. However, it has to be kept in mind that total leaf extracts may not be representative for single organelles, such as the chloroplast. More detailed studies addressing this problem are discussed below. No reports of enhanced nitrite contents in conifers are known to us. As pointed out in section III, recent investigations have questioned the production of nitrite in the apoplast by dissolution of NO and NO$_2$ at moderate mixing ratios for plants with a low apoplastic pH.

NiR activity changes are not linked to the respective nitrite content of different organs. Yoneyama et al. (1979) showed a good correlation between the two for *Helianthus annuus* leaves in light but poor correlation in darkness, and in *Phaseolus vulgaris* and *Zea mays* in both light and darkness. Yu et al. (1988) showed NiR increase in both spinach and kidney bean in darkness after a 6-h exposure to 4000 and 1500 nl l^{-1} NO$_2$ respectively, and in both species an increase in nitrite content was detected at 3500 nl l^{-1} in darkness. No increase in NiR activity was recorded for either species when exposed to 8000 nl l^{-1} in light, but *P. vulgaris* accumulated large amounts of nitrite and showed severe foliar injury within 2 h. Absolute activities were about 2.5 times higher in spinach which proved to be more resistant to visible foliar injury. The authors concluded that in short-term conditions (a few hours), sensitivity to high NO_2 concentrations is determined by the capacity to remove nitrite from the tissue.

Cultivar specific differences in the reaction of NiR activity to both NO and NO_2 fumigations were reported for *Lycopersicon esculentum* (Wellburn et al., 1980; Murray and Wellburn, 1985). In 3-h fumigations with NO concentrations between 400 and 2500 nl l^{-1}, maximum levels of NiR activity were higher in the less sensitive cultivar (Wellburn et al., 1980). In *Capsicum annuum* L. (cv Bell Boy) NiR activity even substantially decreased with both NO and NO_2 treatments, while cv Rumba showed no significant effect when exposed to NO_2 (Murray and Wellburn, 1985). Changes in nitrate supply to the roots did not influence the effect of NO_2 fumigation on *Hordeum vulgare* L. (cv Patty) NiR activity in either shoots or roots (Rowland et al., 1987). Again, in all these studies very high NO_x concentrations were used during short periods of time. In long-term studies NiR activity showed significant responses at much lower pollutant concentrations. The lowest NO_2 concentration causing an increase in NiR activity in plants with at least partial foliar nitrate reduction was 31 nl l^{-1} for 34 days (*Phaseolus vulgaris*; Bender et al., 1991) in the vegetative growth phase; during anthesis a significant decrease in NiR activity was noted. Another study with a realistic exposure regime by Wellburn et al. (1981) also showed a response of NiR activity to low NO_2 levels. In *Triticum aestivum*, Weber et al. (1995) measured a transient increase in short-term fumigation with 60 nl l^{-1}. These examples illustrated that extrapolating results from one time scale or concentration range to another should be done with great caution.

There are a few reports on nitrite contents and nitrite reduction in conifers. In aseptically grown *Picea abies* seedlings, Tischner et al. (1988) observed changes in NiRA similar to those of NRA in needles (increase) and roots (decrease). Thoene et al. (1991) found a consistent, but only in one instance significant increase in NiRA after exposure of twigs to 78 and 125 nl l^{-1} NO_2 for up to 72 h. The flux of NO_2 measured concurrently was identical at both concentrations due to stomatal closure at the higher concentration. NO at 1,400 nl l^{-1} caused no change in NiRA of *Picea abies* seedlings during a 7-d fumigation (Nussbaum et al., 1991).

Judging from the few published studies, foliar NiR activity is affected at much lower pollutant concentrations than nitrite content. Since the potential inducers of NiR are not known in detail, this should not be mistaken as an indication of an altered nitrite influx or a changing nitrite release by NR. However, as indicated by the differential response of NR and NiR activities described by several researchers, the induction signal and/or its transmission appears to be different for each enzyme. No fundamental differences between partial 'foliar nitrate reducers' and 'root nitrate reducers' are apparent. In the first group both NO_2 and NO may affect NiR activity. Given the lower concentrations at which effects were observed and the much lower affinity for absorption, NO was by far more efficient in evoking a response. No reports of effects on NiR activity by NO fumigation for conifers are known to us. Effects on root NiR activity were seldom measured and only Tischner et al. (1988) have reported on this aspect. No effects of HNO_3, HONO or PAN on NiR are reported in the literature.

3. *Incorporation of Oxidized Atmospheric Nitrogen into Organic Compounds*: Using isotope tracer techniques, airborne NO_2-nitrogen has been found to be incorporated into organic compounds within the plants (Durmishidze and Nutsubidze, 1976; Rogers et al., 1979; Yoneyama and Sasakawa, 1979; Okano et al., 1984; Latus et al., 1990; Ammann et al., 1995a; Segschneider et al., 1995). In most cases, 80% or more of the label was recovered in reduced or organic form within hours after commencement of fumigation, and incorporation was linear with time. Kaji et al. (1980) exposing *Helianthus annuus* L. (cv Russian Mammoth) and *Spinacia oleracea* L. (cv New Asia) to 4000 to 6000 nl l^{-1} $^{15}NO_2$ for 20 min recovered 99% of the label in the plant in reduced form when fumigating in light, whereas in darkness 11–15% was present in acidic forms, possibly nitrate and nitrite. Segschneider et al. (1995) found a linear relation between $^{15}NO_2$ concentration and $\delta^{15}N$ in all nitrogen pools analyzed (total N, NO_3-N, soluble amino acids, protein-N) after fumigating *Helianthus annuus* L. (cv Giganteus) for 8 h in light with 5 to 40 nl l^{-1}. More than 95% of the label was detected in reduced form. Similar fractions of labelled NO_2-nitrogen in reduced form were found in *Phaseolus vulgaris* L. (cv Bush Blue Lake 290) exposed to 320 nl l^{-1} NO_2 for 3 h (Rogers et al., 1979). In *Picea abies* (L.) Karst. exposed to 81 nl l^{-1} $^{13}NO_2$ for up to 60 min the proportions of NO_2-nitrogen after one hour in organic and mineral (nitrate) N were 58 and 42% respectively (Ammann et al., 1995a). Because of the stabilization of the absolute amount of NO_2-N in nitrate as well as the linear increase in organic nitrogen from airborne NO_2 over time, this ratio is expected to increase with time.

In higher plants ammonia as the product of nitrite reduction is incorporated into glutamate in the chloroplasts. Glutamine synthetase (GS; E.C. 6.3.1.2) and glutamine oxoglutarate aminotransferase (GOGAT; E.C. 2.6.1.53) are the two enzymes involved in this ferredoxin-dependent process. Subsequent transamination transfers the amino group of glutamate to other carbon skeletons, thereby producing other amino acids. The reaction catalyzed by glutamate dehydrogenase (GDH; E.C.

1.4.1.2.), localized in the mitochondrion, is thought to be almost exclusively the deamination of glutamate, thus having an important regulatory function in carbon and nitrogen catabolism (Robinson et al., 1991). The same reactions occur with nitrogen taken up from the atmosphere. Using $^{15}NO_2$, Kaji et al. (1980) and Yoneyama and Sasakawa (1979) found the highest label in the amide of glutamine followed by glutamate and aspartate; this points to the GS/GOGAT cycle being active in the primary amination in sunflower and spinach. The label was detected in 11 free amino acids. Nussbaum et al. (1993) and Weber et al. (1995), analyzing α-amino-N only, found the highest ^{15}N enrichment in glutamate after fumigating *Picea abies* (L.) Karst. and *Triticum aestivum* L. (cv Obelisk) with 60 nl l^{-1} $^{15}NO_2$ for up to 4 d. The label was found in 13 and 12 soluble amino acids, respectively.

GS/GOGAT activities in leaves measured under optimum conditions are usually high compared to NR activity (Srivastava and Ormrod, 1984; Tischner et al., 1988; Thoene et al., 1991). Hence, an influence from oxidized airborne nitrogen compounds is expected to be smaller than in the case of NR activity. Srivastava and Ormrod (1984) still recorded increasing GOGAT activity at all levels of nitrate applied to the roots when *Phaseolus vulgaris* L. (cv Kinghorn Wax) was exposed to increasing NO_2 concentrations between 20 and 500 nl l^{-1}. GDH decreased progressively with increasing nitrate supply but exhibited no consistent response to NO_2. The authors concluded that at the NO_2 concentrations used the GS/GOGAT system was operational in primary amination. A significant increase in the GDH/GS ratio in different *Lolium perenne* L. clones exposed to 68 nl l^{-1} NO_2 for 11 weeks was found by Wellburn et al. (1981). The same was observed by Bender et al. (1991) during anthesis of *Phaseolus vulgaris* (cv Rintintin) exposed to 31 nl l^{-1} NO_2. In both cases this effect was due to an unchanged GS activity accompanied by enhanced GDH activity. According to Wellburn (1990) the increased activity may serve to remove excess glutamate from NO_2-treated tissues, rather than to assimilate an overload of NH_4^+. *Pisum sativum* L. (cv Feltham First) showed no effect on this ratio after exposure to 100 and 200 nl l^{-1} NO_2 for 6d. The same was reported by Bender et al. (1991) for *Phaseolus vulgaris* in the vegetative state. The developmental stage of the plant, or alternatively the applied dose, may have played an important role in determining the reaction of GDH, since it differed from that measured 14 d later (see above). Investigating the response of tomato to NO concentrations up to 800 nl l^{-1}, Wellburn et al. (1980) found progressive increases in GDH and GOGAT activities with increasing concentrations in both sensitive and resistant cultivars, while the high GS activity remained constant. The increase in GDH was more pronounced in the sensitive cultivar Ailsa Craig compared to the resistant Sonato. Glutamate pyruvate transaminase activity was only enhanced in Ailsa Craig. Yu et al. (1988) exposed leaf discs of spinach in light and in darkness to nitrite solutions (5 and 25 mM, pH 5) and found an inhibition of GS and GOGAT. Together with the results from NO_2 fumigation experiments with kidney bean and spinach they concluded that the resultant NH_4^+ accumulation may be responsible for toxic effect of the pollutant. But, as outlined above, this effect is not expected to be of importance at realistic NO_2 concentrations. In *Picea abies* needles no activity change of GS was found after exposure to 125 nl l^{-1} NO_2 for 72 h (Thoene et al., 1991). On the other hand, Tischner et al. (1988) measured more than doubled activity in shoots, but unaffected activity in roots, after fumigating seedlings for 11 weeks with 500 nl l^{-1} NO_2. GDH activity was not altered by the pollutant treatment in either organ. Nussbaum et al. (1991) found no activity change in the GS/GOGAT system when the same species was exposed to 1400 nl l^{-1} NO for up to 6 d. Again, obvious differences in the reaction to NO_x are observed when time scales change.

We found no studies addressing foliar reduction and incorporation into organic compounds of nitrogen derived from HNO_3, HONO or PAN.

4. *Synthesis*: Changes in foliar enzymes of assimilatory nitrate reduction suggest that nitrogen derived from NO_2, NO and HNO_3 is incorporated into organic compounds by means of the pathway known to be generally operating in plant tissues. We are not aware of similar studies for HONO and PAN. The effect on these enzymes is dependent on species, cultivar, developmental stage of the plant, plant organ, photosynthetic photon flux density, nitrogen-nutrient form and concentration and also on concentration or dose of the pollutant. The relative effect is greater in 'root nitrate reducers' as opposed to partial 'leaf nitrate reducers', whereas the absolute values for enzyme activities are generally lower. Pearson and Soares (1995) hypothesized that this is an expression of the loosely linked carbon-nitrogen acquisition in the leaves of 'root nitrate reducers'; Raven (1988) suggested that it points to limited buffering capacity for incoming acidity from airborne nitrogen compounds. Thus, such plants are at higher risk for suffering from the respective pollutants. Enzyme activities in the roots were affected as well, indicating translocation of a signal. Direct proof of incorporation of airborne nitrogen into plant organic compounds has only been established for NO_2. Judging from enzyme activities, airborne oxidized nitrogen compounds may serve as an alternative nitrogen-source. On the other hand, increase in GDH, a catabolic enzyme, may be indicative of the need to remove excess amino acids. This may become more important when plants are sufficiently supplied with nitrogen to the roots. Accumulation

of toxic amounts of intermediates of assimilatory nitrate reduction (nitrite, ammonium) is only expected at exceedingly high pollutant concentrations.

Clearly more information is needed to assess the effects of airborne oxidized nitrogen compounds on plants using realistic scenarios, i.e., low concentrations persisting for prolonged time (several weeks), combined with typical nutrient relations for each species. From the scarce information available, important differences between short-term and long-term experiments must be expected, which may be related either to the applied dose or to effects caused by different growth stages or seasons.

B. Nitrogen Containing Compounds: Pool Sizes and Cycling

Assimilated nitrogen from compounds taken up by the leaves from the atmosphere may be exported to other parts of the plant, as shown by using tracer techniques (Rogers et al., 1979; Kaji et al., 1980; Okano et al., 1984; Rowland et al., 1987; Latus et al., 1990; Jensen and Pilegaard, 1993; Nussbaum et al., 1993; Segschneider et al., 1995; Weber et al., 1995). In both short-term (a few hours) and long-term (42 d; Jensen and Pilegaard, 1993) experiments, the amounts exported to the roots were usually shown to be between 3 and 9% of the total tracer. Therefore, they might affect the pool size of different nitrogen-containing compounds, and fluxes between them, not only in the leaf but throughout the entire plant. Several workers have determined concentrations of nitrogen-containing compounds and their changes in response to NO_2 in both shoots and roots (Table 4.3).

With one exception total-nitrogen decreased in none of the studies listed in Table 4.3 in either roots or shoots. Okano et al. (1988) only found an increased total-nitrogen content in intact plants of *Helianthus annuus, Raphanus sativus, Lycopersicon esculentum, Nicotiana tabacum, Cucumis sativus, Phaseolus vulgaris* and *Zea mays* and no change in *Sorghum vulgare* when exposing the plants to 500 nl l^{-1} NO_2 for 2 weeks. On the other hand, fumigation of 7 broad-leaved tree species (*Populus, Nerium, Zelkova, Euonymus, Quercus, Viburnum, Cinnamomum*) with 300 nl l^{-1} NO_2 for 30 d caused no increase in total-nitrogen concentrations with the exception of one of three *Populus* cultivars tested (Okano et al., 1989). An interesting differential effect of increasing pollutant concentrations was observed in *Glycine max* (Sabaratnam et al., 1988a). Fumigation with 100, 200, 300 and 500 nl l^{-1} NO_2 for 5 d caused an increase in foliar total-nitrogen at the end of the experiment relative to the unfumigated control at all pollutant levels. This increase correlated linearly with the pollutant concentration up to 300 nl l^{-1} but then dropped sharply to an increase at 500 nl l^{-1}, which was smaller than the one measured at 100 nl l^{-1}. A similar effect of increasing NO_2 concentrations was demonstrated by Srivastava and Ormrod (1984) for ethanol soluble and insoluble nitrogen in leaves of kidney bean grown on nutrient solution with > 5 mM nitrate, where at the highest NO_2 level of 500 nl l^{-1}, even decreases in the respective pools were reported. Below this nutrient concentration, increasing pool sizes were recorded with increasing NO_2 concentrations up to 500 nl l^{-1}. Apparently, there is some regulation of total nitrogen uptake involved. Unfortunately, neither of the research groups measured the absorption of NO_2. This, together with measurements of root nitrogen uptake, would have given some indication about the mechanisms involved.

Some researchers have compared the relative uptake of N from the air and the soil. Okano and Totsuka (1986) exposed *Helianthus annuus* L. (cv Russian Mammoth) growing at 3 different nitrate levels to 300 and 2000 nl l^{-1} NO_2 for 7 d: they found an enhanced total nitrogen uptake per plant during this period in all pollutant treatments. At the lower NO_2 concentration N uptake by the roots was equal in the control and fumigation treatment at all nitrate levels, whereas at 2000 nl l^{-1} NO_2 it decreased in plants exposed to NO_2 at the two higher nitrate levels. Muller et al. (1996) exposed *Picea abies* to 100 nl l^{-1} NO_2 for 48 h and supplied no nitrogen, only NO_3^-, or both NO_3^- and NH_4^+ to the roots. Total nitrogen uptake was increased significantly only with the nitrogen-free solution. This increase was completely due to NO_2-derived nitrogen. In the two other nitrogen treatments the plants absorbed the same amount of NO_2-nitrogen as did the nitrogen-deprived plants, but they reduced root nitrogen uptake, resulting in no significant total nitrogen gain. This regulation of total nitrogen uptake was not found by Rowland et al. (1987) with *Hordeum vulgare*. No impact of NO_2 fumigation was found on root uptake of nitrate, but the ability of the nitrate uptake system to react to changes in nitrate concentration was depressed. Apparently, uptake of airborne nitrogen-containing compounds may exert some feedback regulation on root nitrogen uptake, unless the nitrogen status of the plant drops below a certain threshold. Muller and Touraine (1992) showed that some amino acids may be involved in such a regulation in soybean.

The concentrations of soluble free amino acids (fAA) increased in barley at low levels of nitrate applied to the roots and decreased at higher nitrate levels (Rowland, 1986). An increase in foliar fAA concentrations was also reported by Prasad and Rao (1980) and Ito et al. (1984b and 1986). Weber et al. (1995) measured an increase in total fAA in the leaves of *Triticum aestivum* mainly caused by an increase in glutamate, but concentrations of practically all other fAA species were increased as well. In stalks and roots the concentration increased most for asparagine

Table 4.3. Effects of NO_2 on weight-based concentrations of nitrate, nitrite, ammonium, free amino acids (fAA), protein (Prot.), reduced nitrogen (N_{red}) and total nitrogen (N_{tot}) in plants grown under different conditions. Legend for arrows as given in Table 4.2

Species	Root N-supply form	mM	NO_2 conc. nl l^{-1}	Duration	NO_3^-	NO_2^-	NH_4^+	fAA	Prot.	N_{red}	N_{tot}	Reference
Helianthus annuus	h/NO_3^-/NH_4^+	2.4/0.6	5–73	8 h				→/→	→/→		→/→	Segschneider et al., 1995
Hordeum vulgare	s/NO_3^-		33	42 d							↑/→	Jensen and Pilegaard, 1993
Hordeum vulgare	h/NO_3^-	0	300	9 d	→/↑			↑/↑		↑/↑	↑/↑	Rowland, 1986
		0.01			↑/→			→/↑		↑/→	↑/→	Rowland et al., 1987
		0.1			↑/→			↑/↓		→/↑	→/↑	
		0.5			↑/→			↓/↓		→/↑	→/↑	
		1.0			↑/→			↓/↓		→/↑	→/↑	
		10			↑/→			↓/↓		↓/→	→/→	
Phaseolus vulgaris	s/NO_3^-/NH_4^+		2,000	2 d				↑/→				Ito et al., 1984b *
				4 d				↑/↑				
				7 d				↑/↑				
			4,000	2 d				↑/↑				
				4 d				↑/↑				
				7 d				↑/↑				
Picea abies	h/NO_3^-/NH_4^+	2.0	100	48 h	→/→		→/→				↑/↑	Muller et al., 1996
	h/NO_3^-	2.0			↑/→		↑/→					
	—	0			→/→		↑/→					
	s/NO_3^-/NH_4^+	2.0	100	28 d	→/→		↑/→			→/↑	→/→	
				49 d	↑/↑		→/↑			→/↑	→/↑	

Picea abies	s		60	4 d		↑/→	Nussbaum et al., 1993
							Nussbaum, 1992
Picea abies	h/NO$_3^-$	0.72	500	77 d	↑/→	(↑)/(↑)	Tischner et al., 1988
Pinus sylvestris	h/NH$_4^+$	1.42	20/30	19/39 d	(↓)/→ →/→	→/→	Näsholm et al., 1991
Triticum aestivum	s/NO$_3^-$/NH$_4^+$		60	48 h	↑/↑		Weber et al., 1995

* Significance levels not given by the authors.

and glutamine, in agreement with reports by Prasad and Rao (1980) and Ito et al. (1984b and 1986). Murray and Wellburn (1985) found a 3.5-fold increase in fAA in tomato leaves (cv Ailsa Craig) as opposed to a slight decrease in pepper (cv Bell Boy) after treatment with 1500 nl l^{-1} NO$_2$. They attributed the difference between the two species to differences in their basic nitrogen metabolism. In conifers, Näsholm et al. (1991) measured a decrease in arginine, GABA and proline in the needles of *Pinus sylvestris*. Wingsle et al. (1987) registered a 50% decrease in total amino acids in needles of the same species, while the total-N content increased slightly; arginine was reduced by 84%. Large variations in the concentration of this amino acid are not unusual in conifers (Schmeink and Wild, 1990; Ericsson et al., 1993; Gessler et al., 1998) and concentration changes may be an indicator of changes in mineral nutrient proportions (Edfast et al., 1990) caused by nitrogen uptake by the leaves. Concentrations of single amino acids have been shown to react differently in *Picea abies* at different times of the year (Nussbaum et al., 1993). Thus, besides species specificity, effects of NO$_2$ fumigation on fAA concentrations dependent on nitrogen-nutrient concentrations applied to the roots, nutrient relations (i.e., pedospheric culture conditions), and season/growth stage.

Amino acid concentrations reflect subtle regulation processes in plants, involving different processes such as synthesis, degradation and also translocation. This is illustrated by the analysis of xylem sap exudates presented by Muller et al. (1996) and Rowland et al. (1987) for *Picea abies* and *Hordeum vulgare*, respectively. Muller et al. (1996) found a decrease in aspartate and the most abundant amino acid glutamine, while a number of other amino acids became more abundant when NO$_2$ was present. No assessment of total flux was made. In barley the change in amino acid composition in response to NO$_2$ fumigation was most pronounced at low nitrate supply (0.1 mM) wherein the fraction containing serine, asparagine and glutamine increased in a decreased total flux of reduced nitrogen. The nitrate flux in the xylem increased nine-fold. It is noteworthy that the reaction of glutamine in leaves, roots and xylem is the opposite in angiosperms compared to conifers.

Translocation of NO$_2$-nitrogen to the roots was higher in nitrogen-deficient barley plants and allocation of root-derived nitrate-nitrogen to the roots showed a tendency to decrease at low nitrate supply (Rowland et al., 1987). This together with the analysis of the xylem flux (see above), demonstrated an increased cycling of nitrogen between shoot and root of exposed plants growing at low nitrate supply, compared to those with high nitrate supply (Fig. 4.4). This may reduce nitrogen-shortage, although the authors mentioned that the proportion of additional nitrogen was marginal compared to the amount necessary to reach normal nitrogen status. The decrease in distribution of nitrogen-

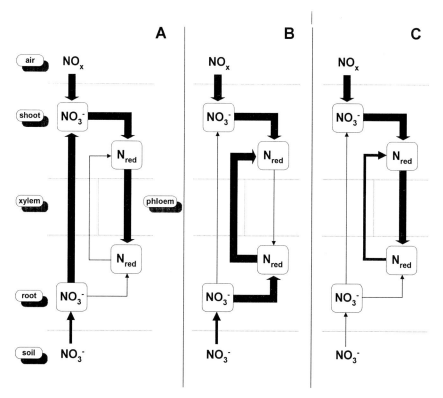

Fig. 4.4. Changes in fluxes between nitrate-nitrogen and reduced nitrogen pools in leaves and roots of plants exposed to airborne NO_x, compared to clean air controls. Narrow, wide and intermediate arrows indicate reduced, increased or unchanged fluxes, respectively. **A**: 'Leaf nitrate reducers' growing at low nitrate supply to the roots have increased nitrogen cycling between root and shoot. The relative contribution of the shoot to total nitrate reduction is increased and reduced nitrogen (N_{red}) is preferentially allocated to roots. Nitrate uptake by the roots is not affected. **B**: In 'leaf nitrate reducers' growing at high nitrate supply, nitrate translocation from roots to shoots is reduced. At the same time nitrate reduction in the roots is increased with a similar increase in the proportion of nitrogen translocated as reduced nitrogen in the xylem. Total nitrogen flux in the xylem is reduced, indicating that each plant part is sufficiently supplied with nitrogen. Nitrate uptake by the roots is not affected. **C**: 'Root nitrate reducers' keep the total nitrogen influx into the plant constant. Hence, nitrate uptake by the roots is reduced and the nitrogen supply to the roots occurs from the shoots.

containing compounds at high nitrate supply was explained by a decrease in sink strength by well-supplied tissues. In such plants uptake of airborne nitrogen may be seen as a forced nutrition with no particular benefit but with extra energy costs for assimilation/detoxification.

Muller et al. (1996), presenting a balance sheet for a 7-week fumigation, calculated enhanced pool sizes of nitrate and ammonium, together with a decreased pool of organic nitrogen (N_{org}), in roots of fumigated *Picea abies*. In the shoot only the nitrate pool increased while a decrease was registered in the other two pools. In shoots the nitrogen flux from nitrate to N_{org} increased, whereas the opposite was observed in roots. The already small nitrate flux in the xylem was reduced whereas the N_{org} flux increased significantly by 5%. The downward transport of N_{org} on the other hand increased by 240%. These flux changes were similar to the low nitrogen-supply case presented by Rowland et al. (1987), but the two species differed in the fate of nitrate in the roots. In barley nitrate reduction in the roots was only enhanced in the high nitrate supply case when sink strength of the leaves was decreased; in the low nitrogen-supply case surplus nitrate was translocated to the leaves where reduction increased. In *Picea abies* this pathway was practically absent, and this species reacted by reducing nitrate uptake by the roots and translocating less to the shoot. Again, this probably reflects fundamental differences in handling additional nitrogen input from the atmosphere between species with preferential nitrate reduction in the shoot or in the root. Similar experiments using a variety of species differing in nitrate reduction distribution and/or nitrogen acquisition could help to better understand the regulating mechanisms.

C. Effects on Chloroplasts

1. *Photosynthesis*: As outlined above, a large proportion of the absorbed atmospheric oxidized nitrogen enters the chloroplast in the form of nitrite. The highly reactive and toxic nature of this compound may be expected to have impacts on physiological processes associated with this organelle, such as photosynthesis. Responses of net photosynthesis (P_n) to NO_x were investigated using chamber techniques (Table 4.4). Hill and Bennett (1970) found a threshold concentration for P_n inhibition between 500 and 700 nl l^{-1} for both NO and NO_2. Above this concentration a linear reduction with increasing NO_x concentrations was observed. NO_2 was more effective than NO, and the effect of the combination of the two was additive. Additivity of the effects of NO and NO_2 on P_n was also reported by Capron and Mansfield (1977). In *Medicago sativa* and *Hordeum vulgare* greater toxicity of NO_2 towards P_n was found compared to NO after fumigation for 2 h (Bennett and Hill, 1973). Contrary to this, Saxe (1986a) stated that at 1000 nl l^{-1}, on average NO was four times more inhibitory to photosynthesis than NO_2. Taking into account the clearly lower deposition rate (Saxe, 1986b), absorbed NO was even 22 times more effective.

Two studies with *Pinus sylvestris* yielded contrasting results. In detached twigs no reaction of P_n was observed in any of the 3 needle age classes measured (Oleksyn, 1984). With the same fumigation protocol during 3 d with progenies differing in susceptibility to NO_2, P_n was depressed in susceptible seedlings 30 min after the end of fumigation, which was reversible within 24 h (Lorenc-Plucinska, 1988). This confirmed earlier reports that inhibition of P_n by NO_2 is reversible (Hill and Bennett, 1970), but other researchers have found no sign of reversibility (Sabaratnam et al., 1988b). Short-term experiments may not produce the same effect as experiments involving entire life cycles or growing seasons. P_n inhibition in *Phaseolus vulgaris* (cv Pure Gold Wax) was dependent on both pollutant concentration and duration of exposure (Srivastava et al., 1975a), and dose-dependence was also observed by Bull and Mansfield (1974). The observed effects on P_n are strongly influenced by experimental conditions. Most of the negative effects in short-term fumigations were found at NO_x concentrations above 500 nl l^{-1}. Only in one case was an increase in P_n observed (Sabaratnam et al., 1988a — 200 nl l^{-1}). Incorporation studies with $^{13}CO_2$ revealed enhanced incorporation with 2000 nl l^{-1} NO_2 compared to control (Okano et al., 1985a).

Ito et al. (1984a and 1985) studied the partitioning of assimilates in *Phaseolus vulgaris* exposed to 2000 and 4000 nl l^{-1} NO_2 for 2, 4, and 7 d. They found a decreased sugar content and a shift in proportions between the different sugars in the roots, resulting in negative effects on root respiration at the end of the experiment. In a long-term experiment with *Pseudotsuga menziesii* Mirb. water-use efficiency in shoots exposed to 50 nl l^{-1} NO_2 for 5 months was reduced (van Hove et al., 1992). This was due to increased stomatal conductance over the entire range of the light response curve. Contrary to short-term experiments with high concentrations (Hill and Bennett, 1970; Srivastava et al., 1975 a, b; Saxe, 1986a); the main effect of NO_2 on gas exchange in this case was on guard cells and not due to an increased mesophyll resistance. Differences between long-term and short-term experiments must be taken into account in risk assessments under ambient conditions. Extrapolating effects observed in short-term experiments at high concentrations to chronic low concentrations appears to be prone to errors due to nonlinearity of plant response. The database for long-term exposures is narrow. Integrating measurements such as gas exchange may mask effects in the contributing reactions (van Hove et al., 1992); hence, more detailed mechanistic studies at the cellular level are needed in order to understand the impact of airborne oxidized nitrogen on photosynthesis.

Import of nitrite into the chloroplast is a prerequisite for detoxification by reduction to ammonium. This is associated with acidification of the stroma (Heber and Purczeld, 1978) and may reduce the transthylakoid proton gradient driving ATP synthesis. In thylakoid preparations, quenching of a 9-amino-acridine probe, a measure for the transthylakoid

Table 4.4. Effects of NO_x on net photosynthesis (P_n), dark respiration (DR), stomatal conductance (g_s), chlorophyll content (Chl.) and carotenoid content in leaves of plants grown under different conditions. Legend for arrows as given in Table 4.2.

Species	Root N-supply form	mM	Pollutant	NO_2 conc nl l^{-1}	Duration	P_n	DR	g_s	Chl.	Carot.	Reference
Araucaria cunninghamii	s		NO_2	76	112 d/ 2.5 h d^{-1}				→		Murray et al., 1992a
Ficus, Hedera, Hibiscus, Dieffenbachia and *Nephrolepis*	s		NO NO_2	1,000 1,000	3 d	→ ↑	↑ ←	↑ ↑			Saxe, 1986a
Glycine max	s/NO_3^-/NH_4^+		NO_2	100 200 300 500	5 d/7 h d^{-1}	↑ ← ↑ →	→ ↑ ← ↑ →	↑ ↑ ↑ ↑	↑ ↑ ↑ →		Sabaratnam et al., 1988a/b Sabaratnam and Gupta, 1988
Glycine max	s		NO_2	≤600	2 h	↑					Carlson, 1983
Glycine max	s		NO_2	50 100	5 d	↑ ←		↑ ↑	↑ ←		Gupta and Narayanan, 1992 Gupta et al., 1992
Helianthus annuus	h/NO_3^-/NH_4^+	2.4/ 0.6	NO_2	5–73	8 h	↑		↑			Segschneider et al., 1995
Helianthus annuus	s		NO_2	200–1,000	2 h	↑					Furukawa and Totsuka, 1979
Lycopersicon esculentum	s		NO	400	28 d				↑		Wellburn et al., 1980
Medicago sativa and *Avena sativa*	s		NO/ NO_2/ $NO \cdot NO_2$	<500–700 >500–700	45–90 min	↑ →		(↓)			Hill and Bennett, 1970

Species			Concentration	Exposure	Effects					Reference
Medicago sativa	s		NO$_2$ 250–400 / NO 100–150 / NO$_2$+NO	1–2 h	↑↑↑					White et al., 1974
Phaseolus vulgaris	s/NO$_3^-$		NO$_2$ 1,000–7,000	2 h	→					Srivastava et al., 1975a
Phaseolus vulgaris	s		NO$_2$ 100	5 d		(↓)				Ashenden, 1979a
Phaseolus vulgaris	s		NO$_2$ 2000	8 h	↑*	↑				Okano et al., 1985
Phaseolus vulgaris	s/NO$_3^-$/NH$_4^+$		NO$_2$ 25 / 50 / 100	15 d/7h d^{-1}	↑↑←	←←←				Sandhu and Gupta, 1989
Phaseolus vulgaris	s/NO$_3^-$/NH$_4^+$		NO$_2$ 30–40	34 d / 48 d		↑→				Bender et al., 1991
Phaseolus vulgaris	s/NO$_3^-$	0–50	NO$_2$ 3,000	5 h	→	↑↑←				Srivastava et al., 1975c
Phaseolus vulgaris	h/NO$_3$	<5 / ≥5	NO$_2$ 20–500	5 d		(↓)	↑→←			Srivastava and Ormrod, 1984
Picea abies	h/NO$_3^-$	0.72	NO$_2$ 500	77 d	↑→→→					Tischner et al., 1988
Pinus sylvestris tolerant / susceptible	s		NO$_2$ 500 / ≥1,000 / 500 / ≥1,000	3 × 6 h	↑→→→ ↑	→↑→↑ ↑				Lorenc-Plucinska, 1988
Pinus sylvestris	h		NO$_2$ 500/1,000	2 × 6 h	↑					Oleksyn, 1984
Pisum sativum	s		NO$_2$ ≤100 / 250	28 d (Wellburn, 1990)	↑→					Bull and Mansfield, 1974
Pisum sativum	s		NO$_2$ 1,000	6 d		(↓)	(↑)			Horsman and Wellburn, 1975
Pseudotsuga menziesii	s		NO$_2$ 50	5 months	↑	←	←			van Hove et al., 1992

*Result of $^{13}CO_2$ incorporation study.

proton gradient, was reduced at nitrite concentrations of around 0.2 mM in *Avena sativa* (Robinson and Wellburn, 1983). Wellburn (1985) fumigating *Hordeum vulgare* (cv Patty) with 280 nl l^{-1} NO$_2$ for up to 3 d measured a transient 50% increase in the nitrite concentration in chloroplast preparations, reaching a maximum of 0.15 mM. Hitherto no corresponding experiments with NO, HONO or HNO$_3$ have been published. The transient nature of this increase may indicate that plants are physiologically adapted to deal with nitrite, but stress may occur when changes are quick or when the incoming flux exceeds the capacity for removal of protons by nitrite reduction (Zeevaart, 1976). As already outlined above, this may be expected to happen only at exceedingly high NO$_x$ concentrations. Thus the manganese release from thylakoids at 2.0 mM nitrite shown by Wellburn (1984) may not be of importance at realistic NO$_x$ concentrations. Enzymatic reduction of enhanced stromal nitrite concentrations may compete for NADPH used for CO$_2$ fixation and thus impair photosynthetic performance especially at low light levels and high nitrite concentrations (Pierson and Elliot, 1988). Wellburn (1990) pointed out that these are the conditions prevailing in plants exposed to NO and NO$_2$ at high latitudes. Interactions with other air pollutants such as SO$_2$ may aggravate this situation (Wolfenden et al., 1991).

Inhibition of photosynthesis by PAN was reported for *Phaseolus vulgaris* and *Petunia hybrida* when visible injury developed on the leaves (Dugger et al., 1963a; Mayumi and Yamazoe, 1983); this appears to be associated with destruction of the respective cellular structures.

2. *Pigments*: Most of the studies shown in Table 4.4 report no reaction or an increase in the chlorophyll content of plants exposed to moderate NO$_2$ concentrations. With the exceptions of van Hove et al. (1992) and Murray et al. (1992a) the effects on chlorophyll *a* were greater than those on chlorophyll *b*. Bender et al. (1991) mentioned changing responses depending on the duration of the fumigation. It is not clear whether this was a dose-dependent effect or caused by differences in response depending on the developmental stage. Sabaratnam et al. (1988a) also noted different responses of chlorophyll at different pollutant concentrations. In conifers only results of long-term fumigations have been published. They show both increases (Tischner et al., 1988; van Hove et al., 1992) and decreases (Murray et al., 1992a) in chlorophyll content. Nutrient nitrate supply to the roots of *Phaseolus vulgaris* determined the NO$_2$ effect on chlorophyll content (Srivastava and Ormrod, 1984). The increase at nitrate concentrations below 5 mM was attributed to a general stimulation of chloroplast biogenesis rather than to specific incorporation of NO$_2$ into nitrogenous chlorophyll precursors. Likewise, Srivastava et al. (1994a) found that in greening *Phaseolus vulgaris* seed-

lings, even though nitrogen was not the limiting factor for chlorophyll synthesis, it was promoted by 300 nl l^{-1} NO$_2$. The only published result involving NO showed no change in chlorophyll content of *Lycopersicon esculentum* during a 4-week fumigation with 400 nl l^{-1} (Wellburn et al., 1980).

Of the other plant pigments, only carotenoid responses to NO$_2$ have been measured in two instances (Srivastava and Ormrod, 1984; Tischner et al., 1988). The direction of the observed changes was the same as for chlorophyll but differed in magnitude of the effect. Dugger et al. (1963b) showed that PAN damage to *Phaseolus vulgaris* was mediated through a photochemical reaction with some plant component. The action spectrum of PAN damage revealed that carotenoids rather than chlorophyll may be involved in this reaction. We do not know of any study on pigments with HONO and HNO$_3$.

D. Other Effects on Plants

1. *Dark Respiration*: Inhibition of dark respiration (DR; Table 4.4) at concentrations above 500 nl l^{-1} NO$_2$ have been noted by Carlson (1983) and Srivastava et al. (1975a), and Lorenc-Plucinska (1988) who showed that this effect was reversible. Saxe (1986a) registered an insignificant reduction in DR by 1000 nl l^{-1} NO but an enhanced DR by the same concentration of NO$_2$. At lower concentrations, reversibility of an effect on DR (increase) was found (Sandhu and Gupta, 1989). Like Lorenc-Plucinska (1988), Sabaratnam et al. (1988b) observed clear differences in the reaction of DR immediately after termination of their fumigation depending on the NO$_2$ concentration. During a 24-h recovery period effects disappeared. However, the insignificant effect at 200 nl l^{-1} (+12 %) became significant (+44%). Overall, the observed effects of NO$_x$ on DR were small, appeared only at relatively high pollutant concentrations, and tended to be reversible. It is thus unlikely that DR is seriously affected with realistic pollutant regimes (van Hove et al., 1992). No report is available on the effects of HONO, HNO$_3$ and PAN on DR.

2. *Lipids and Fatty Acids*: Nouchi and Toyama (1988) exposed *Pharbitis nil* Choisy (morning glory) and *Phaseolus vulgaris* to 100 nl l^{-1} PAN for up to 8 h. Subsequent analysis of the total fatty acid content revealed a significant decrease at the end of the fumigation, which became more pronounced 24 h later. The relative contribution of 18:3 and 18:2 acids was decreased as well, accompanied by an increase in the malondialdehyde (MDA) content. MDA is formed by peroxidation of unsaturated fatty acids. After 4 h of exposure, glycolipids, sulpholipids and all the phospholipids started to decline, with the exception of phosphatidic acid (PA), a breakdown product of phospholipase D, which

showed a significant increase. These results suggest that PAN may attack unsaturated fatty acids by peroxidation, which likely disturbs functions associated with membranes. Unlike in the case of ozone where visible damage developed well after changes in fatty acid composition were detected, water-soaked lesions developed on the leaves simultaneously in response to PAN. Likewise, as mentioned above, inhibitory effects on photosynthesis were only detected when visible symptoms developed. Thus, it is possible that this effect is mainly due to alterations in the thylakoid membranes.

The free radical nature of NO_2 and NO may lead to oxidation of bisallylic groups of polyunsaturated fatty acids, thereby initiating a radical chain reaction eventually leading to membrane damage (Kunert, 1987). However, Mudd et al. (1984) concluded that the concentrations of NO_2 are usually too low to produce such effects. Inhibition of the biosynthesis of phospholipids and galactolipids was observed in *Pinus banksiana* exposed to 2000 nl l^{-1} NO_2 for 2 days (Malhotra and Khan, 1984). This may be due to high levels of nitrite (Yung and Mudd, 1966) caused by the extremely high concentration, and probably is of no importance at realistic concentrations.

An alternative mechanism triggered by NO_x and possibly damaging membranes was proposed by Mehlhorn and Wellburn (1987). They showed that 150 nl l^{-1} NO or NO_2 causes stress ethylene formation in *Phaseolus vulgaris* leaves and enhanced ozone toxicity. They suggested that the reaction of ozone with ethylene produced free radicals which, in turn, attacked biomolecules, such as unsaturated fatty acids. Since NO_x pollution is often accompanied by enhanced ozone levels, this chain of reactions may be realistic, but further evidence is needed.

3. *Enzymes*: Only a few reports exist on the effects of NO_x on enzymes other than those involved in reductive nitrogen assimilation or primary carbon assimilation. Tischner et al. (1988) measured increased glucose 6 phosphate-dehydrogenase activity in shoots but not in roots of *Picea abies* exposed to 500 nl l^{-1} NO_2 for 11 weeks. Malate dehydrogenase activity decreased in the shoots and increased in the roots. Horsman and Wellburn (1975) found no increase in peroxidase activity in *Pisum sativum* (cv Feltham first) when they exposed the plants to 100 and 1000 nl l^{-1} NO_2 for 6 d. For PAN there is some evidence that damage to enzymes is associated with oxidation of sulphhydryl groups (Dugger et al., 1966; Dugger and Ting, 1968; Taylor, 1969; Dugger and Ting, 1970). Mudd (1963) noted that enzymes without SH-groups were not affected by PAN. However, these studies were conducted with plant extracts and very high pollutant concentrations. It remains to be shown that these effects are relevant with realistic pollution scenarios.

4. *Symbiotic Nitrogen Fixation*: Srivastava and Ormrod (1986) fumigated *Phaseolus vulgaris* seedlings grown with different nutrient nitrate levels for 15 d (6 h d^{-1}) with 20–500 nl l^{-1} NO$_2$. They observed the typical decrease in nodule number and total nodule fresh weight with increasing nitrate supply. In response to NO$_2$ exposure, the nodule number was increased at ≤ 5 mM nitrate and decreased at higher nitrate concentrations. Total nodule fresh weight decreased at 0 and 1 mM nitrate and increased at higher concentrations when NO$_2$ was introduced. These effects were most pronounced at 20 nl l^{-1} NO$_2$; increasing pollutant concentrations reduced number and fresh weight. Specific nitrogenase activity was generally reduced by NO$_2$ fumigation at all concentrations applied. A contradictory response of nitrogenase activity has been reported at the flowering stage in *Glycine max* exposed for 5 d/7 h d^{-1} to 50 and 100 nl l^{-1} NO$_2$ (Gupta and Narayanan, 1992). Like Srivastava and Ormrod (1986) with > 5 mM nitrate in the nutrient solution, Klarer et al. (1984) noted an increased total nodule weight and a decreased nodule number of *Glycine max* roots when the plants were exposed 15 times for 3 h every other day to 200 nl l^{-1} NO$_2$. These plants were grown with N-free solutions. Documentation of the effect of NO$_2$ on nodulation is scarce and no definite conclusions can be drawn.

5. *Visible Damage*: PAN is by far the most effective oxidized atmospheric nitrogen compound producing visible injury. Bobrov (1955) described the typical symptoms as abaxial glazing, bronzing or silvering on younger leaves. Taylor (1969) noted that damage was practically limited to rapidly expanding leaves. In contrast, ozone injury was typically found on the adaxial surface of mature leaves. PAN has been found to damage *Phaseolus vulgaris* only when applied during illumination (Dugger et al., 1963c). In contrast to other air pollutants, sensitivity to PAN did not correlate with stomatal conductance despite the fact that deposition to leaves is largely under stomatal control (Dugger et al., 1963c; Okano et al., 1990). The reason for this discrepancy is still largely unknown. A stronger relation between PAN damage and applied dose was observed by Nouchi and Toyama (1988) but appearance of symptoms was delayed by several hours. They exposed *Phaseolus vulgaris* to 100 nl l^{-1} PAN for up to 8 h and found water-soaked lesions after 6 h, an often reported symptom of PAN damage. During the 24 h following exposure, 70% of the leaves exposed for only 1 h developed lower surface bronzing. This percentage increased to 100 % after 2 h of exposure. In treatments ≥ 2 h bifacial interstitial bleached necrosis developed. Extremely low doses of PAN causing visible foliar injury have been reported by Sun and Huang (1995) for *Lactuca sativa* L., *Solanum nigrum* L. and *Galinsoga parviflora* Canv. Injury appeared after 3h at 15

nl l^{-1} PAN under controlled conditions, and from their field observations the authors suggest an even lower threshold concentration for damage in the range of 4–5 nl l^{-1}.

NO_2 also caused visible injury in controlled experiments. Symptoms included tanned to necrotic leaf margins, brownish rust-like lesions, yellow chlorotic lesions and tiny colourless or yellow spots on leaves of *Phaseolus vulgaris* (Srivastava and Ormrod, 1984; Srivastava et al., 1990). Okano et al. (1985b) observed a slight crinkling or puckering and a darkening of the green colour of the newly expanding leaves of *Helianthus annuus* in response to 500 nl l^{-1} NO_2 or more. Wild type *Hordeum vulgare* exhibited whitening of the leaves from the tip downward after exposure to the same NO_2 concentration for 3 d (Srivastava et al., 1994b). Srivastava and Ormrod (1984) and Srivastava et al. (1990) observed foliar injury in *Phaseolus vulgaris* only, when plants were grown with sufficient nitrate (20 mM and 5 mM respectively). To our knowledge, 250 nl l^{-1} NO_2 applied for 10 d for 6 h d^{-1} is the lowest concentration/dose for which visible injury has been reported for NO_2 (Srivastava et al., 1990). These researchers also mention that damage was reduced at low photosynthetic flux densities (PPFD), possibly because of different uptake rates of NO_2 at the different PPFD.

At NO_x concentrations above 1000 nl l^{-1}, necrosis and wilting become the most often described symptoms in many different species (Zeevaart, 1976), sometimes appearing within a few hours after commencement of fumigation. In some cases this was the reason for limiting the duration of an experiment (Yoneyama et al., 1979; Yu et al., 1988). Accumulation of nitrite in the leaves was thought to be responsible for the observed damage, but this is not a general rule (see section IV.1). Zeevaart (1976) described an interstitial chlorosis on leaves of *Leguminosae* depending on light. At 3000 nl l^{-1} NO_2, Srivastava et al. (1975c) noted a protection against foliar damage with increasing nitrate concentration applied to the roots. The same was reported by Okano and Totsuka (1986) for *Helianthus annuus* exposed to 2000 nl l^{-1} NO_2.

In *Picea abies* exposed to 71 and 9 nl l^{-1} HNO_3 and in *Picea rubens* exposed to 75 nl l^{-1} HNO_3, necrotic needle tips appeared after only 2 d (Cadle et al., 1991; Norby et al., 1989). The same concentration of NO_2 evoked no symptoms (Norby et al., 1989). Due to the very limited number of reports on effects of HNO_3 vapour on plants, it is not known whether these observations indicate higher toxicity of HNO_3 compared to NO_2.

6. *Growth Responses*: Growth responses of plants exposed to atmospheric nitrogen oxides are listed in Tab. 4.5. Increased shoot growth is limited to plants grown on low root nitrogen supply, fumigated with low to moderate NO_2 concentrations. Likewise, negative growth responses appear only when root nitrogen supply is increased. For

instance, Srivastava and Ormrod (1984) working with *Phaseolus vulgaris* at 20 to 500 nl l⁻¹ NO_2, observed a decrease in leaf weight and specific leaf area with increasing NO_2 concentration only when nitrate concentrations fed to the roots were $\geq$ 5 mM. As in greening bean seedlings (Srivastava et al., 1994a), dry weight was more affected than fresh weight. In two long-term studies with low NO_2 concentrations an initial growth increase had reversed to a decrease at the time of fruit production (Bender et al., 1991—*Phaseolus vulgaris*; Whitmore and Mansfield, 1983— *Poa pratensis*). It is not known whether this effect is dose-dependent or due to differences in sensitivity during different growth stages.

The shoot:root ratio proved the most sensitive growth parameter in plants exposed to NO_2. Typically it was enhanced, and Srivastava et al. (1994b) showed that the most striking feature of NO_2-resistant *Hordeum vulgare* mutants was their reduced root growth when fumigated with the pollutant. It is not known whether this is a common stress tolerance reaction of plants exposed to NO_2, and there is no mechanistic explanation for this reaction. In trees, such an effect may be of particular importance since it may affect mechanical stability. This may be aggravated by increased growth due to atmospheric input of otherwise growth-limiting nutrients. It may also have important implications for drought and cold tolerance because it affects the balance between water uptake by roots and its loss by transpiration. Okano et al. (1985b) and Sabaratnam and Gupta (1988) have reported that the net assimilation rate (NAR) reduced by NO_2 was partially compensated for by an increase in leaf area ratio (LAR) but could not prevent a reduction in the relative growth rate (RGR). On the other hand, Sandhu and Gupta (1989) measured increased NAR and RGR without a change in LAR. Hence these parameters respond differently to NO_2 exposure in different species.

There are only a few reports concerning effects on the economic yield of annual crops. In *Glycine max* exposed to 50 nl l⁻¹ NO_2 seed number remained unchanged, and increased pod number was counterbalanced by a decreased seed weight resulting in a slight reduction in yield (Gupta and Narayanan, 1992). At 100 nl l⁻¹ NO_2, both pod number and seed number were increased, while seed weight was the same as in control, resulting in a substantial yield increase. For *Phaseolus vulgaris*, all of these parameters were increased with increasing NO_2 concentrations up to 100 nl l⁻¹ (Sandhu and Gupta, 1989). The weight of seeds was increased already at 25 nl l⁻¹. Bender et al. (1991) found a slight reduction in both pod number and pod fresh weight in the same species. An increase in ear weight, grain number and weight was also registered in *Hordeum vulgare* fumigated for 108 d/4h d⁻¹ with 170 nl l⁻¹ NO_2 (Murray et al., 1992b). Pande and Mansfield (1985) reported increased 1000-grain weight for the same species after exposure to 93 nl l⁻¹ NO_2 for

Table 4.5. Effects of NO_x on weight (fresh weight or dry weight), shoot: root ratio, total leaf area, relative growth rate (RGR, biomass gain per biomass and time unit), net assimilation rate (NAR, biomass gain per unit leaf area and time unit), leaf area ratio (LAR, leaf area per plant biomass) and leaf weight ratio (LWR, leaf weight per plant biomass) of plants grown under different conditions. Legend for arrows as given in Table 4.2

Species	Root N-supply form	mM	Pollutant	NO_2 conc. nl l^{-1}	Duration	Weight	Shoot : root	Leaf area	RGR	NAR	LAR	LWR	Reference
Dactylis glomerata	s		NO_2	110	140 d / 15 h d^{-1}	↓/		↑					Ashenden and Mansfield, 1978
Lolium multiflorum						→/		↑					Ashenden, 1979b
Phleum pratense						↑/		↑					
Poa pratensis						↓/		↑					
Glycine max	s/NO_3^-/NH_4^+		NO_2	100	5 d / 7 h d^{-1}		↑		↑↑↑→	↑↑↑→	←		Sabaratnam and Gupta, 1988
				200			↑		↑↑↑→	↑↑↑→	↑↑→		
				300			↑		↑↑↑→	↑↑↑→	↑↑→		
				500			↑						
Helianthus annuus	s/NO_3^-/NH_4^+		NO_2	200	14 d	→/↑	↑	→	←	←	→	→	Okano et al., 1985b
				500		↑/↑	↑←	↑	↑→	↑→	↑←	←	
				1,000		(↓)/↓	←	→					
Zea mays				200		→/↑	↑	↑	↑↑↑	↑↑↑	↑↑↑	↑↑↑	
				500		→/↑	↑	↑	↑↑↑	↑↑↑	↑↑↑		
				1,000		→/↑	↑	↑					
Helianthus annuus	s/NO_3^-/NH_4^+		NO_2	500	14 d							←←→	Okano et al., 1988
Raphanus sativus												→→	
Lycopersicon esculentum													
Nicotiana tabacum													
Cucumis sativus													
Phaseolus vulgaris												↑	

Species		Concentration	Duration			Reference
Zea mays						
Sorghum vulgare					→ ↑	
Helianthus annuus	h/NO$_3^-$	0 NO$_2$ 5 15 0 5 15	300	7 d	↑/→ →/→ (↑)/→ →/↓ (↓)/↓	↑ ↑ ↑ ← ↑ →	Okano and Totsuka, 1986
Hordeum vulgare	s/NO$_3^-$	1 g N NO$_2$ 3 g N	33	42 d	→/→ →/→	↑ ↑	Jensen and Pilegaard, 1993
Hordeum vulgare cv Zephyr (wild type) B1 (NO$_2$ resistant) W3 (NO$_2$ resistant) cv Zephyr (wild type) B1 (NO$_2$ resistant) W3 (NO$_2$ resistant)	s/NO$_3^-$	0 NO$_2$ 5	500	3 d	→/→ →/↓ →/↓ →/↑ →/↓	(→) (∈) (↑) (∈) (∈)	Srivastava et al., 1994b
Hordeum vulgare Trifolium subterraneum	s	NO$_2$	170	108 d 4 h d^{-1}	↑/ ↓/		Murray et al., 1992b
Hordeum vulgare	h/NO$_3^-$	0 NO$_2$ 0.01 0.1 0.5 1.0 10.0	300	9 d	↑/↑ ↑/↑ →/↑ ↑/↑ ↑/↑		Rowland, 1986
Lycopersicon esculentum	s	NO$_2$ NO NO$_2$+NO	100 400 100/400	19 d	→/ ↓/ ↓/	↑ → →	Capron and Mansfield, 1977

(contd.)

Table 4.5 (contd.).

Species	Root N-supply form	mM	Pollu-tant	NO$_2$ conc. nl l^{-1}	Duration	Wei-ght	Shoot : root	Leaf area	RGR	NAR	LAR	LWR	Reference
Phaseolus vulgaris	s/ NO$_3^-$/NH$_4^+$		NO$_2$	39	34 d ('88)	→ /		↑					Bender et al., 1991
					48 d	→ /		↑					
					62 d	↑ /		↑ ←					
				31	34 d ('89)	↑ /							
					48 d	↑ /		↑ →					
					62 d	↓ /							
Phaseolus vulgaris	s/ NO$_3^-$/NH$_4^+$		NO$_2$	2,000	2 d		↑ ←						Ito et al. 1984a
					4 d		↑ ← ← ←						Ito et al. 1985
					7 d								
				4,000	2 d		↑ ← ← ←						
					4 d								
					7 d		←						
Phaseolus vulgaris	h/NO$_3^-$	0	NO$_2$	250	10 d/		↓/↓						Srivastava et al., 1990
		5			6 h d^{-1}		↓/↓						
Phaseolus vulgaris	s/ NO$_3^-$/NH$_4^+$		NO$_2$	25	15 d/	→/↑	↑		↑	↑	↑	↑	Sandhu and Gupta, 1989
				50	7 h d^{-1}	↑/↑	↑		←	↑	↑	↑	
				100		↑/↑	↑		←	←	↑	↑	
Picea abies	s/ NO$_3^-$/NH$_4^+$	2.0	NO$_2$	100	28 d	→/↓	←						Muller et al., 1996
					49 d	→/↓	←						
Picea abies	h/NO$_3^-$	0.72	NO$_2$	500	77 d	→/→							Tischner et al., 1988
Pinus sylvestris	h/NH$_4^+$	1.42	NO$_2$	20/30	19/39 d	→/→	↑	↑					Näsholm et al., 1991
Populus, Nerium, Zelkova, Euonymus, Quercus, Viburnum, Cinnamomum	s/ NO$_3^-$/NH$_4^+$		NO$_2$	300	30 d			↑					Okano et al., 1989

8 h d^{-1} and 5 d week^{-1}. Spierings (1971) reported yield reductions in *Lycopersicon esculentum* fumigated for 4 months with 250 nl l^{-1} NO$_2$. Yield always reacted comparable to above-ground biomass. Wellburn (1990) showed an example of growth inhibition in *Lactuca sativa* at 2000 nl l^{-1} NO, and growth reductions were published for exposures of *Lycopersicon esculentum* with $\geq$ 400 nl l^{-1} NO with concomitantly enhanced CO$_2$ concentrations (Capron and Mansfield, 1977; Anderson and Mansfield, 1979). Saxe and Christensen (1985) found that the beneficial effect of CO$_2$ on 5 commercial indoor crop genera was eliminated by 1000 nl l^{-1} NO.

Reports on changes in growth in response to oxidized atmospheric nitrogen compounds other than NO$_2$ and NO were not found in the published literature.

V. DAMAGE OR BENEFIT?

It is well established that atmospheric oxides of nitrogen are absorbed by plant leaves. The relative contribution of cuticular and stomatal absorption differ for each molecule. Uptake and incorporation into plant organic matter of the nitrogen moiety has only been shown conclusively for NO$_2$. However, there is experimental evidence that nitrogen from NO and HNO$_3$ vapour is incorporated as well. Judging from their deposition behaviour this may also be expected for HONO vapour and PAN. While for NO$_2$, and to a lesser extent for NO, the proportion of the deposition flux that is actually taken up is known to a certain extent, there are large uncertainties about the other compounds. This is an important gap in our knowledge when it comes to assessing total foliar nitrogen uptake from the atmosphere, especially in the case of HNO$_3$. Due to its large dry deposition velocity, and the suggested uptake pathway through the cuticle, it may contribute considerably to total foliar plant N uptake. The fate of HNO$_3$ and HONO nitrogen in wet and occult deposition is not well known either. While for HNO$_3$ some studies have addressed the question of reversibility of absorption, none have assessed uptake and incorporation. Thus, the uncertainties in the quantification of the proportion of the deposition flux that is used by the plants are still quite large. In addition, uptake of solutes in trees through the bark may add to the above-ground uptake of nitrogen (Katz et al., 1989) with possible physiological implications.

In spite of the possible fertilizing effect of airborne oxidized compounds, an increase in growth is rare, or growth reductions may even occur. Increased total-nitrogen and growth are usually limited to short-term experiments and to plants with a low nitrogen supply from the roots. In plants with a high nitrogen status or in legumes (Murray et al., 1992b), negative effects of the additional nitrogen input from the atmosphere are a common response. However, nitrogen supply alone is not the only parameter resulting in differences in susceptibility to NO_y absorption. Differences in the basic nitrogen metabolism may explain species-specific responses as discussed in several studies (Murray and Wellburn, 1985; Murray et al., 1992b; Pearson and Soares, 1995; Soares et al., 1995). Pearson and Soares (1995) have pointed out the importance of the differences between species in their ability to utilize nitrate. They argue that species which easily take up nitrate through the roots and translocate it to the shoots for reduction (mostly pioneer and ruderal species) are more efficient at inducing foliar NR activity when nitrate is imported from the air. According to Raven (1988) this may serve to cope with the concomitant increase in acidity which may cause physiological disturbances within the plant. 'Root nitrate reducers' (mostly climax species) may thus encounter more problems from atmospheric nitrogen input. Many of these plants have a low capacity for utilizing nitrate efficiently and their low nitrate uptake rate is coupled to a low uptake of base cations with buffering properties within the plant (Soares et al., 1995). In addition, they usually grow on acid soils low in base cations. Schulze (1989) and Slovik (1996) have discussed the possibility of the generation of nutrient imbalances by atmospheric nitrogen input, which would increase nitrogen: base cation ratios in plants on acid soils. Such imbalances are suggested by changes in amino acid contents as discussed in section IV. This may be aggravated i) by cation leaching from the soil induced by acidic nitrogen-containing deposition (Schulze, 1989; Durka et al., 1994; Likens et al., 1996), ii) by cation leaching from the foliage, and iii) if the feedback regulation on root nitrogen uptake occurs (Muller et al., 1996). Hence, for calcifuge 'root nitrate reducers' acidity generated by uptake of oxidized nitrogen compounds may offset the benefits of the additional nitrogen supply. Similarly, the impaired nitrate reducing capacity observed in leaves of 'leaf nitrate reducers' well supplied with nitrate (see Fig. 4.3) could result in increased acidification. However, for these plants, in ecological terms often pioneer species growing in disturbed soils, base cation shortage is usually of no importance and hence buffering by this means may be sufficient. However, proof for this hypothesis is lacking to date.

Morgan et al. (1992) also found differences between calcicole and calcifuge bryophytes, with the former being better adapted to high nitrogen supply and thus more resistant to NO_x toxicity. They assumed that the increased nitrite content caused problems for these plants. Missing sinks for excess nitrogen, such as vacuoles and long range transport to other organs, may render bryophytes more susceptible for nitrite accumulation. Wellburn (1990) has extensively discussed nitrite toxicity as a possible cause for NO_x toxicity in higher vascular plants. Evidence for nitrite accumulation in NO_x exposed leaves is limited to concentrations not usually found in the open field (see section IV). However, concentrations may be increased in organelles, such as the chloroplast. The few mechanistic studies addressing possibly deleterious effects of such an accumulation (see section IV) have shown serious impairments, but the nitrite concentrations applied were in excess of those expected in plants exposed to realistic pollutant concentrations. In view of the large phytotoxic potential of nitrite (Wellburn, 1990), clearly more research with realistic scenarios is needed. In this context, knowledge of the dissolution reaction of NO_x in the apoplast (disproportionation, nitrate only, nitrite only) is of considerable importance. The acidity hypothesis and the nitrite hypothesis for plant damage by NO_y are not mutually exclusive and a complete separation of the two effects appears to be difficult. In both cases, plants reducing the most nitrate in the roots would be disturbed more.

Differences in nitrogen metabolism and cycling may determine susceptibility to foliar nitrogen-uptake. During the growing season or life cycle, plants undergo distinct changes in these properties. Hence, different growth stages may differ in susceptibility. Some long-term studies appear to support this (Whitmore and Mansfield, 1983; Bender et al., 1991), though a dose-related effect is possible as well. These results suggest that results from short-term fumigations should be extrapolated to long-term exposures very carefully. Since their reaction with plant structures or with volatile substances emitted by the plants, such as unsaturated hydrocarbons, may produce highly reactive radicals at the atmosphere-plant interface, exposure of plants to oxidized nitrogen species may impose oxidative stress (see section IV), which in turn would increase energy demand for the plant defence and repair system. When nitrogen supply by the roots is sufficient, as is usually the case for agricultural crops, the observed nitrogen assimilation may be mostly an energy-consuming detoxification process without a nutritional benefit for plant. This is supported by the repeatedly reported GDH activity increase. According to Wellburn (1990), the resultant imbalance between nitrogen-containing and non-nitrogen-containing carbon skeletons and their allocation may impose an additional strain on the plants. Thus the forced nitrogen nutrition together with the undesirable chemical reactions triggered by atmospheric NO_y could impose increased energy demand. Most reports show that plants do not react to this by increasing photosynthetic activity. In some cases DR was shown to be enhanced at moderate NO_2 levels. Sandhu and Gupta (1989) noted that this may be a reflection of increased cellular activity to transform the pollutant into a non-toxic form. Thus the increased energy demand may be sufficient to prevent growth increases or even result in negative growth responses.

A large body of literature exists on the combined effect of different air pollutants on plants and their review is beyond the scope of this article. Many synergistic negative effects of the combination SO_2/NO_2 on plant growth or physiological parameters have been observed (e.g. Horsman and Wellburn, 1975; Whitmore and Mansfield, 1983; Freer-Smith, 1984, 1985; Freer-Smith and Mansfield, 1987; Murray et al., 1992; van Hove et al., 1992; for review: Darrall, 1989; Wolfenden and Mansfield, 1991). In the context of deposition the effect of SO_2 on stomatal conductance is especially important. The careful analysis by Darrall (1989) reveals that at low SO_2 concentrations usually stomatal opening is stimulated. At the same time stomatal response to environmental changes is impaired. Thus aerial nitrogen uptake may be increased in combination with SO_2, and a change in water relations may occur. Interactions of the two pollutants may also occur because the two assimilation pathways are interconnected (Murray et al., 1992b; Kast et al., 1995). Wellburn et al. (1981) showed that SO_2 inhibited nitrate reductase activity and thereby increased nitrite content of leaves in the presence of NO_2. This may have been the reason for an observed increase in nitrite content in needles of *Picea rubens* during an overwinter SO_2+NO_2 fumigation (Wolfenden et al., 1991). Since nitrite accumulation is discussed as one of the reasons for NO_y toxicity, evergreens could be at risk during winter dormancy when both NO_2 and SO_2 are present. This possibility should be investigated further, since most of the economically important evergreens of the northern hemisphere are 'root nitrate reducers', which may be especially susceptible (see above).

Some studies have looked at interactions between ozone and NO_x (Furukawa et al., 1984; Okano et al., 1985a). Mehlhorn and Wellburn (1987) presented the hypothesis that NO_x-induced stress ethylene production may determine plant sensitivity to ozone (Mehlhorn et al., 1991). After exposure to ozone, frost injury of *Picea rubens* during the winter could be prevented by removing NO with Purafil® (Neighbour et al., 1990). Therefore they concluded that the presence of NO in charcoal-filtered air in the low nl l^{-1} range was responsible for the observed differences. Nussbaum et al. (1995) found that 8 nl l^{-1} NO had an effect on ascorbate peroxidase and water-use efficiency in *Triticum aestivum* growing in open-top chambers at low ozone concentrations. The nature of this interaction is not yet understood. In view of these results, the point made by Wellburn (1990) that many of the observed effects of NO_2 in earlier studies may have included effects of NO, or that controls were not free of NO, is important, and should be carefully observed in future.

Likewise, reports of NO emission by plant leaves or extracts may become important in toxicity considerations. NO may evolve both enzymatically or non-enzymatically from nitrite (Dean and Harper, 1988) or result from the reduction of NO_2 by a not further characterized low-molecular weight polysaccharide (Nishimura et al., 1986). It may be dissolved in fluids as •N=O known for its hormone-like nature in the human blood circulation. Wellburn (1990) has discussed the potentially deleterious effect of binding to transition element-containing proteins (e.g. reduced ferredoxin). He argues that protective regulatory reactions could be prevented from taking place. Hence, the toxicity of NO in plants may need re-evaluation.

Another factor which could lead to changes in plant performance is the change in susceptibility to pests. Masters and McNeill (1996) reported an increased feeding of aphids on *Vicia faba* exposed to NO_2, which was cultivar specific. Changes in insect interactions after exposure to combinations of air pollutants have been reported earlier (Dohmen, 1988; Flückiger et al., 1988; Houlden et al., 1992).

VI. CONCLUSIONS AND FUTURE PROSPECTS

Foliar absorption, uptake and assimilation of nitrogen moiety are suggested for all the atmospheric nitrogen oxides discussed in this article. The forced nitrogen-nutrition without balanced nutrition provided by root-uptake systems, and the reactive nature of these compounds, appear to eliminate a possible, positive fertilizing effect. NO_y input from the air is beneficial in special cases only (e.g. N-starved 'leaf nitrate reducers'). Judging from the literature reviewed in this article, the following toxicity classification for dry deposition to leaves appears to be justified: PAN > NO > HNO_3 ≥ NO_2. Nitrogen nutrition, nitrogen metabolism and nitrogen cycling, in this order, are important factors in determining the susceptibility of single plant species to each compound. Due to the high acute toxicity, this may not be the case for PAN, but there are no studies with long-term low concentration exposures addressing this issue. The fate of occult and wet deposition absorbed by leaves is known to a far lesser extent but the contribution to foliar nitrogen uptake may be important.

Assessment of the consequences of foliar nitrogen uptake for plants clearly has to include the nitrogen economy of the whole plant. Hence, knowledge of atmospheric and pedospheric contributions to nitrogen nutrition and its mutual regulation under realistic scenarios is imperative. In this context, absorbed doses rather than applied concentrations are important. This aspect has been neglected in chamber studies in the past but has gained attention during the last decade. It would facilitate establishing links to the micrometeorological deposition measurements and thus create a broader database for risk assessment for natural vegetation and agricultural crops. Long-term studies covering entire life cycles or growing seasons are scarce, but the data provided suggest that susceptibility to pollutants may change considerably within long time spans.

Our knowledge of the behaviour of the various oxidized nitrogen species in deposition, absorption, uptake and metabolism is not well balanced. Of the different oxidized nitrogen species, NO_2 is the best studied in all processes, followed by NO. There is a clear gap in knowledge concerning absorption, uptake and assimilation of HNO_3. Given the rather high amount of nitrogen deposited to leaves in this form, these uncertainties may become important for the total nitrogen-balance. Except for deposition data, virtually nothing is known about the behaviour of HONO towards plants. Since the contribution to total-N deposition is usually very low, its impact on plants may be assumed to be negligible. But since nitrite and its undissociated acid are assumed to be the products of the absorption process, it may be overproportionally toxic. Research on effects of PAN was done mostly in the 60s and 70s, and the deposition behaviour, acute visible damage, and some damaging processes are known. However, practically nothing is known about the effect of concentrations below the threshold for visible injury, as found over large parts of industrialized areas.

Overall, in polluted air, toxicity of foliar absorption of oxidized nitrogen compounds in terrestrial plants overrides nutritional benefits. But at the ecosystem level they appear to be important and should be considered together with the often negative effects of the deposition to the soil (see *Water, Air and Soil Pollution*, vol. 85, 1995). However, more detailed knowledge about the effects of realistic exposures and increased attention to the set-up of chamber experiments with respect to compatibility with field deposition measurements appear to be essential in order to arrive at valid assessments of ecosystem impacts.

LITERATURE CITED

Adams, M.A. and Attiwill, P.M. 1982. Nitrate reductase activity and growth response of forest species to ammonium and nitrate sources of nitrogen. *Plant and Soil* 66: 373–381.

Ammann, M., Stalder, M., Suter, M., Brunold, C., Baltensperger, U., Jost, D.T., Türler, A. and Gäggeler, H.W. 1995a. Tracing uptake and assimilation of NO_2 in spruce needles with ^{13}N. *J. Exp. Bot.* 46: 1685–1691.

Ammann, M., von Ballmoos, P., Stalder, M., Suter, M. and Brunold, C. 1995b. Uptake and assimilation of atmospheric NO_2-N by spruce needles (*Picea abies*): a field study. *Water, Air, Soil Pollut.* 85: 1497–1502.
Anderson, L.S. and Mansfield, T.A. 1979. The effects of nitric oxide pollution on the growth of tomato. *Environ. Pollut.* 20: 113–121.
Andrews, M. 1986. The partitioning of nitrate assimilation between root and shoot of higher plants. *Plant, Cell Environ.* 9: 511–519.
Ashenden, T.W. 1979a. Effects of SO_2 and NO_2 pollution on transpiration in *Phaseolus vulgaris*. *Environ. Pollut.* 18: 45–50.
Ashenden, T.W. 1979b. The effects of long-term exposures to SO_2 and NO_2 pollution on the growth of *Dactylis glomerata* L. and *Poa pratensis* L. *Environ. Pollut.* 18: 249–258.
Ashenden, T.W. and Mansfield, T.A. 1978. Extreme pollution sensitivity of grasses when SO_2 and NO_2 are present in the atmosphere together. *Nature* 273: 142–143.
Baldocchi, D. 1988. A multi layer model for estimating sulfur dioxide deposition to a deciduous oak forest canopy. *Atm. Environ.* 22: 869–884.
Baldocchi, D., Hicks, B.B. and Camara, P. 1987. A canopy stomatal resistance model for gaseous deposition to vegetated surfaces. *Atm. Environ.* 21: 91–101.
Bambauer, A., Brantner, B., Paige, M. and Novakov, T. 1994. Laboratory study of NO_2 reaction with dispersed and bulk liquid water. *Atm. Environ.* 28: 3225–3232.
Beevers, L. and Hageman, R.H. 1980. Nitrate and nitrite reduction. In: *The Biochemistry of Plants, Vol. 5* (B.J. Miflin, ed.). Academic Press, New York, pp. 115–168.
Bender, J., Weigel, H.J. and Jäger, H.J. 1991. Response of nitrogen metabolism in beans (*Phaseolus vulgaris* L.) after exposure to ozone and nitrogen dioxide, alone and in sequence. *New Phytol.* 119: 261–267.
Bennett, J.H. and Hill, A.C. 1973. Inhibition of apparent photosynthesis by air pollutants. *J. Environ. Qual.* 2: 526–530.
Beswick, K.M., Gallagher, M.W., Hummelshoj, P., Pilegaard, K., Jensen, N.O. and Duyzer, J. 1994. Aerosol exchange to Speulder forest. In: *Proc. EUROTRAC Symp. '94 Garmisch-Partenkirchen, 11th–15th April 1994* (P.M. Borrell, P. Borrell, T. Cvitas and W. Seiler, eds.). SPB Academic Publ., Den Haag, The Netherlands, pp. 683–688.
Bobrov, R.A. 1955. The leaf structure of *Poa annua* with observations on its smog sensitivity in Los Angeles County. *Am. J. Bot.* 42: 467–474.
Bruckner, G., Schulze, E-D. and Gebauer, G. 1993. ^{15}N labelled NH_3 uptake experiments and their relation to natural conditions. In: *General Assessment of Biogenic Emissions and Depositions of Nitrogen Compounds, Sulphur, Compounds and Oxidants in Europe* (J. Slanina, G. Angeletti, and S. Beilke, eds.). *Air Pollut. Res. Rep.* 47: 305–311.
Burnold, C., von Ballmoos, P., Nussbaum, S., Ammann, M., Stalder, M., Schlunegger, U.P. and Gfeller H. 1997. Uptake and distribution of NO_2 in Norway spruce (*Picea abies* [L.] Karst.). In: *Biosphere-Atmosphere Exchange of Pollutants and Trace Substances* (S. Slanina ed.) Springer, Berlin, Germany, pp. 413–419.
Bull, J.N. and Mansfield, T.A. 1974. Photosynthesis in leaves exposed to SO_2 and NO_2. *Nature* 250: 443–444.
Cadle, S.H., Marshall, J.D. and Mulawa, P.A. 1991. A laboratory investigation of the routes of HNO_3 dry deposition to coniferous seedlings. *Environ. Pollut.* 72: 287–305.
Capron, A.C. and Mansfield, T.A. 1977. Inhibition of growth in tomato by air polluted with nitrogen oxides. *J. Exp. Bot.* 28: 112–116.
Carlson, R.W. 1983. Interaction between SO_2 and NO_2 and their effects on photosynthetic properties of soybean. *Glycine max. Environ. Pollut.* 32: 11–38.
Choularton, T.W., Coe, H., Walton, S., Gallagher, M.W., Beswick, K.M., Dore, C., Duyzer, J., Westrate, H., Pilegaard, K., Jensen, N.O. and Hummelshoj, P. 1994. Photochemical modification of ozone and NO_x deposition to forests: results from the Speulderbos and Rivox experiments. In: *Proc. EUROTRAC Symp. '94 Garmisch-Partenkirchen, 11th–15th April 1994* (P.M. Borrell, P. Borrell, T. Cvitas and W. Seiler, eds.). SPB Academic Publ. Den Haag, The Netherlands, pp. 645–650.

Darrall, N.M. 1989. The effect of air pollutants on physiological processes in plants. *Plant, Cell Environ.* 12: 1–30.
Dasch, J.M. 1989. Dry deposition of sulfur dioxide or nitric acid to oak, elm and pine leaves. *Environ. Pollut.* 59: 1–16.
Dasch, J.M. and Cadle, S.H. 1986. Dry deposition to snow in an urban area. *Water, Air, Soil Pollut.* 9: 297–308.
Davidson, E.A. 1991. Fluxes of nitrous oxide and nitric oxide from terrestrial ecosystems. In: *Microbial Production and Consumption of Greenhouse Gases: Methane, Nitrogen Oxides, and Halomethanes* (J.E. Rogers and W.B. Whitman, eds.). Amer. Soc. Microbiology, Washington D.C., USA, pp. 219–235.
Dean, J.V. and Harper, J.E. 1988. The conversion of nitrite to nitrogen oxide(s) by the constitutive NAD(P)H-nitrate reductase enzyme from soybean. *Plant Physiol.* 88: 389–395.
Demerjian, K.L. 1986. Atmospheric chemistry of ozone and nitrogen oxides. In: *Air Pollutants and Their Effects on the Terrestrial Ecosystem* (A.H. Legge and S.V. Krupa eds.). John Wiley & Sons, New York, USA, pp. 105–127.
Dohmen, G.P. 1988. Indirect effects of air pollutants: changes in plant parasite interactions. *Environ. Pollut.* 53: 197–207.
Dollard, G.J., Atkins, D.H.F., Davies, T.J. and Healy, C. 1987. Concentrations and dry deposition velocities of nitric acid. *Nature* 326: 481–483.
Draaijers, G.P.J., van Ek, R., Bleuten, W. and Meijers, R. 1992. Measuring and modelling atmospheric dry deposition in complex forest terrain. In: *Acidification Research, Evaluation and Policy Applications* (T. Schneider, ed.). Elsevier Science Publishers, The Netherlands, pp. 285–294.
Dugger, W.M.Jr., Koukol, J. and Palmer, R.L. 1966. Physiological and biochemical effects of atmospheric oxidants on plants. *J. Air Pollut. Contr. Ass.* 16: 467–471.
Dugger, W.M.Jr. and Ting, I.P. 1968. The effect of peroxyacetyl nitrate on plants: photoreductive reactions and susceptibility of bean plants to PAN. *Phytopathology* 58: 1102–1107.
Dugger, W.M.Jr. and Ting, I.P. 1970. Air pollution oxidants—their effects on metabolic processes in plants. *Ann. Rev. Plant Physiol.* 21: 215–234.
Dugger, W.M.Jr., Koukol, J., Reed, W.D. and Palmer, R.L. 1963a. Effects of peroxyacetyl nitrate on $^{14}CO_2$ fixation by spinach chloroplasts and pinto bean plants. *Plant Physiol.* 38: 468–472.
Dugger, W.M.Jr., Taylor, O.C., Klein, W.H. and Shropshire, W.Jr. 1963b. Action spectrum of peroxyacetyl nitrate damage to bean plants. *Nature* 198: 75–76.
Dugger, W.M.Jr., Taylor, O.C., Thompson, C.R. and Cardiff, E. 1963c. The effect of light on predisposing plants to ozone and PAN damage. *J. Air Pollut. Contr. Ass.* 13: 423–428.
Durka, W., Schulze, E.-D., Gebauer, G. and Voerkelius, S. 1994. Effects of forest decline on uptake and leaching of deposited nitrate determined from ^{15}N and ^{18}O measurements. *Nature* 372: 765–767.
Durmishidze, S.V. and Nutsubidze, N.N. 1976. Absorption and conversion of nitrogen dioxide by higher plants. *Dokl. Biochem.* 277: 104–107.
Duyzer, J. and Fowler, D. 1994. Modelling land atmosphere exchange of gaseous oxides of nitrogen in Europe. *Tellus* 46B: 353–372.
Duyzer, J.H., Deinum, G. and Baak, J. 1995. The interpretation of measurements of surface exchange of nitrogen oxides: correction for chemical reactions. *Phil. Trans. R. Soc. Lond. A* 351: 231–248.
Edfast, A.-B., Näsholm, T. and Ericsson, A. 1990. Free amino acid concentrations in needles of Norway spruce and Scots pine trees on different sites in areas with two levels of nitrogen deposition. *Can. J. For. Res.* 20: 1132–1136.
Egger, A., Landolt, W. and Brunold, C. 1988. Effects of NO_2 on assimilatory nitrate and sulfate reduction in needles from spruce trees (*Picea abies* L.). In: *Air Pollution and*

Forest Decline (J.B. Bucher and I. Bucher-Wallin, eds.). Proc. 14th Intern. Meeting Specialists in Air Pollution Effects on Forest Ecosystems, Interlaken, Switzerland, pp. 401–403.

Emes, M.J. and Fowler, M.W. 1979. The intracellular location of the enzymes of nitrate assimilation in the apices of pea seedling roots. *Planta* 144: 249–253.

Ericsson, A., Nordén, L.-G., Näsholm, T. and Walheim, M. 1993. Mineral nutrient imbalances and arginine concentrations in needles of *Picea abies* (L.) Karst. from two areas with different levels of airborne deposition. *Trees* 8: 67–74.

Erisman, J.W., Beier, C., Draaijers, G. and Lindberg, S. 1994. Review of deposition monitoring methods. *Tellus* 46B: 79–93.

Eugster, W. and Hesterberg, R. 1996. Transfer resistances of NO_2 determined from eddy correlation flux measurements over a litter meadow at a rural site on the Swiss Plateau. *Atm. Environ.* 30: 1247–1254.

Evans, L.S., Canada, D.C. and Santucci, K.A. 1986. Foliar uptake of ^{15}N from rain. *Environ. Exp. Bot.* 26: 143–146.

Faller, N. 1972. Schwefeldioxid, Schwefelwasserstoff, Nitrose Gase und Ammoniak als ausschliessliche S-bzw. N-Quellen der höheren Pflanzen. *Z. Pfl. Ernähr. Düng. Bodenk.* 131: 120–130.

Febo, A., and Perrino, C. 1991. Prediction and experimental evidence for high air concentration of nitrous acid in indoor environments. *Atm. Environ.* 25A: 1065–1061.

Flückiger, W., Braun, S. and Bolsinger, M. 1988. Air pollution: effect on host plant-insect relationships. In: *Air Pollution and Plant Metabolism* (S. Schulte-Hostede, N.M. Darrall and L.W. Blank, eds.). Elsevier Applied Science, London-New York, pp. 366–381.

Fowler, D., Duyzer, J.H. and Baldocchi, D.D. 1991. Inputs of trace gases, particles and cloud droplets to terrestrial surfaces. *Proc. Royal Soc. Edinburgh* 97B: 35–39.

Freer-Smith, P.H. 1984. The response of six broad-leaved trees during long-term exposure to SO_2 and NO_2. *New Phytol.* 97: 49–61.

Freer-Smith, P.H. 1985. The influence of SO_2 and NO_2 on the growth, development and gas exchange of *Betula pendula* Roth. *New Phytol.* 99: 417–430.

Freer-Smith, P.H. and Mansfield, T.A. 1987. The combined effects of low temperature and $SO_2 + NO_2$ pollution on the new season's growth and water relations of *Picea sitchensis*. *New Phytol.* 106: 237–250.

Fuhrer, J. and Erismann, K.H. 1980. Uptake of NO_2 by plants grown at different salinity levels. *Experientia* 36: 409–410.

Furukawa, A. and Totsuka, T. 1979. Effect of NO_2, SO_2 and O_3 alone and in combinations on net photosynthesis in sunflower. *Environ. Control in Biology* 17: 161–166.

Furukawa, A., Yokoyama, M., Ushijima T. and Totsuka T. 1984. The effects of NO_2 and/or O_3 on photosynthesis of sunflower leaves. *Res. Rep. Natl. Inst. Environ. Stud. Jpn.* 65: 89–98.

Galangau, R. Daniel-Vedèle, F., Moureaux, T., Dorbe, M.F., Leydecker, M.T. and Caboche, M. 1988. Expression of leaf nitrate reductase genes from tomato and tobacco in relation to light-dark regimes and nitrate supply. *Plant Physiol.* 88: 383–388.

Garland, J.A. and Penkett, S.A. 1976. Absorption of peroxyacetyl nitrate and ozone by natural surfaces. *Atm. Environ.* 10: 1127–1131.

Garten, C.T. and Hanson, P.J. 1990. Foliar retention of ^{15}N-nitrate and ^{15}N-ammonium by red maple (*Acer rubrum*) and white oak (*Quercus alba*) leaves from simulated rain. *Environ. Exp. Bot.* 30: 333–342.

Gessler, A., Schneider, S., Weber, P., Hanemann, U. and Rennenberg, H. 1998. Soluble N compounds in trees exposed to high loads of N: a comparison between the roots of Norway spruce (*Picea abies*) and beech (*Fagus sylvatica*) trees grown under field conditions. *New Phytol.* 138: 385–399.

Grace, J., Malcolm, D.C. and Bradbury, I.K. 1975. The effect of wind and humidity on leaf diffusive resistance in Sitka spruce seedlings. *J. Appl. Ecol.* 12: 931–940.

Guenthardt, M.S. 1984. Epicuticular wax of *Picea abies* needles. In: *Structure, Function and Metabolism of Plant Lipids* (P.-A. Siegenthaler and W. Eichenberger, eds.). Elsevier Science Publishers, The Netherlands, pp. 499–502.

Gupta, G. and Narayanan, R. 1992. Nitrogen fixation in soybean treated with nitrogen dioxide and molybdenum. *J. Environ. Qual.* 21: 46–49.

Gupta, G., Li, Y. and Sandhu, R. 1992. Photosynthesis and nitrogen fixation in soybean exposed to nitrogen dioxide and carbon dioxide. *J. Environ. Qual.* 21: 624–626.

Gupta, S.C. and Beevers, L. 1983. Environmental influences on nitrate reductase in *Pisum sativum* L. seedlings. *J. Exp. Bot.* 34: 1455–1462.

Gutknecht, J. and Walter, A. 1981. Hydrofluoric and nitric acid transport through lipid bilayer membranes. *Biochim. Biophys. Acta* 644: 153–156.

Hanson, P.J. and Lindberg, S.E. 1991. Dry deposition of reactive nitrogen compounds: a review of leaf, canopy and non-foliar measurements. *Atm. Environ.* 25A: 1615–1634.

Hanson, P.J. and Garten, C.T. 1992. Deposition of $H^{15}NO_3$ vapour to white oak, red maple and loblolly pine foliage: experimental observations and a generalized model. *New Phytol.* 122: 329–337.

Hargreaves, K.J., Fowler, D., Storeton-West, R.L. and Duyzer, J.H. 1992. The exchange of nitric oxide, nitrogen dioxide and ozone between pasture and the atmosphere. *Environ. Pollut.* 75: 53–59.

Harrison, R.M., Rapsomanikis, S. and Turnbull, A. 1989. Land-surface exchange in a chemically reactive system; surface fluxes of HNO_3, HCl and NH_3. *Atm. Environ.* 23: 1795–1800.

Hartung, W., Radin, J.W. and Hendrix, D.L. 1988. Abcissic acid movement into apoplastic solution of water-stressed cotton leaves: role of apoplastic pH. *Plant Physiol.* 86: 908–913.

Heber, U. and Purczeld, P. 1978. Substrate and product fluxes across the chloroplast envelop during bicarbonate and nitrite reduction. In: *Proc. 4th Internat. Cong. Photosynthesis*. (O.D. Hall, J. Coombs and T.W. Goodwin, eds.). Biochemical Society, London, UK, pp. 197–218.

Hidy, G.M. and Mueller, P.K. 1986. The sulfur oxide particulate matter complex. In: *Air Pollutants and Their Effects on the Terrestrial Ecosystem* (H.A. Legge and S.V. Krupa, eds.). John Wiley & Sons, Inc., New York, pp. 51–104.

Hill, A.C. and Bennett, J.H. 1970. Inhibition of apparent photosynthesis by nitrogen oxides. *Atm. Environ.* 4: 341–348.

Hisamatsu, S., Nihira, J., Takeuchi, Y., Satoh, S. and Kondo, N. 1988. NO_2 suppression of light-induced nitrate reductase in squash cotyledons. *Plant Cell Physiol.* 29: 395–401.

Horsman, D.C. and Wellburn, A.R. 1975. Synergistic effects of SO_2 and NO_2 polluted air upon enzyme activity in pea seedlings. *Environ. Pollut.* 8: 123–133.

Houlden, G., McNeill, S., Craske, S. and Bell, J.N.B. 1992. Air pollution and agricultural aphid pests. II: Chamber filtration experiments. *Environ. Pollut.* 72: 45–55.

Huebert, B.J. and Robert, C.H. 1985. The dry deposition of nitric acid to grass. *J. Geophys. Res.* 90: 2085–2090.

Hummelshoj, P., Pilegaard, K. and Jensen, N.O. 1994. Flux measurements of O_3 and NO_2 over a regrown wheat field. In: *Proc. EUROTRAC Symp. '94 Garmisch-Partenkirchen, 11th-15th April 1994* (P.M. Borrell, P. Borrell, T. Cvitas and W. Seiler, eds.). SPB Academic Publ., Den Haag, The Netherlands, pp. 544–548.

Intergovernmental Panel on Climate Change. 1990. Climate Change: The IPCC Scientific Assessment (J.T. Houghton, G.J. Jenkins, and J.J. Ephraums, eds.). Cambridge University Press, New York.

Ito, O., Okano, K., Kuroiwa, M. and Totsuka, T. 1984a. Effects of NO_2 and O_3 alone or in combination on kidney bean plants: I. Growth, partitioning of assimilates and root activities. *Res. Rep. Natl. Inst. Environ. Stud. Jpn.* 66: 1–12.

Ito, O., Okano, K. and Totsuka, T. 1984b. Effects of NO_2 and O_3 alone or in combination on kidney bean plants: II. Amino acid pool size and composition. *Res. Rep. Natl. Inst.*

Environ. Stud. Jpn. 66: 15–24.
Ito, O., Okano, K. and Totsuka, T. 1986. Effects of NO_2 and O_3 alone or in combination on kidney bean plants: II. Amino acid pool size and composition. Soil Sci. Plant Nutr. 32: 351–363.
Ito, O., Okano, K., Kuroiwa, M. and Totsuka, T. 1985. Effects of NO_2 and O_3 alone or in combination on kidney bean plants (Phaseolus vulgaris L.): Growth, partitioning of assimilates and root activities. J. Exp. Bot. 36: 652–662.
Jarvis, P.G. 1976. The interpretation of the variations in leaf water potential and stomatal conductance found in canopies in the field. Phil. Trans. R. Soc. Lond. B 273: 593–610.
Jensen, E.S. and Pilegaard, K. 1993. Absorption of nitrogen dioxide by barley in open-top chambers. New Phytol. 123: 359–364.
Johansson, C. 1987. Pine forest: a negligible sink for atmospheric NO_x in rural Sweden. Tellus 39B: 426–438.
Johansson, C. and Granat, L. 1986. An experimental study of the dry deposition of gaseous nitric acid to snow. Atm. Environ. 20: 1165–1170.
Kaiser, W.M., Spill, D. and Glaab, J. 1993. Rapid modulation of nitrate reductase in leaves and roots: Indirect evidence for the involvement of protein phosphorylation/dephosphorylation. Physiol. Plant. 89: 557–562.
Kaji, M., Yoneyama, T., Totsuka, T. and Iwaki, H. 1980. Absorption of atmospheric NO_2 by plants and soils. VI. Transformation of NO_2 absorbed in the leaves and transfer of nitrogen through the plants. Res. Rep. Natl. Inst. Env. Stud. Jpn. 11: 51–58.
Kast, D., Stalder, M., Rüegsegger, A., Galli, U. and Brunold, C. 1995. Effects of NO_2 and nitrate on sulfate assimilation in maize. J. Plant Physiol. 147: 9–14.
Katz, C., Oren, R., Schulze, E.-D. and Milburn, J.A. 1989. Uptake of water and solutes through twigs of Picea abies (L.) Karst. Trees 3: 33–37.
Kerstiens, G., Federholzner, R. and Lendzian, K.J. 1992. Dry deposition and cuticular uptake of pollutant gases. Agriculture, Ecosystems and Environment, 42: 239–253.
Kisser-Priesack, G.M. and Gebauer, G. 1991. Kinetics of $^{15}NO_x$ uptake by plant cuticles. In: Stable Isotopes in Plant Nutrition, Soil Fertility and Environmental Studies. International Atomic Energy Agency, Vienna, Austria, pp. 619–625.
Kisser-Priesack, G.M., Scheunert, I. and Gnatz, G. 1987. Uptake of $^{15}NO_2$ and ^{15}NO by plant cuticles. Naturwissenschaften 74: 550–551.
Klarer, C.I., Reinert, R.A. and Huang, J.S. 1984. Effects of sulfur dioxide and nitrogen dioxide on vegetative growth of soybeans. Phytopathology 74: 1104–1106.
Kramm, G., Dlugi, R., Dollard, G.J., Foken, T., Mölders, N., Müller, H., Seiler, W. and Sievering, H. 1995. On the dry deposition of ozone and reactive nitrogen species. Atm. Environ. 29: 3209–3231.
Kunert, K.J. 1987. Lipidperoxidation als phytotoxische Folge atmosphärischer Schadstoffwirkung. In: Projekt Europäisches Forschungszentrum für Massnahmen zur Luftreinhaltung PEF. PEF-Bericht 34. Kernforschungsanlage Karlsruhe, Germany.
Latus, C., Förstel, H., Führ, F., Neubert, A. and Kley, D. 1990. Quantitative measurements of NO_2 uptake and metabolism by sunflower plants. Naturwissenschaften 77: 283–285.
Lee, Y.-N. and Schwartz, S.E. 1981. Evaluation of the kinetics of uptake of nitrogen dioxide by atmospheric and surface liquid water. J. Geophys. Res. 86: 11971–11973.
Leighton, P.A. 1961. Photochemistry of Air Pollution. Academic Press, New York.
Lendzian, K.J. and Kahlert, J. 1988. Interactions between nitrogen dioxide and plant cuticles: binding to cuticular structures. Plant Physiol. 7: 197–207.
Lendzian, K.J. and Kerstiens, G. 1991. Sorption and transport of gases and vapors in plant cuticles. Rev. Environ. Contamination and Toxicology 121: 65–128.
L'Hirondelle, S.J., Jacobson, J.S. and Lassoie, J.P. 1992. Acidic mist and nitrogen fertilization effects on growth, nitrate reductase activity, gas exchange, and frost hardiness of red spruce seedlings. New Phytol. 121: 611–622.
Likens, G.E., Driscoll, C.T. and Buso, D.C. 1996. Long-term effects of acid rain: response and recovery of a forest ecosystem. Science 272: 244–246.

Lindberg, S.E. and Lovett, G.M. 1985. Field measurements of particle dry deposition rates to foliage and inert surfaces in a forest canopy. *Environ. Sci. Technol.* 19: 238–244.

Lindberg, S.E., Lovett, G.M., Richter, D.D. and Johnson, D.W. 1986. Atmospheric deposition and canopy interactions of major ions in a forest. *Science* 231: 141–145.

Lindberg, S.E., Lovett, G.M., Schaefer, D.A. and Bredemeier, M. 1988. Coarse aerosol deposition velocities and surface-to-canopy scaling factors from forest canopy throughfall. *J. Aerosol Sci.* 19: 1187–1190.

Lindberg, S.E., Bredemeier, M., Schaefer, D.A. and Qi, L. 1990. Atmospheric concentrations and deposition of nitrogen and major ions in conifer forests in the United States and Federal Republic of Germany. *Atm. Environ.* 24A: 2207–2220.

Lindberg, S.E., Cape, J.N., Garten, C.T.Jr. and Ivens, W. 1992. Can sulphate fluxes in forest canopy throughfall be used to estimate atmospheric sulphur deposition? In: *Proc. 5th IPSASEP Conf., Richland, 15–19 June 1991, USA* (S.E. Schwartz, and W.G. Sklinn, eds.). Hemisphere Publishing Corporation, Washington, USA, pp. 1367–1378.

Logan, J.A. 1983. Nitrogen oxides in the troposphere: global and regional budgets. *J. Geophys. Res.* 88: 785–807.

Lorenc-Plucinska, G. 1988. Effect of nitrogen dioxide on CO_2 exchange in Scots pine seedlings. *Photosynthetica* 22: 108–111.

Lovett, G.M. and Kinsman, J.D. 1990. Atmospheric pollutant deposition to high-elevation ecosystems. *Atm. Environ.* 24A: 2767–2786.

Lovett, G.M. and Lindberg, S.E. 1993. Atmospheric deposition and canopy interactions of nitrogen in forests. *Can. J. For. Res.* 23: 1603–1616.

Malhotra, S.S. and Khan, A.A. 1984. Biochemical and physiological impact of major pollutants. In: *Air Pollution and Plant Life* (M. Treshow, ed.). John Wiley & Sons, Chichester, UK, pp. 113–157.

Masters, G.J. and McNeill, S. 1996. Evidence that plant varieties respond differently to NO_2 pollution as indicated by resistance to insect herbivores. *Environ. Pollut.* 91: 351–354.

Mayumi, H. and Yamazoe, F. 1983. Effects of photochemical air pollution on plants. *Bull. Natl. Inst. Agric. Sci.* B. 35: 1–71.

McMillen, R.T. 1990. Estimating the spatial variability of trace gas deposition velocities. *NOAA Technical Memorandum,* ERL ARL 181, Oak Ridge.

Mehlhorn, H. and Wellburn, A.R. 1987. Stress ethylene formation determines plant sensitivity to ozone, *Nature* 327: 193–197.

Mehlhorn, H., O'Shea, J.M. and Wellburn, A.R. 1991. Atmospheric ozone interacts with stress ethylene formation by plants to cause visible plant injury. *J. Exp. Bot.* 42: 17–24.

Meixner, F.X., Franken, H.H., Duyzer, J.H. and van Aalst, R.M. 1988. Dry deposition of gaseous HNO_3 to a pine forest. In: *Air Pollution Modeling and its Application, VI* (H. van Dop, ed.). Plenum, New York, pp. 23–35.

Meyers, T.P., Huebert, B.J. and Hicks, B.B. 1989. HNO_3 deposition to a deciduous forest. *Boundary Layer Met.* 49: 395-410.

Momen, B. and Helms, J.A. 1996. Effects of simulated acid rain and ozone on foliar chemistry of field grown *Pinus ponderosa* seedlings and mature trees. *Environ. Pollut.* 91: 105–111.

Morgan, S.M., Lee, J.A. and Ashenden, T.W. 1992. Effects of nitrogen oxides on nitrate assimilation in bryophytes. *New Phytol.* 120: 89–97.

Mudd, J.B. 1963. Enzyme inactivation by peroxyacetyl nitrate. *Arch. Biochem. Biophys.* 102: 59–65.

Mudd, J.B., Banerjee, S.K., Dooley, M.M. and Knight, K.L. 1984. Pollutants and plant cells. In: *Gaseous Air Pollutants and Plant Metabolism.* Proc. 1st Internat. Symp. Air Pollution and Plant Metabolism, Oxford, 1982 (M.J. Koziol and F.R. Whatley eds.). Butterworths Scientific Press, London, UK, pp. 105–116.

Muller, B. and Touraine, B. 1992. Inhibition of NO_3^- uptake by various phloem translocated amino acids in soybean seedlings. *J. Exp. Bot.* 43: 617–623.
Muller, B., Touraine, B. and Rennenberg, H. 1996. Interaction between atmospheric and pedospheric nitrogen nutrition in spruce (*Picea abies* L. Karst.) seedlings. *Plant, Cell Environ.* 19: 345–355.
Murray, A.J.S. and Wellburn, A.R. 1985. Differences in nitrogen metabolism between cultivars of tomato and pepper during exposure to glasshouse atmospheres containing oxides of nitrogen. *Environ. Pollut.* 39: 303–316.
Murray, F., Clarke, K. and Wilson, S. 1992a. Effects of NO_2 on hoop pine can be counteracted by SO_2. *Europ. J. For. Path.* 22: 403–409.
Murray, F., Wilson, S. and Monk, R. 1992b. NO_2 and SO_2 mixtures stimulate barley grain production but depress clover growth. *Environ. Exp. Bot.* 32: 185–192.
Näsholm, T., Högberg, P. and Edfast, A.-B. 1991. Uptake of NO_x by mycorrhizal and nonmycorrhizal Scots pine seedlings: quantities and effects on amino acid and protein concentrations. *New Phytol.* 119: 83–92.
Neighbour, E.A., Person, M. and Mehlhorn, H. 1990. Purafil-filtration prevents the development of ozone-induced frost injury: a potential role for nitric oxide. *Atm. Environ.* 24A: 711–715.
Neubert, A., Kley, D., Wildt, J., Segschneider, H.J. and Förstel, H. 1993. Uptake of NO, NO_2 and O_3 by sunflower (*Helianthus annuus* L.) and tobacco plants (*Nicotiana tabacum* L.): Dependence on stomatal conductivity. *Atm. Environ.* 27A: 2137–2145.
Nishimura, H., Hayamizu, T. and Yanagisawa, Y. 1986. Reduction of NO_2 to NO by rush and other plants. *Environ. Sci. Technol.* 20: 413–416.
Norby, R.J., Weerasuriya, Y. and Hanson, P.J. 1989. Induction of nitrate reductase activity in red spruce needles by NO_2 and HNO_3 vapor. *Can. J. For. Res.* 19: 889–896.
Nouchi, I. and Toyama, S. 1988. Effects of ozone and peroxyacetyl nitrate on polar lipids and fatty acids in leaves of morning glory and kidney bean. *Plant Physiol.* 87: 638–646.
Nussbaum, S. 1992. Einbau von atmosphärischem $^{15}NO_2$-Stickstoff in freie Aminosäuren durch die Fichte (*Picea abies* [L.] Karst.). Ph.D. thesis, University of Berne, Switzerland.
Nussbaum, S., Geissmann, M. and Fuhrer, J. 1995. Effects of nitric oxide and ozone on spring wheat (*Triticum aestivum*). *Water, Air, Soil Pollut.* 85: 1449–1454.
Nussbaum, S., von Ballmoos, P., Guggisberg, R. and Brunold, C. 1991. Effects of NO_x fumigation on spruce (*Picea abies* [L.] Karst.). In: *Air Pollution Research Report 39* (G. Angeletti, S. Beilke and J. Slanina, eds.). Commission European Communities, Brussels, Belgium, pp. 203–210.
Nussbaum, S., von Ballmoos, P., Gfeller, H., Schlunegger, U.P., Fuhrer, J., Rhodes, D. and Brunold, C. 1993. Incorporation of atmospheric $^{15}NO_2$-nitrogen into free amino acids by Norway spruce *Picea abies* (L.) Karst. *Oecologia* 94: 408–414.
Okano, K. and Totsuka, T. 1986. Absorption of nitrogen dioxide by sunflower plants grown at various levels of nitrate. *New Phytol.* 102: 551–562.
Okano, K., Machida, T. and Totsuka, T. 1988. Absorption of atmospheric NO_2 by several herbaceous species: estimation by the ^{15}N dilution method. *New Phytol.* 109: 203–210
Okano, K., Machida, T. and Totsuka, T. 1989. Differences in ability of NO_2 absorption in various broad-leaved tree species. *Environ. Pollut.* 58: 1–17.
Okano, K., Tobe, K. and Furukawa, A. 1990. Foliar uptake of peroxyacetyl nitrate (PAN) by herbaceous species varying in susceptibility to this pollutant. *New Phytol.* 114: 139–145.
Okano, K., Tatsasmi, J., Yoneyama, T., Kono, Y. and Totsuka, T. 1984. Comparison of the fates of $^{15}NO_2$ and $^{13}CO_2$ absorbed through a leaf of rice plants. *Res. Rep. Natl. Inst. Environ. Stud. Jpn.* 66: 59–67.
Okano, K., Ito, O., Takeba, G., Shimizu, A. and Totsuka, T. 1985a. Effects of O_3 and NO_2 alone or in combination on the distribution of ^{13}C-assimilate in kidney bean plants. *Jap. J. Crop Sci.* 54: 152–159.

Okano, K., Totsuka, T., Fukuzawa, T. and Tazaki, T. 1985b. Growth responses of plants to various concentrations of nitrogen dioxide. *Environ. Pollut.* 38: 361–373.

Oleksyn, J. 1984. Effects of SO_2, HF and NO_2 on net photosynthetic and dark respiration rates of Scots pine needles of various ages. *Photosynthetica* 18: 259–262.

Pande, P.C. and Mansfield, T.A. 1985. Responses of winter barley to SO_2 and NO_2 alone and in combination. *Environ. Pollut.* 39: 281–291.

Panofsky, H.A. and Dutton, J.A. 1984. *Atmospheric Turbulence*. John Wiley, & Sons, New York, 397 pp.

Pearson, J. and Soares, A. 1995. A hypothesis of plant susceptibility to atmospheric pollution based on intrinsic nitrogen metabolism: why acidity really is the problem. *Water, Air Soil Pollut.* 85: 1227–1232.

Percy, K.E., Jagels, R., Marden, S., McLaughlin, C.K. and Carlisle, J. 1993. Quantity, chemistry and wettability of epicuticular waxes on needles of red spruce along a fog-acidity gradient. *Can. J. For. Res.* 23: 1472–1479.

Peters, K. and Eiden, R. 1992. Modelling the dry deposition velocity of aerosol particles to a spruce forest. *Atm. Environ.* 26A: 2555–2564.

Peters, K. and Bruckner-Schatt, G. 1995. The dry deposition of gaseous and particulate nitrogen compounds to a spruce stand. *Water, Air, Soil Pollut.* 85: 2217–2222.

Pfanz, H. and Dietz, K.-J. 1987. A fluorescence method for the determination of the apoplastic proton concentration in intact leaf tissues. *J. Plant Physiol.* 129: 41–48.

Pierson, D. and Elliott, J.R. 1988. Effect of nitrite and bicarbonate on nitrite utilization in leaf tissue of bush bean (*Phaseolus vulgaris*). *J. Plant Physiol.* 133: 425–429.

Prasad, B.J. and Rao, D.N. 1980. Alterations in metabolic pools of nitrogen dioxide exposed wheat plants. *Ind. J. Exp. Bot.* 18: 879–882.

Radin, R.W. 1975. Differential regulation of nitrate reductase induction in roots and shoots of cotton plants. *Physiol. Plant* 71: 517–521.

Rajasekhar, V.K. and Oelmüller, R. 1987. Regulation of induction of nitrate reductase and nitrite reductase in higher plants. *Physiol. Plant.* 71: 517–521.

Ramge, P., Bedeck, F.-W., Plöchl, M. and Kohlmaier, G.H. 1993. Apoplastic antioxidants as decisive elimination factors within the uptake process of nitrogen dioxide into leaf tissues. *New Phytol.* 125: 771–785.

Raven, J.A. 1988. Acquisition of nitrogen by the shoots of land plants: its occurrence and implications for acid-base regulation. *New Phytol.* 109: 1–20.

Reich, P.B., Schoettle, A.W., Stroo, H.F. and Amundson, R.G. 1988. Effects of ozone and acid rain on white pine (*Pinus strobus*) seedlings grown in five soils. III. Nutrient relations. *Can. J. Bot.* 66: 1517–1531.

Reynhardt, E.C. and Riederer, M. 1994. Structure and molecular dynamics of plant waxes. II: Cuticular waxes from leaves of *Fagus sylvatica* L. and *Hordeum vulgare* L. *European Biophys. J.* 23: 59–70.

Roberts, J.M. 1995. Reactive odd-nitrogen (NO_y) in the atmosphere. In: *Composition, Chemistry and Climate of the Atmosphere* (H.B. Singh, ed.). Van Nostrand Reinhold, New York, pp. 176–215.

Robertson, G.P. 1993. Fluxes of nitrous oxide and other nitrogen trace gases from intensively managed landscapes: a global perspective. In: *Agricultural Ecosystem Effects on Trace Gases and Global Climate Change* (L.A. Harper, A.R. Mosier, J.M. Duxbury and D.E. Rolston, eds.). Amer. Soc. Agronomy, Madison Wisconsin, USA, pp. 95–108.

Robinson, D.C. and Wellburn, A.R. 1983. Light-induced changes in the quenching of 9-amino-acridine fluorescence by photosynthetic membranes due to atmospheric pollutants and their products. *Environ. Pollut.* 32: 109–120.

Robinson, S.A., Slade, A.P., Fox, G.G., Phillips, R., Ratcliffe, R.G. and Steward, G.R. 1991. The role of glutamate dehydrogenase in plant nitrogen metabolism. *Plant Physiol.* 95: 509–516.

Rogers, H.H., Campbell, J.C. and Volk, R.J. 1979. Nitrogen-15 dioxide uptake and incorporation by *Phaseolus vulgaris* (L.). *Science* 206: 333–335.

Rondon, A. and Granat, L. 1994. Studies on the dry deposition of NO_2 to coniferous species at low NO_2 concentrations. *Tellus* 46B: 339–352.
Rondon, A., Johansson, C. and Granat, L. 1993. Dry deposition of nitrogen dioxide and ozone to coniferous forests. *J. Geophys. Res.* 98: 5159–5172.
Rowland, A.J. 1986. Nitrogen uptake, assimilation and transport in barley in the presence of atmospheric nitrogen dioxide. *Plant and Soil* 91: 353–356.
Rowland, A.J., Drew, M.C. and Wellburn, A.R. 1987. Foliar entry and incorporation of atmospheric nitrogen dioxide into barley plants of different nitrogen status. *New Phytol.* 107. 357–371.
Rowland-Bamford, A.J., Lea, P.J. and Wellburn, A.R. 1989. NO_2 flux into leaves of nitrate reductase-deficient barley mutants and corresponding changes in nitrate reductase activity. *Environ. Exp. Bot.* 29: 439–444.
Sabaratnam, S. and Gupta, G. 1988. Effects of nitrogen dioxide on biochemical and physiological characteristics of soybean. *Environ. Pollut.* 55: 149–158.
Sabaratnam, S., Gupta, G. and Mulchi, C. 1988a. Effects of nitrogen dioxide on leaf chlorophyll and nitrogen content of soybean. *Environ. Pollut.* 51: 113–120.
Sabaratnam, S., Gupta, G. and Mulchi, C. 1988b. Nitrogen dioxide effects on photosynthesis in soybean. *J. Environ. Qual.* 17: 143–146.
Sakamaki, F., Hatakeyama, S. and Akimoto, H. 1983. Formation of nitrous acid and nitric oxide in the heterogeneous dark reaction of nitrogen dioxide and water vapor in a smog chamber. *Intern. J. Chem. Kin.* 15: 1013–1029.
Sandhu, R. and Gupta, G. 1989. Effects of nitrogen dioxide on growth and yield of black turtle bean (*Phaseolus vulgaris* L.) cv. *Domino. Environ. Pollut.* 59: 337–344.
Sandness, H. and Styve, H. 1992. Calculated budgets for airborne acidifying components in Europe, 1985, 1987, 1988, 1989, 1990, 1991. EMEP/MSC-W Report 1/92.
Saxe, H. 1986a. Effects of NO, NO_2 and CO_2 on net photosynthesis, dark respiration and transpiration of pot plants. *New Phytol.* 103: 185–197.
Saxe, H. 1986b. Stomatal-dependent and stomatal-independent uptake of NO_2. *New Phytol.* 103: 199–205.
Saxe, H. and Christensen, O.V. 1985. Effects of carbon dioxide with and without nitric oxide pollution on growth, morphogenesis and production time of pot plants. *Environ. Pollut.* 38: 159–169.
Schmeink, B. and Wild, A. 1990. Studies on the content of free amino acids in needles of undamaged and damaged spruce trees at a natural habitat. *J. Plant Physiol.* 136: 66–71.
Schulze, E.-D. 1989. Air pollution and forest decline in a spruce (*Picea abies*) forest. *Science* 244: 776–783.
Segschneider, H.-J., Wildt, J. and Förstel, H. 1995. Uptake of $^{15}NO_2$ by sunflower (*Helianthus annuus*) during exposures in light and darkness: quantities, relationship to stomatal aperture and incorporation into different nitrogen pools within the plant. *New Phytol.* 131: 109–119.
Seinfeld, J.H. 1986. Atmospheric Chemistry and Physics of Air Pollution. John Wiley & Sons Inc., New York.
Shen, T.-L., Wooldridge, P.J. and Molina, M.J. 1995. Stratospheric pollution and ozone depletion. In: *Composition, Chemistry and Climate of the Atmosphere* (H.B. Singh, ed.). Van Nostrand Reinhold, New York, pp. 394–442.
Shimazaki, K., Yu, S.-W., Sakaki, T. and Tanaka, K. 1992. Differences between spinach and kidney bean plants in terms of sensitivity to fumigation with NO_2. *Plant Cell Physiol.* 33: 267–273.
Sievering, H., Enders, G., Kins, L., Kramm, G., Ruoss, K., Roider, G., Zelger, M., Anderson, L. and Dlugi, R. 1993. Nitric acid, particulate nitrate and ammonium profiles at the Bayerischer Wald: Evidence for large deposition rates of total nitrate. *Atm. Environ.* 28: 311–315.

Skärby, L., Bengtson, C., Boström, C.-Å., Grennfelt, P. and Troeng, E. 1981. Uptake of NO_x in Scots pine. *Silva Fennica* 15: 396–398.
Slovik, S. 1996. Chronic SO_2- and NO_x-pollution interferes with the K^+ and Mg^{2+} budget of Norway spruce trees. *J. Plant Physiol.* 148: 276–286.
Slovik, S., Siegmund, A., Führer, H.-W. and Heber, U. 1996. Stomatal uptake of SO_2, NO_x and O_3 by spruce crowns (*Picea abies*) and canopy damage in Central Europe. *New Phytol.* 132: 661–676.
Smirnoff, N., Todd, P. and Stewart, G.R. 1984. The occurrence of nitrate reduction in the leaves of woody plants. *Ann. Bot.* 54: 363–374.
Soares, A., Ming, J.Y. and Pearson, J. 1995. Physiological indicators and susceptibility of plants of acidifying atmospheric pollution: a multivariate approach. *Environ. Pollut.* 87: 159–166.
Spierings, F. 1971. Influence of fumigations with NO_2 on growth and yield of tomato plants. *Neth. J. Plant Pathol.* 77: 194–200.
Srivastava, H.S. and Ormrod, D.P. 1984. Effects of nitrogen dioxide and nitrate nutrition on growth and nitrate assimilation in bean leaves. *Plant Physiol.* 76: 418–423.
Srivastava, H.S. and Ormrod, D.P. 1986. Effects of nitrogen dioxide and nitrate nutrition on nodulation nitrogenase activity, growth and nitrogen content of bean plants. *Plant Physiol.* 81: 737–741.
Srivastava, H.S. and Ormrod, D.P. 1989. Nitrogen dioxide and nitrate nutrition effects on nitrate reductase activity and nitrate content of bean leaves. *Environ. Exp. Bot.* 29: 433–438.
Srivastava, H.S., Joliffe, P.A. and Runeckles, V.C. 1975a. Inhibition of gas exchange in bean leaves by NO_2. *Can. J. Bot.* 53: 466–474.
Srivastava, H.S., Joliffe, P.A. and Runeckles, V.C. 1975b. The effects of environmental conditions on the inhibition of leaf gas exchange by NO_2. *Can. J. Bot.* 53: 475–482.
Srivastava, H.S., Joliffe, P.A. and Runeckles, V.C. 1975c. The influence of nitrogen supply during growth on the inhibition of gas exchange and visible damage to leaves by NO_2. *Environ. Pollut.* 9: 35–47.
Srivastava, H.S., Ormrod, D.P. and Hale Marie, B. 1990. Photosynthetic photon flux effects on bean response to nitrogen dioxide. *Environ. Exp. Bot.* 30: 463–467.
Srivastava, H.S., Ormrod, D.P. and Hale, B.A. 1994a. Responses of greening bean seedling leaves to nitrogen dioxide and nutrient supply. *Environ. Pollut.* 86: 37–42.
Srivastava, H.S., Wolfenden, J., Lea, P.J. and Wellburn, A.R. 1994b. Differential responses of growth and nitrate reductase activity in wild type and NO_2-tolerant barley mutants to atmospheric NO_2 and nutrient nitrate. *J. Plant Physiol.* 143: 738–743.
Srivastava, H.S., Ormrod, D.P. and Hale, B. 1995. Assimilation of nitrogen dioxide by plants and its effects on nitrogen metabolism. In: *Nitrogen Nutrition in Higher Plants* (H.S. Srivastava and R.P. Singh, eds.). Assoc. Publ. Co. New Delhi, India, pp. 417–430.
Stocker, D.W., Stedman, D.H., Zeller, K.F., Massman, W.J. and Fox, D.G. 1993. Fluxes of nitrogen oxides and ozone measured by eddy correlation over a shortgrass prairie. *J. Geophys. Res.* 98: 12619–12630.
Sun E.-J. and Huang, M.-H. 1995. Detection of peroxyacetyl nitrate at phytotoxic level and its effects on vegetation in Taiwan. *Atm. Env.* 29: 2899–2904.
Takeuchi, Y., Nihira, J., Kondo, N. and Tezuka, T. 1985. Change in nitrate reducing activity in squash seedlings with NO_2 fumigation. *Plant. Cell Physiol.* 26: 1027–1035.
Tanner, C.B. 1960. Energy balance approach to evapotranspiration from crops. *Soil Sci. Soc. Amer. Proc.* 24: 1–9.
Taylor, O.C. 1969. Importance of peroxyacetyl nitrate (PAN) as a phytotoxic air pollutant. *J. Air Pollut. Contr. Assoc.* 19: 347–351.
Thiemens, M.H. and Trogler, W.C. 1991. Nylon production: An unknown source of atmospheric nitrous oxide. *Science* 251: 932–934.

Thoene, B., Schröder, P., Papen, H., Egger, A. and Rennenberg, H. 1991. Absorption of atmospheric NO_2 by spruce (*Picea abies* L. (Karst.)) trees. I. NO_2 influx and its correlation with nitrate reduction. *New Phytol.* 117: 575–585.
Tischner, R., Peuke, A., Godbold, D.L., Feig, R., Merg, G. and Huttermann, A. 1988. The effect of NO_2- fumigation on aseptically grown spruce seedlings. *J. Plant Physiol.* 133: 243–246.
Turner, N.C. and Begg, J.E. 1973. Stomatal behaviour and water status of maize, sorghum, and tobacco under field conditions. I. At high soil water potential. *Plant Physiol.* 51: 31–36.
van Hove, L.W.A., Bossen, M.E., Mensink, M.G.J. and van Kooten, O. 1992. Physiological effects of a long term exposure to low concentrations of NH_3, NO_2 and SO_2 onDouglas fir (*Pseudotsuga menziesii*). *Physiol. Plant.* 86: 559–567.
Vincentz, M., Moureaux, M., Leydecker, T., Vaucheret, H. and Caboche, M. 1993. Regulation of nitrate and nitrite reductase expression in *Nicotiana plumbaginifolia* leaves by nitrogen and carbon metabolites. *Plant J.* 3: 315–324.
von Ballmoos, P. 1995. Wirkung von atmosphärischem NO_2 auf die Aktivität der Nitratreduktase und die Stickstoffassimilation von Fichten. Ph.D. thesis, University of Berne, Switzerland.
von Ballmoos, P., Nussbaum, S. and Brunold, C. 1993. The relationship of nitrate reductase activity to uptake and assimilation of atmospheric $^{15}NO_2$-nitrogen in needles of Norway spruce (*Picea abies* [L.] Karst.). *Isotopenpraxis Environ. Health Stud.* 29: 59–70.
Vose, J.M. and Swank, W.T. 1990. Preliminary estimates of foliar absorption of ^{15}N-labeled nitric acid vapor (HNO_3) by mature eastern white pine (*Pinus strobus*). *Can. J. For. Res.* 20: 857–860.
Walcek, D.J., Brost, R.A., Chang, J.S. and Wesely, M.L. 1986. SO_2, sulfate and HNO_3 deposition velocities computed using regional land use and meteorological data. *Atm. Environ.* 20: 1165–1170.
Wallsgrove, R.M., Lea, P.L. and Miflin, B.J. 1979. Distribution of the enzymes of nitrogen assimilation within the leaf cell. *Plant Physiol.* 63: 232–236.
Watson, R.T., Rodhe, H., Oeschger, H. and Siegenthaler, U. 1990. Greenhouse gases and aerosols. In: *Climate Change: The IPCC Scientific Assessment* (J.T. Houghton, G.J. Jenkins and J.J. Ephraums, eds.). Cambridge University Press, New York, pp. 1–40.
Weber, P. and Rennenburg, H. 1996. Dependency of nitrogen dioxide (NO_2) fluxes to wheat (*Triticum aestivum* L.) leaves from NO_2 concentration, light intensity, temperature and relative humidity determined from controlled dynamic chamber experiments. *Atm. Environ.* 30: 3001–3009.
Weber, P., Nussbaum, S., Fuhrer, J., Gfeller, H., Schlunegger, U.P., Brunold, C. and Rennenberg, H. 1995. Uptake of atmospheric $^{15}NO_2$ and its incorporation into free amino acids in wheat (*Triticum aestivum*). *Physiol. Plant.* 94: 71–77.
Wellburn, A.R. 1984. The influence of atmospheric pollutants and their cellular products upon photophosphorylation and related events. In: *Gaseous Air Pollutants and Plant Metabolism* (M.J. Koziol and F.R. Whatley, eds.). *Proc. 1st Internat. Symp. Air Pollution and Plant Metabolism*, Oxford 1982. Butterworths Scientific. Publ., London, pp. 203–221.
Wellburn, A.R. 1985. Ion chromatographic determination of levels of anions in plastids from fumigated and non-fumigated barley seedlings. *New Phytol.* 100: 329–339.
Wellburn, A.R. 1990. Why are atmospheric oxides of nitrogen usually phytotoxic and not alternative fertilizers? *New Phytol.* 115: 395–429.
Wellburn, A.R., Wilson, J. and Aldridge, P.H. 1980. Biochemical responses to nitric oxide polluted atmospheres. *Environ. Pollut.* 22A: 219–228.
Wellburn, A.R. Higginson, C., Robinson, D. and Walmsley, C. 1981. Biochemical explanations of more than additive inhibitory effects of low atmospheric levels of sulphur dioxide plus nitrogen dioxide upon plants. *New Phytol.* 88: 223–237.

Wesely, M.L. 1989. Parameterization of surface resistance to gaseous dry deposition in regional scale numerical models. *Atm. Environ.* 23: 1293–1304.
Wesely, M.L., Eastman, J.A., Stedman, D.H. and Yalvac, E.D. 1982. An eddy-correlation measurement of NO_2 flux to vegetation and comparison to O_3 flux. *Atm. Environ.* 16: 815–820.
White, K.L., Hill, A.C. and Bennett, J.H. 1974. Synergistic inhibition of apparent photosynthesis rate of alfalfa by combinations of sulfur dioxide and nitrogen dioxide. *Environ. Sci. Technol.* 8: 574–576.
White, M.C., Decker, A.M. and Chaney, R.L. 1981. Metal complexation in xylem fluid. 1. Chemical composition of tomato and soybean stem exudate. *Annal. Physiol.* 67; 292–300.
Whitmore, M.E. and Mansfield, T.E. 1983. Effects of long-term exposure to SO_2 and NO_2 on *Poa pratensis* and other gases. *Environ. Pollut.* 31: 217–235.
Williams, E.J. 1992. NO_x and N_2O emission from soil. *Global Biogeochem. Cycles* 6: 351–388.
Wingsle, G., Näsholm, T., Lundmark, T. and Ericsson A. 1987. Induction of nitrate reductase in Scots pine seedlings by NO_x and NO_3^-. *Physiol. Plant.* 70: 399–403.
Wolfenden, J. and Mansfield, T.A. 1991. Physiological disturbances in plants caused by air pollutants. *Proc. Royal Soc. Edinburgh.* 97B: 117–138.
Wolfenden, J., Pearson, M. and Francis, B.J. 1991. Effects of over-winter fumigation with sulphur and nitrogen dioxides on biochemical parameters and spring growth in red spruce (*Picea rubens* Sarg.). *Plant, Cell Environ.* 14: 35–45.
Wright, B.C., Lockaby, B.G., Meldahl, R., Thornton, F. and Chappelka, A.H. 1990. The influence of acid precipitation and ozone on nitrogen nutrition of young loblolly pine. *Water, Air, Soil Pollut.* 54: 135–142.
Wyers, G.P., Geusebroek, M., Wayers, A., Möls, J.J. and Veltkamp, A.C. 1994. Dry deposition of submicron aerosol on a coniferous forest. In: *Proc. EUROTRAC Symp. '94 Garmisch Partenkirchen, 11th–15th April 1994* (P.M. Borrell, P. Borrell, T. Cvitas and W. Seiler, eds.). SPB Acad. Publ. Den Haag, The Netherlands, pp. 712–715.
Yoneyama, T. and Sasakawa, H. 1979. Transformation of atmospheric NO_2 absorbed in spinach leaves. *Plant Cell Physiol.* 20: 263–266.
Yoneyama, T., Sasakawa, H., Ishizuka, S. and Totsuka, T. 1979. Absorption of atmospheric NO_2 by plants and soils. (II) Nitrite accumulation, nitrite reductase activity and diurnal change of NO_2 absorption in leaves. *Soil Sci. Plant Nutr.* 25: 267–275.
Yu, S., Li, L. and Shimazaki, K. 1988. Response of spinach and kidney bean plants to nitrogen dioxide. *Environ. Pollut.* 55: 1–13.
Yung, K.H. and Mudd, J.B. 1966. Lipid synthesis in the presence of nitrogenous compounds in *Chlorella pyrenoidosa*. *Plant Physiol.* 41: 506–509.
Zeevaart, A.J. 1974. Induction of nitrate reductase by NO_2. *Acta Bot. Neerl.* 23: 345–346.
Zeevaart, A.J. 1976. Some effects of fumigation plants for short periods with NO_2. *Environ. Pollut.* 11: 97–108.

Chapter 5

PHYSIOLOGY OF NITROGEN-FIXING *CASUARINA-FRANKIA* SYMBIOTIC ASSOCIATION

Anita Sellstedt

I. Introduction

II. Host Plants: Casuarinaceae

III. Symbiotic Microbes of Casuarinaceae
   A. Frankia
   B. Mycorrhizal Fungi

IV. Infection Process, Formation of Nodules, Cellular Modification and Specificity in Root Nodulation
   A. Infection Process
   B. Formation of Nodules
   C. Cellular Modifications as a Result of Infection
   D. Specificity in Root Nodulation

V. Nitrogen Metabolism
   A. Nitrogen Fixation
   B. Nitrogen Assimilation

VI. Hydrogen Metabolism

VII. Environmental Effects on Growth, Nitrogen Fixation and Biomass Production

VIII. Plant Selection and Improvements for Increased Symbiotic Performance

IX. Current Uses of Casuarinaceae
   A. Rehabilitation of Degraded Lands
   B. Casuarina as a Source of Fuelwood
   C. Casuarina as an Intercrop in Agroforestry Systems
   D. Casuarina as a Source for Fodder

X. Conclusions and Future Prospects

Literature Cited

I. INTRODUCTION

All living organisms need nitrogen. The atmosphere consists of 78% nitrogen gas, a huge resource of nitrogen, which most organisms cannot use. There are, however, some bacteria and archaebacteria that have the capability to reduce nitrogen gas into ammonia, a biologically useful form. Also, some nitrogen-fixing bacteria can live in symbiosis with plants and the reduced nitrogen thus becomes available for the plants.

Actinorhizal species, i.e., plants in symbiosis with the actinomycete *Frankia*, belonging to 8 plant families and 25 genera, fix more nitrogen on a global basis than do legumes, yet remain less known. These species are generally found on nitrogen-poor sites and often serve many purposes, as mentioned in section IX. Most actinorhizal species appear as pioneer plants at early succession on disturbed sites such as landslides and volcanic eruptions. There are, however, species capable of dominance among species of stable plant communities, i.e., species of Casuarinaceae.

Although taxonomically diverse, actinorhizal species have some features in common. All are perennial dicotyledons (Bond, 1983) and all except *Datisca*, which has herbaceous shoots, are woody trees or shrubs (Tjepkema et al., 1986). Actinorhizal plants are primarily found in temperate climates, except tropical and subtropical species of *Myrica* and Casuarinaceae, the latter being native to the Southern Hemisphere. When species of *Casuarina* live in symbiosis with *Frankia*, they form nitrogen-fixing root nodules. In addition, *Casuarina* and *Gymnostoma* form upward growing root hairs. *Frankia* strains are readily recognized in free-living cultures in which nitrogen is absent by their formation of specific cell types. *Casuarina* also form symbiotic associations with other partners; the three generic associations have been found on the same plant in some situations (Torrey, 1981). Recent studies have dealt with various aspects of actinorhizal plants such as infection processes (Berry and Sunell, 1990), root nodule physiology (Tjepkema et al., 1986; Silvester and Winship, 1990); general biology (Benson and Silvester, 1993) and morphology (Newcomb and Wood, 1987). However, these studies deal with the actinorhizal plant group as a whole. This chapter concentrates solely on nitrogen-fixing *Casuarina-Frankia* symbiotic associations and in particular their physiology. Firstly, the host plants, secondly, the microbes and thirdly, the special events occurring when the symbiosis is formed are presented. Next the processes of nitrogen fixation and hydrogen metabolism as well as the factors affecting nitrogen fixation and growth are discussed. Since the species of Casuarinaceae could be important in nitrogen cycling, plant selection and breeding for increased symbiotic performance and current use of Casuarinaceae are also discussed.

II. HOST PLANTS: CASUARINACEAE

The family Casuarinaceae consists of four genera: *Allocasuarina, Casuarina, Ceuthostoma* and *Gymnostoma* (Johnson, 1988). Some economically promising species, their natural climates and suitability are presented in Table 5.1. There are some 90 species within the family ranging from Australia and the islands of the Western Pacific to the shores of the Indian Ocean. Some 43 species are indigenous to Australia (Bowen and Reddell, 1986). The representatives of Casuarinaceae are trees capable of occupying a diversity of sites ranging from sandy coastal dunes to tropical rain forests and arid deserts. In Australia *C. cunninghamiana* grows along fresh river banks while *C. equisetifolia* grows along the coastal areas. Another species, *C. glauca* grows well in waterlogged swampy soils. *A. torulosa* thrives in the arid inland, but nodulation has been reported to be variable and sometimes absent (Bowen and Reddell, 1986). Species of *Gymnostoma*, however, occur in the north of Australia and can also occur in Malaysia.

The plants of the family are leafless, evergreen trees or shrubs with regularly spaced nodes on the branchlets. At each of these nodes there is a ring of teeth, which are tips of the reduced leaves. All members of the family have photosynthetic deciduous branchlets, so called assimilatory branchlets (Flores, 1978; Johnson, 1982).

III. SYMBIOTIC MICROBES OF CASUARINACEAE

A. *Frankia*

Frankia belong to the actinomycetous group of bacteria. Actinomycetes are filamentous, branching, gram-positive bacteria. Most are free-living saprophytes living on dead organic matter. They are found in soil, compost, and plant litter, as well as in air and water. In general, *Frankia* differentiates into three different cell types, both in free-living culture and in symbiosis: 1) vegetative cells that are filamentous and form mycelia, 2) sporangia containing spores that can be found in free-living cells if the nutritional requirements are met, but can only be found in some symbioses, and 3) vesicles that are structures surrounded by lipids and produced on lateral branches of the vegetative mycelium. *Frankia* are heterotrophic, areobic, sometimes microaerophilic. It was

Table 5.1. Some economically promising species of Casuarinaceae[a]

Genus and species	Climate	Altitude(m)	Rainfall[b] (mm)	Soil	Suitability
Casuarina cunninghamiana	Warm subhumid Warm semi-arid, 50 days frost	0–1000	500–1500	Acid	Very suitable as shelter and wind-breaks, and firewood
C. equisetifolia var. *equisetifolia*	Hot humid to hot subhumid	0–100	1000–2150	Tolerates salt, calcareous soil	Can be used as shelter and wind-breaks, and firewood
C. equisetifolia var. *incana*	Warm subhumid	0–100	1000–1500		Very suitable as ornament and firewood
C. glauca	Warm humid Warm Subhumid	0–30	900–1150	Tolerates salt, calcium clay	Very suitable as ornament and firewood
C. obesa	Warm subhumid warm semi-arid warm arid	0–300	250–500	Various	Very suitable as ornament and firewood
C. cristata	Warm arid Warm semi-arid	25–400	175–275	Tolerates alkalinity, clay	Very suitable as turnery and special items, and firewood
Allocasuarina littoralis	Humid/warm humid Warm subhumid Cool subhumid	0–1200	650–1250	Various, well-drained, acid	Very suitable as shelter and wind-breaks, and firewood
A. decaisneana	Arid	250–700	200–250	Deep, sandy	Very suitable as turnery and special items, and firewood
A. torulosa	Warm subhumid Warm humid	0–1100	950–1500	Various	Very suitable as ornament, turnery and special items, and firewood
A. luehmania	Warm subhumid	0–800	425–600	Various, tolerates salt, clay	Very suitable as ornament, turnel and special items, and firewood

[a] After Doran and Hall (1983)
[b] Fifty percentile values probability of one year in two.

not until 1978 that *Frankia* from *Comptonia peregrina*, an actinorhizal species, was successfully isolated and grown in culture (Callaham et al., 1978). However, in only two genera within the actinorhiza is the vesicle not formed and the bacteria remain filamentous within the host plant (see section VI), that is *Casuarina* and *Allocasuarina* (Torrey, 1976, 1978; Torrey and Callaham, 1979; Berg, 1983). In *Gymnostoma* vesicles are spherical to clavate and sephtated (Newcomb et al., 1990), while the situation in *Ceuthostoma* is not known. Isolation of *Frankia* from *C. equisetifolia* was successfully performed is 1981 (Gauthier et al., 1981). *Frankia* strains isolated from *Casuarina* do form vesicles when grown without nitrogen, and vesicles developed from hyphae were observed in *Frankia* KB isolated from *C. equisetifolia* (Fig. 5.1).

Sporulation of *Frankia* has been reported in quite some strains isolated from *Casuarina* (Tzean and Torrey, 1989; Zhang and Torrey, 1985). Interesting results were found by Tzean and Torrey (1989), showing that *Frankia* strains having spontaneous release of spores germinated better than strains that did not release spores spontaneously. *Frankia* strain Ce15 had a 75% germination of spores and hyphal outgrowth was

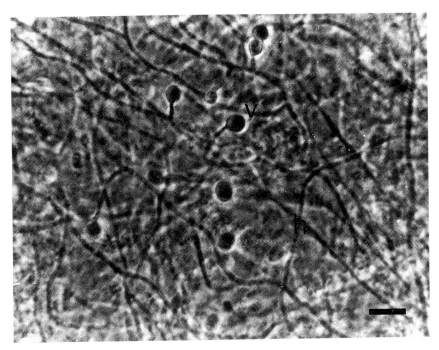

Fig. 5.1. Light micrograph of *Frankia* KB5 isolated from *C. equisetifolia* and grown in medium without addition of nitrogen. Abbreviations: h, hyphae and v, vesicle. Bar = 1 μm.

observed after two days. The effect of sporulation on *Frankia* growth has yet to be completely elucidated, however.

1. *Isolation of Frankia*: Frankia are considered relatively difficult to isolate from *Casuarina* (Berg, 1983; Diem and Dommergues, 1983; Diem et al., 1983) and are slow growing compared to other *Frankia* strains (Nazaret et al., 1989). Recommendation for isolation includes: 1) surface sterilization of the nodules, 2) use of hyphal suspensions rather than surface-sterilized nodules in an antibiotic mixture and 3) choosing the medium carefully in order to obtain a successful isolation. Surface sterilization has mainly been performed using diluted sodium hypoclorite or osmium tetroxide (Lechevalier and Lechevalier, 1990). Mostly nodule pieces have been used for isolation but it is preferable to use hyphal cluster suspensions. The advantage of using hyphal suspensions and plating them out is that they can be monitored by microscopy and any contaminant can be removed by hand. Several media have been reported to date as successful in the isolation of *Frankia* in general (Lechevalier and Lechevalier, 1990), but only a few media have been reported for *Frankia* from *Casuarina*. Gauthier et al. (1981) used the medium QMOD, a rich medium, when they first succeeded in isolation of *Frankia* from *Casuarina*. However, rich media tend to increase the growth of contaminants. Antifungal agents, e.g. cycloheximide, are therefore added. In spite of the fact that *Frankia* from *Casuarina* are relatively difficult to isolate, several strains have now been isolated from *Casuarinaceae* throughout the world since 1981 (Gauthier et al., 1981).

Growth kinetics of *Frankia* isolated from *Casuarinaceae* are complex due to the fact that they are filamentous but usually consist of an exponential phase, then a stationary phase (Fig. 5.2). The stationary phase is characterized by a slower growth. The pattern of growth varies with subculturing frequency, inoculum density, medium and degree of stirring. The doubling time was decreased when the above-mentioned factors were optimized (Fig. 5.2). A rapid growth was observed when phosphatidyl choline was added and the culture was grown under stirring conditions (Schwenke, 1991). Exponential growth was shown to require supplementation of long-chain fatty acids such as palmitic acid (Selim and Schwenke, 1994). Also, factors such as salt concentration have an effect on *in-vitro* growth of *Frankia* (Dawson and Gibson, 1987; Burleigh and Dawson, 1991).

2. *Carbon Metabolism*: Information on the media requirements of *Frankia* strains isolated from *Casuarina* is increasing (Diem and Dommergues, 1983; Shipton and Burggraaf, 1983; Zhang et al., 1984). However, most studies to date have focussed on the requirements of *Frankia* from Northern Hemisphere plant species (e.g. Benson and Schultz,

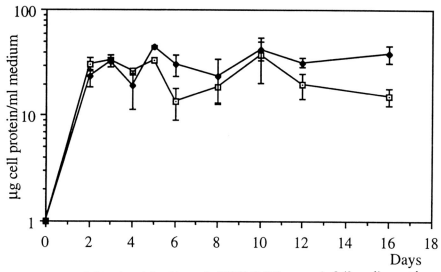

Fig. 5.2. Growth kinetics of *Frankia* strain HFPCci3 R43 grown in L/2 medium under stirring conditions after subsequent subculturing without (□) and with (♦) addition of proteins secreted from *Frankia* strain R43. x ± SE (when bars not seen, SE is within the limit of the symbol).

1990). In general, it is believed that *Frankia* strains isolated from *Casuarina* differ from *Frankia* isolated from nodules of other plant genera in their physiology. *Frankia* in general are known to grow well on propionate (Benson and Schultz, 1990). Interestingly, Shipton and Burggraaf (1982, 1983) showed that a *Frankia* strain from *Casuarina* can also use propionate to support growth. This was recently confirmed by Sellstedt et al. (1994) and Sempavalan et al. (1996). The ability of three *Frankia* isolates from *Casuarina* sp. to utilize nine different carbon sources when supplied singly in a commonly used basal medium was studied by Sellstedt et al. (1994). Utilization of the carbon source(s) was evaluated initially by analysis of protein content and subsequently by measurement of nitrogenase activity. For each of the isolates, JCT287 and KB5 isolated from *C. equisetifolia* and HFPCcI3 Isolated from *C. cunninghamiana*, growth was greatest on the carbon sources pyruvate and propionate. In general the carbon sources which gave the greatest growth gave the highest levels of nitrogenase activity. It has also been shown that succinate could support growth (Shipton and Burggraaf, 1983; Zhang et al., 1984, 1986), an interesting finding which shows that *Frankia* strains isolated from Casuarinaceae resemble other *Frankia* strains in demand of carbon sources.

3. *Secretion of Proteins*: Secretion of proteins has been observed in some *Frankia* strains isolated from *Casuarina* (Safo-Sampah and Torrey,

1988; Seguin and Lalonde, 1989; Benoist and Schwenke, 1990; Müller et al., 1991; Benoist et al., 1992; Tavares et al., 1994). Characterization of these proteins revealed cellulase activities in HFPCcI2, HFPCcI3 and HFPGpI1. Müller et al. (1991) showed that an increase occurs in extracellular proteins, aminopeptidases and also in proteinase activities in *Frankia* isolate BR with age of the culture. Benoist et al. (1993) characterized a high-molecular-mass proteinase complex of 1300 kDa, and concluded that this complex is an important component of the *Frankia* proteolytic system. Tavares et al. (1994) investigated protein excretion from HFPCcI3 and R43, both isolated from *C. cunninghamiana* and found that the amount of protein excreted from *Frankia* R43 exceeded that of *Frankia* HFPCcI3. Also, growth of *Frankia* strain HFPCcI3 increased after addition of proteins secreted from *Frankia* strain R43 (Fig. 5.2). One possible function of these excreted proteins was suggested to be nuclease activity (Tavares et al., 1997). This suggestion is attractive since *Frankia* are soil bacteria and this would then be a way of defending themselves from invasion of DNA from other bacteria. Other roles of these proteins are suggested to be in the infection process, whereby the bacteria get inside the host plant cell wall. This last hypothesis is interesting but not yet verified.

4. *Identification of, and Relationship Between Frankia Strains*: Quite a number of *Frankia* strains from *Casuarina* have been shown not to reinfect the host after isolation. These strains are included in Genomic species 9 (Fernandez et al., 1989). In order to verify whether these strains are *Frankia* or not, techniques have been developed to distinguish between *Frankia* isolated from *Casuarina*, *Frankia* isolated from other actinorhiza and other actinomycetes. Restriction pattern analysis (An et al., 1985a; Dobritsa, 1985), restriction fragment length polymorphism analysis (RFLP) (Normand et al. 1988), DNA homology testing (An et al., 1985b; Fernandez et al., 1989), analysis of enzymatic patterns (Benoist and Schwenke, 1990; Girgis and Schwenke, 1993; Maggia et al., 1993; Puppo et al., 1985), analysis of random PCR-products (Sellstedt et al., 1992), 16S rDNA sequence analysis (Niner et al., 1996) and PCR combined with RFLP (Rouvier et al., 1996) have been used to distinguish between *Frankia* strains. These methods have all been very valuable, showing, for example, that two isolates from different geographic origin had a high level of homology (97%), as found between *Casuarina* strains D11 and G2 (An et al., 1985b). Analysis of strains from different *Casuarina* species has also been done. Nazaret et al. (1989) tested various *Frankia* strains isolated from the nodules of *C. equisetifolia* and *C. glauca*. They found that the isolates could be separated into two groups: one that effectively nodulated the original *Casuarina* plants and another that was unable to reinfect the original host but infective on *Hippophae rhamnoides*. In gen-

eral, the results show that there is a genetic homogeneity among *Frankia* strains infective on Casuarinaceae. However, a recent study by Rouvier et al. (1996) confirms the indicative results by Sellstedt (1995) showing that there is a genetic diversity among these *Frankia* strains. The reason for the discrepancy between earlier and more recent results might be that the former studies focussed on *Frankia* isolated from areas outside the natural geographic range of Casuarinaceae, while the latter studies also included those from natural geographic areas as well.

A restriction map of cloned *Frankia* DNA containing nif hybridizing sequences from CeD, a *C. equisetifolia* isolate, showed that *nif* H, K and D were clustered within a 5-kb region (Normand et al., 1988). They also showed that the *Casuarina*-compatible strains had a similar hybridization pattern. Using a 10-mer primer amplifying random parts of *Frankia* DNA, Sellstedt et al. (1992) showed a distinct pattern between different *Frankia* isolated from *Casuarina*.

Also, differentiation of *Frankia* strains by their electrophoretic patterns of esterases and aminopeptidases has been used (Girgis and Schwenke, 1993; Maggia et al., 1993) and strains isolated from the same genera distinguished. Analysis of partial 16S rDNA sequences revealed that isolates from *C. equisetifolia* in Mexico are members of a novel clade (Niner et al., 1996). A recent publication by Normand et al. (1996) suggests the emendation of the family Frankiaceae so that the other genera, *Geodermatophilus* and *Blastococcus*, are excluded and only *Frankia* is retained. The suggestion was proposed after comparison of complete 16S ribosomal sequences.

B. Mycorrhizal Fungi

In addition to forming symbiosis with *Frankia*, *Casuarina* can form symbioses with endomycorrhizal fungi (AM) and ectomycorrhizal fungi (EM) (Torrey, 1981). Together with the fact that nodulation of *Casuarina* increases the species capability to inhabit nitrogen-poor soils, it was claimed by Rose (1980) and Gardner, (1986) that this capability of the species was greatly enhanced by the presence of mycorrhiza on the root systems. Endomycorrhiza are more ubiquitous and not so host specific but still capable of improving the nutrient status of the host plant that is infected. Diem and Dommergues (1982) found that inoculation of *C. equisetifolia* with *Frankia* and *Glomus mossae* improved plant growth and nodulation, and indeed mycorrhizal inoculated plants contained twice as much nitrogen as seedlings inoculated with *Frankia* solely. Also, ectomycorrhiza have been observed on species of Casuarinaceae, particularly on *Allocasuarina* (Reddell et al., 1986). The benefits of having two or three associations are quite clear. But is there a competition between two associations during the colonization process? This question was addressed by Sempavalan et al. (1995) using *C. equisetifolia* inoculated with *Frankia* and *Glomus*. They found a linear relation between nodule number per plant based on percentage *Glomus* root colonization and the time after *Frankia* and *Glomus* inoculation. The data indicated co-operation but not competition between *Frankia* and *Glomus* for nodulation and mycorrhizal colonization of *Casuarina*. It has also been claimed by Diem and Dommergues (1990) that the networks of the endomycorrhizal hyphae can intermesh sand particles and form aggregates which help to stabilize the sand (Koske and Halvorson, 1981).

IV. INFECTION PROCESS, FORMATION OF NODULES, CELLULAR MODIFICATION AND SPECIFICITY IN ROOT NODULATION

A. Infection Process

The infection process of *Casuarina* roots by *Frankia* has mainly been studied by Torrey (1976) and Callaham et al. (1979) and reviewed by Newcomb and Wood (1987). Uninfected roots are single-celled and straight, while roots exposed to nodule suspensions showed regions of deformation (Callaham et al., 1979; Torrey, 1976). Newly developed root hairs were believed to become curled or club-shaped and not to elongate normally (Torrey, 1976). Root hairs that had completed elongation did not become affected by the nodule extract. The response occurred within 24 h according to Callaham et al. (1979). Invasion of the root by the bacteria occurred exclusively by root hair infection (Callaham et al., 1979; Torrey, 1976). These events were followed by an intracellular infection of the cortex cells (Torrey, 1976; Callaham et al., 1979). An intimate association between *Frankia* growing in the root environment allowed the cell-to-cell contact necessary to effect chemical dissolution of the root hair. Root hair deformation involves changes in host cells i.e., wall deformation of expanding root hair. Responding to unknown signals growing root hairs initiate multiple tips and tip growth is supposedly modified. The site where *Frankia* penetrate the hyphae are continuous with the hyphae of *Frankia* in the soil or medium where the root is growing. Progression of the infection from cell to cell occurs by means of invasive hyphal filaments (Berg and McDowell, 1988). The hyphal filaments penetrate the host cell wall and traverse a longitudinal path through cortical tissues toward the nodule lobe apex. The septate invasive hyphae grow through host cell walls and nonseptate branched intracellular hyphae develop within the host cytoplasm. The direct involvement of root hair deformation in infection of actinorhiza was earlier questioned by Quispel (1954), who concluded that the actinomycete was incapable of growing outside the host cell and therefore could not be responsible for root hair deformation. Recently, it was also questioned by Sellstedt et al. (unpubl. results), and the questioning appears justified since the experiments by Torrey (1976) and Callaham et al. (1979) were not carried out under strictly aseptic conditions. Sellstedt et al. (unpubl. results) kept seedlings of *C. cunninghamiana* under very strict aseptic conditions and inoculated them with KB5 and no root hair deformation was found after 4 days of observations. However, as pointed out earlier, one root hair deformation suffices for infection, and there may be technical difficulties in observations of root hair deformation.

Inoculation of plants is a delicate matter and the inoculation procedure should be performed under optimal conditions. But first we must ask the question, what is the infective unit? Burleigh and Torrey (1990) tried to answer that question by testing the effectiveness of different *Frankia* cell types as inocula for *C. cunninghamiana*. They found that fresh hydrated preparations of whole cells, hyphae and spores were all infective. Dehydrated hyphae had no infection capacity, while dehydrated spores remained infective at a reduced level. Interestingly, it was shown that spore suspensions were three times more infective than hyphae. Crushed nodule inoculum has been widely used. However, as pointed out by Reddell and Bowen (1985a) there might be more than one *Frankia* strain in the nodule. So crushed nodule inoculum should be used with precautions to ensure that *Frankia* with a certain characteristic is introduced. It is also important to screen for tolerance of temperature and desiccation of inoculum when selecting strains for inoculations in different climates. Sayed et al. (1997) showed that infectivity of three *Frankia* strains isolated from *C. equisetifolia* showed no change in infectivity after storage at 25°C in wet sand, but infectivity was lowered by more than 50% after 12 weeks in moderate and dry conditions. Also, some researchers have tried to grow *Frankia* in alginate beads in order to obtain inoculum (Burleigh et al., 1988; Diem et al., 1988; Frioni et al. 1994). *Frankia* entrapped inside the alginate beads are able to grow and form sporangia. After drying, it is possible to store the inoculum for years.

B. Formation of Nodules

The root nodule of *Casuarina* is a perennial, coralloid structure consisting of nodule lobes. Each nodule lobe is a modified lateral root (Berry and Sunell, 1990). It was shown by Torrey (1976) that every nodule is initiated from a lateral root and in his experiments the initiation points were always sites of deformed root hairs. Torrey (1976) also showed that in the early nodule initiation stage bacteria are present in the cortical cells of the lateral root in the same area as the deformed root hair. The formation of the root nodule in *Casuarina* has been reviewed by Newcomb and Wood (1987) and shall therefore not be detailed here.

An indeterminate way of growth is found in *Casuarina* and hence nodular roots from the apex of mature nodule lobes are observed. Soon after formation of the zone forming the young nodule, primordia are the first sign of root nodules (Torrey, 1976). Initiation is in the pericycle, i.e., nodule roots arise in the same manner as do lateral roots. Nodule roots are the result of continuous activity of the nodule lobe meristem (Bond, 1957), are negatively geotrophic and develop from most nodule lobes. Zhang and Torrey (1985) used a strain isolated from *Allocasuarina lehmaniana* producing nodules but no nodule roots and inoculated that onto *Casuarina* species. The nodules of the *Casuarina* species produced nodule roots, which led to the conclusion that the host genome determines the morphological expression of the nodule. Nodule roots do not contain *Frankia* cells, so they do not contribute to nitrogen fixation. Instead they contain large air spaces, an anatomical character that facilitates oxygen diffusion. It was shown for *Myrica gale*, another actinorhizal species, that nodule roots was the major pathway for oxygen diffusion (Tjepkema, 1978).

What controls the growth of a nodule? There is of course more than one answer to this question, but Mansour (1994) showed that cytokinin level was higher in *C. cunninghamiana* root nodules inoculated with a pure *Frankia* culture than in nodules inoculated with a crushed nodule inoculum. The occurrence of the growth hormones cytokinin and auxin in the root nodule supports the idea that the balance between the hormones controls the growth of the mature nodule (Wheeler et al., 1979). Another interesting finding was reported by Prin et al. (1991). They found aerial nodules in *C. cunninghamiana* actively reducing acetylene. The aerial nodules were similar to root nodules in anatomy.

C. Cellular Modification as a Result of Infection and as a Protection Against O_2

As mentioned earlier, it is only in two genera within actinorhizal plants that *Frankia* fail to form vesicles in the nodule, namely *Casuarina* and *Allocasuarina*, in which *Frankia* remains filamentous in the nodule, in contrast to *Frankia* in nodules of *Elaeagnus* in which vesicles occur and house the nitrogen-fixing enzyme nitrogenase (Sasakawa et al., 1988). Under aerobic conditions the vesicles possess a multilaminate envelop which provides protection of the nitrogen-fixation enzyme from access of molecular oxygen (Torrey and Callaham, 1982). One could then speculate how can nitrogen fixation commence when there is no vesicle to provide oxygen protection? This interesting topic has been intensively studied by Berg (1983), Berg and McDowell (1987), and they found that the absence of vesicles in nodule of *Casuarina* correlates with the presence of modified cell walls. Berg (1983) used enzymes that degrade cell walls and found that walls of infected cells did not degrade while walls of non-infected cells did. Histological analysis showed that the cell walls were autofluorescent and stained with phloroglucinol (Berg, 1983; Berg and McDowell, 1988). Permanganate stained the infected cell walls and together with the other evidence this indicated the presence of lignin in the infected cell walls. Thus, cells of *Casuarina* infected by *Frankia* are modified by cell wall thickenings of lignin-like materials not found in non-infected cells (Berg and McDowell, 1988). In addition, the lignin-like material was concluded to be of a hydrofobic nature (Berg and McDowell, 1988). In a study by Sellstedt et al. (1991b) it was shown that six symbioses with a wide range in symbiotic performance all had lignin-like compounds in their cell walls. This points to the fact of lignin-like compounds as a response to infection and in that case represent a very special response of a host plant-microbe interaction.

These lignin-like substances in the cell walls are believed to present a physical barrier to gaseous diffusion into the infected cells and the site of nitrogenase. It was shown by Zeng et al. (1989) that a diffusion barrier for gases into the infected cells and into the site where nitrogenase is localized actually exists. Vesicle formation seems to be linked to oxygen tension (Murray et al., 1985), as is the thickness of the vesicle wall (Parsons et al., 1987). The gas diffusion barrier represented by the lignin-like hydrofobic wall in infected cells of *Casuarina* would explain the lower oxygen tension within the infected cell without the vesicle. *Casuarina* also has haemoglobin in high concentrations (see section V:3).

D. Specificity in Root Nodulation

Studies on specificity in root nodulation have mostly concentrated on temperate species (Huang et al., 1985; e.g. Baker, 1987). However, both Baker (1987) and Huang et al. (1985) included *Casuarina* sp. Baker (1987) identified four host infectivity groups among *Frankia* strains: (i) strains which nodulate *Alnus* and *Myrica*, (ii) strains which nodulate *Casuarina* and *Myrica*, (iii) strains which nodulate Elaeagnaceae and *Myrica* and (iv) strains that nodulate only Elaeagnaceae. Some studies have been performed on Southern Hemisphere species (Jiabin et al., 1985; Reddell and Bowen, 1986; Fernandez et al., 1989; Nazaret et al., 1989; Sellstedt, 1995). Jiabin et al. (1985) found that *C. equisetifolia* seedling was nodulated by *Frankia* isolated from its own host. Interestingly, Reddell and Bowen (1985a) found more than one strain in one nodule.

These nodules could contain both *Frankia* strains with 'typical' genetic information to form an effective nodule, and 'atypical' strains such as those belonging to the Elaeagnaceae compatibility group, not being able to form an effective nodule but able to coexist in the nodule. Fernandez et al. (1989) confirmed the results of Normand et al. (1988) showing that the Elaeagnaceae compatible strains are highly heterologous. They also showed that the *Casuarina*-compatible strains are homologous.

Zhang et al. (1984) also found that *Frankia* strains isolated from *Casuarina* species nodulated *Elaeagnus*. Torrey and Racette (1989) tested three genera of the host family Casuarinaceae for nodulation by *Frankia*. The most widely infective strain was HFPCcI3 and AlII1, while the *Frankia* strains with the narrowest host range within the Casuarinaceae were CcI2 and GpI1. Intrafamily cross-inoculations were uncommon. *Elaeagnus* sp. were, however, not included in that study. Torrey and Racette (1989) also reported that the most broadly receptive host species was *G. papuanum*. In a study by Sellstedt (1995) also species of Elaeagnaceae were included. Compared with the host plants included in this study, Sellstedt (1995) could not conclude *G. australiensis* to be a receptive host species. In addition, *Frankia* strains HFPCcI3, EL1, A.T. and G.A. were tested for infectivity on *A. torulosa, C. cunninghamiana, G. australianum* and *E. triflora*. It was shown that *A. torulosa* and *C. cunninghamiana, G. australianum* and *E. triflora*. It was shown that *A. torulosa* and *C. cunninghamiana* formed nodules only with the *Frankia* isolated from nodules from their own host plant, while *E. triflora* formed nodules with all four *Frankia* sources. All nodules formed effectively fixed nitrogen, as tested by the acetylene reduction assay.

It can be concluded that *Allocasuarina* is the most specific in infection (Sellstedt, 1995; Torrey and Racette, 1989) but Sellstedt (1995) would also like to add *Casuarina* to that group. Torrey and Racette (1989) concluded that *Gymnostoma* offer the broadest tolerance, but Sellstedt (1995) would rather refer to species of *Gymnostoma* as intermediate, and assign to *E. triflora* the broadest tolerance to infectants.

V. NITROGEN METABOLISM

A. Nitrogen Fixation

The atmosphere consists of 78% dinitrogen, which is not available to most organisms on the earth. However, nitrogen-fixing organisms can use this N_2 and reduce it to NH_3, which is immediately converted to NH_4^+ when it comes into contact with the slightly acidic content of the cell. The reduction requires 6 electrons and 12 ATP. It has been shown, however, that 16 ATP are required for reduction of one molecule of dinitrogen. The reason for this was revealed in the 1980s, when it was shown that the reduction of one N_2 is accompanied with the formation of one molecule of H_2 and therefore the enzyme reaction should be written:

$$N_2 + 8e^- + 8H^+ + 16ATP \rightarrow 2NH_3 + H_2 + 16ADP + 16P_i.$$

Nitrogen fixation is catalyzed by the enzyme nitrogenase. In summary, it can be said that when the enzyme is active electrons are transferred from the Fe-protein to the Mo-Fe-protein. During the first stage of this process the Fe-protein forms a complex with two Mg-ATP molecules. The reduced Fe-protein forms a complex with the Mo-Fe-protein. As one electron is transferred to the Mo-Fe-protein, 2Mg-ADP and 2 inorganic phosphates are released. The enzyme complex only actively reduces dinitrogen when associated.

1. *Measurements of Nitrogenase Activities and Nitrogen Fixation*: Measurements of nitrogen fixation may be performed either by direct or by indirect methods. The indirect methods involve acetylene reduction assay (ARA), while the direct methods consist of ^{15}N enrichment, ^{15}N dilution, natural abundance dilution, hydrogen evolution and nitrogen accumulation. Methods described here are ARA, hydrogen evolution, nitrogen accumulation and the ^{15}N dilution technique. These and other methods for measurements of nitrogenase activities are nicely reviewed by (Silvester and Winship, 1990; Winship and Tjepkema, 1990).

Most studies on nitrogen fixation involve use of (ARA) performed in a closed container, where acetylene is reduced via nitrogenase instead of atmospheric nitrogen, according to the reaction described below:

$$C_2H_2 + 2e^- + 2H^+ \rightarrow C_2H_4.$$

The assay is based on the fact that the enzyme nitrogenase not only reduces atmospheric nitrogen but also is capable of reducing other substrates, e.g. acetylene, which is reduced to ethylene (Hardy et al., 1968). The ethylene can be sampled and quantified by use of a gas chromatograph equipped with a flame ionization detector. ARA is fast, inexpensive and allows repeated measurements of the same material. Therefore ARA has been widely used for the last 30 years. However, there were several reports in the 1980s on limitations in the use of ARA (e.g. Minchin et al. 1983; Witty et al., 1986). Nitrogenase has been shown in many legumes to be affected by acetylene, demonstrating what is now called the acetylene-induced decline (Witty et al., 1986), whereby the nitrogenase activity could be underestimated. Several actinorhizal species have also been investigated for acetylene-induced decline (Tjepkema et al., 1988; Silvester and Winship, 1990; Tjepkema and

Schwintzer, 1992). It was shown that the pattern of response in actinorhizal species for acetylene is quite different from that in legumes. Tjepkema et al. (1988) showed that the decline was small in some actinorhizal nodules and in many nodules the activity recovered to predecline values. Tjepkema and Schwintzer (1992) suggested that the acetylene-induced decline is not due to an increase in nodule diffusion resistance, as proposed for legume nodules. Silvester and Winship (1990) stated that the situation in actinorhiza bears little resemblance to the reported acetylene-induced decline in legumes. The acetylene reduction rate shows an increase within some minutes, which has been shown by use of an open-flow acetylene reduction assay (OARA). OARA includes pumping acetylene through a container in which the root system is placed and subsequently measuring the production of ethylene. Open-flow-through systems have been applied on *Casuarina* plants (Tjepkema et al., 1988; Silvester and Winship, 1990). Typical nitrogenase responses to acetylene by individual plants of *C. equisetifolia* and *G. papuanum* have been described by Silvester and Winship (1990). *Casuarina* and *Gymnostoma* plants always showed a small acetylene-induced decline within 15 min. They proposed introduction of the term acetylene-induced transients, given the fact of recovery in some of the actinorhiza.

Unfortunately, ARA involves also overestimation of nitrogen fixation due to additional causes. As stated above, there is always a hydrogen evolution from nitrogenase resulting in loss of energy. In order to quantify nitrogen fixation by use of ARA a conversion factor has to be used. This factor can be determined experimentally if nitrogen fixation is measured with ^{15}N and the conversion factor calculated after comparison with ARA. Only a few determinations of the conversion factor between acetylene reduction and nitrogen fixation have been made for the actinorhizal species of *Casuarinaceae* (Sellstedt et al., 1993). The theoretical value of the conversion factor is 4; however, Sellstedt et al. (1993) found a much higher conversion factor. The reason for this high value could not be explained since even a diurnal variation in acetylene reduction and hydrogen evolution were taken into consideration (Sellstedt et al., 1989; Sellstedt et al., 1993).

As mentioned above, there is always an evolution of hydrogen concomitant with reduction in atmospheric nitrogen. This hydrogen can be measured by a gas chromatograph equipped with a thermal conductivity detector or with an oxygen electrode converted to measurements of hydrogen (Wang et al., 1971). However, it has been indicated that most *Casuarina* plants have an uptake hydrogenase (Sellstedt et al., 1991a; section VI) recycling the hydrogen that is evolved by nitrogenase. The absolute requirement for this method is to have symbiosis lacking uptake hydrogenase. In addition C_2H_2 inhibits hydrogenase in *Frankia* (Sellstedt and Winship, 1990).

Measurements of nitrogen fixation can also be made by studies of nitrogen accumulation. Accumulation of nitrogen was determined for *Casuarina* symbioses grown in a greenhouse for 165 days (Sellstedt et al., 1991b). The plants were given a small addition of nitrogen before nodulation but after that no nitrogen was given. At the end of the experimental period the plants were harvested and the nitrogen content determined by use of Kjeldahl digestion (Allen, 1974). Important aspects of this method are that the plants have to grow in a substrate without nitrogen, and the amount of nitrogen given initially has to be subtracted. The disadvantage of this method is that the plants have to be harvested at the end and it may be difficult to keep the growth substrate nitrogen-free. Alternatively, a reference plant system can also be grown in vicinity to the plot with the experimental plant system. The nitrogen content of the reference plants would then have to be subtracted.

Estimation by use of N^{15} dilution is a method widely used, and experiments conducted on *Casuarina* are most interesting (Gauthier et al., 1985; Sougoufara et al., 1990; Sellstedt et al., 1993). In Gauthier et al. (1985) three isotope methods were compared in studies of N_2 fixation; N^{15} dilution technique, A-value method and the accumulation of N. Values of 3.27, 2.31 and 3.07 g N_2 were fixed per tree for each method respectively. Three methods were likewise compared by Sellstedt et al. (1993) in studies of N_2 fixation and effects of water deficit on N_2 fixation: ARA, N^{15} dilution technique and accumulated N content. It was concluded that the N^{15} dilution technique could be used for studies of effects of water deficit on N_2 fixation.

2. *Localization of Nitrogenase*: Localization of nitrogenase was first shown in a *Frankia* isolated from *Colletia cruciata* and *Comptonia peregrina* (Meesters et al., 1985). Since members of *Casuarina* do not have vesicles, it was interesting to analyze the exact localization of nitrogenase in symbiosis. It was shown by Sellstedt and Mattsson (1994) that nitrogenase localized in hyphae of *Casuarina cunninghamiana* in the symbiosis, but localized in the vesicles in free-living *Frankia* KB grown in an N-free medium. This, to my knowledge, is the first exact evidence that nitrogenase is localized in the hyphae of symbiotic *Casuarina*.

3. *Haemoglobin*: It is well known that haemoglobins occur in all legume root nodules shown to be effectively fixing nitrogen and play a role as an oxygen carrier (Appleby, 1984). Haemoglobin was first reported in the non-legume *Casuarina* 1960 by Davenport (1960). There have been some contradictory reports, however, on the presence/absence of haemoglobin in actinorhizal systems. Moore (1964) found no

haemoglobin in nodule slices of *Alnus incana* (L. Moench) subsp. *rugosa* (Duroi) Clausen, *Elaeagnus communata* Bernh., *Shepherdia canadensis* (L.) Nutt. nor in *Hippophae rhamnoides* L.. Neither Smith (1949) nor Becking (1970) found any haemoglobin in *Alnus* nodules. In addition, Bond (1974) reported no visible haemoglobin in any of the actinorhizal systems, with the possible exception of *Casuarina* nodules. That led Tjepkema to make a thorough investigation of haemoglobins in actinorhizal symbioses (Tjepkema, 1983). Using spectrophotometry he analyzed segments of actinorhizal nodules of seven species and was able to show high amounts of haemoglobin in *M. gale* and *C. cunninghamiana*. However, only low concentrations were found in *Comptonia peregrina* (L.) Coult., *A. rubra* Bong. and *E. angustifolia* L., and very low to zero amounts in *Ceanothus americanus* L. and *D. glomerata* (Presl.) Baill. CO-reactive haem, which gives the upper limit of haemoglobin concentration, was measured in actinorhizal nodules from six different genera by Tjepkema and Asa (1987). They found the concentrations of haemoglobin in *C. cunninghamiana* and *M. gale* to be approximately half that found in legumes, while *Comptonia peregrina, A. rubra* and *Ceanothus americanus* had concentrations approximately 10 times lower than in legume nodules. The presence of a membrane-bound haemoglobin in root nodules of *Casuarina glauca* was confirmed by Fleming et al. (1987). He found that this haemoglobin was similar to other plant haemoglobins in respect of molecular weight. Also, there was a close immunological relation between *Casuarina* and *Parasponia* haemoglobins. This haemoglobin protein was later characterized by amino sequence analysis (Kortt et al., 1988) and was found to consist of 151 amino acids including cystein, an amino acid not found in the leghaemoglobins. *C. glauca* haemoglobin showed extensive homology (43–52%) with other plant haemoglobins.

Sellstedt et al. (1991a) studied three *Casuarina* species in symbiosis with four *Frankia* sources. There was a marked difference in symbiotic performance measured as growth and nitrogen content in the branches. Symbiosis with high haemoglobin concentration also had a high symbiotic performance. Interestingly, two symbioses having no haemoglobin at all also showed low nitrogen content (Sellstedt et al., 1991a). However, there was only poor correlation between nitrogen concentration in the nodules and haemoglobin concentration. So there must be factors involved in the success of symbiotic performance other than haemoglobin concentration. In a more thorough study by Sellstedt et al. (1991b) a positive correlation was shown between haemoglobin concentration and acetylene reduction. The highest level of cumulative nitrogen fixation was always associated with the highest level of haemoglobin concentration and the highest rate of acetylene reduction. In addition, a symbiosis with a small measurable amount of haemoglobin had no nitrogenase activity. Christensen et al. (1991) stated that *C. glauca* as an actinorhiza was unique due to the fact that it has at least two haemoglobin genes. The degree of homology between these two genes was found to be as homologous as that found in legumes (Kórtt et al., 1988).

In a recent study by Jacobsen-Lyon et al. (1995) it was shown that *Casuarina glauca* has a gene-encoding non-symbiotic haemoglobin (*cashb-nonsym*). That gene appeared to be expressed in a number of plant tissues. *Casuarina* also has a second gene family of haemoglobin genes expressed at high levels in the nodules. Both the symbiotic and the non-symbiotic genes retained their specific patterns when introduced into the legume *Lotus corniculatus*. Jacobsen-Lyon et al. (1995) interpreted their results to mean that control of expression of symbiotic *Casuarina* genes is similar to control of expression found in legume-rhizobia nodules. Thus, it is nowadays stated that *Casuarina* plants do have haemoglobin at levels comparable to those of legumes.

B. Nitrogen Assimilation

Nitrogenous solutes in xylem exudates formed from fixed nitrogen have been measured in a variety of actinorhizal plants (Miettinen and Virtanen, 1952; Bollard, 1957; Leaf et al., 1958; Wheeler and Bond, 1970; Walsh et al. 1984; Sellstedt and Atkins, 1991). However, the reports on *Casuarina* (Bollard, 1957; Wheeler and Bond, 1970; Walsh et al., 1984; Sellstedt and Atkins, 1991) are contradictory. Neither Bollard (1957) nor Wheeler and Bond (1970) found any citrulline at all in *Casuarina*. However, Walsh et al. (1984) found that citrulline was the major component transported in *C. equisetifolia*, irrespective of the nitrogen nutrition. The occurrence of citrulline in non-nodulated *Casuarina* suggested that the formation of the compound is not confined to root nodules. In addition, Sellstedt and Atkins (1991) found a variety of nitrogenous compounds transported in the xylem. The proportional composition of the solutes varied between the species, while the same range of amino acids was found in the xylem in all of the symbioses studied. For example, in *C. glauca* the major amino acid in the root sap was asparagine followed by proline, while in symbiotic *C. equisetifolia* it was citrulline followed by asparagine. However, it was concluded by both Walsh et al. (1984) and Sellstedt and Atkins (1991) that relative citrulline content of the xylem could not be used as a measure of nitrogen fixation in *Casuarina*.

VI. HYDROGEN METABOLISM

All nitrogen-fixing organisms evolve hydrogen concomitant with reduction of atmospheric nitrogen (see section V:A). Some organisms have the ability to utilize this evolved hydrogen, by use of an enzyme, hydrogenase, which in organisms such as *Frankia* aerobically catalyzes oxidation of hydrogen when O_2 is an e^- acceptor.

$$2H_2 + O_2 \rightarrow 2H_2O.$$

Uptake hydrogenase is an enzyme believed to be beneficial for nitrogen-fixing organism (Dixon, 1972). Not only energy is gained, but also O_2 and H_2 is scavenged from the O_2 and H_2 sensitive nitrogenase (Dixon, 1972).

Uptake hydrogenase activity is believed to be a common feature in *Alnus-Frankia* symbioses (Sellstedt, 1989). Also, *Frankia* in symbiosis with *Casuarina* showed the presence of an active uptake hydrogenase (Sellstedt and Winship, 1987, 1990 Winship et al., 1987; Sellstedt et al., 1991a).

In a study by Sellstedt and Smith (1990), five free-living *Frankia* strains isolated from *Casuarina* were investigated for occurrence of hydrogenase activity, nitrogenase activity (acetylene reduction) and hydrogen evolution. Acetylene reduction was recorded in all *Frankia* strains. In spite of this, none of the *Frankia* strains had any hydrogenase activity when grown on nickel-depleted medium and they released hydrogen in atmospheric air. After addition of nickel to the medium, the *Frankia* strains were shown to possess an active hydrogenase, which resulted in hydrogen uptake but no hydrogen evolution. The hydrogenase activity in *Frankia* strain KB5 increased from zero to 3.86 µmol H_2 (mg protein)$^{-1}$ h^{-1} after addition of up to 1.0 µM Ni. It is likely that the hydrogenase activity could be enhanced even more as a response on further addition of Ni. It is indicated that absence of hydrogenase activity in free-living *Frankia* isolated from *Casuarina* spp. was due to nickel deficiency. This, to my knowledge, is the first report on the absence of hydrogenase activity in the free-living form of a micro-organism which has been shown to have activity in the symbiotic state. The situation in symbiosis is different, however, from that in the free-living state. In a symbiosis the plant may have efficient ways of Ni uptake so that the microbe can get enough Ni in order to have an active hydrogenase. Sellstedt and Smith (1990) suggested that most isolates of *Frankia* from *Casuarina* sp. have an uptake hydrogenase, but little is known about other factors affecting its activity. The precise conditions required for its expression have not been studied in detail in *Frankia*, but it has been shown that in addition to nickel being important, low concentrations of oxygen, and also the presence of hydrogen were required for expression of measurable uptake hydrogenase activities in static cultures (Sellstedt and Smith, 1990). Manipulation of the type and concentration of carbon source has been suggested as a way of regulating uptake hydrogenase activity in other micro-organisms (Doyle and Arp, 1987). It was shown that uptake hydrogenase activity was induced when cells of *Alcaligenes latus* were transferred from heterotrophic conditions to chemolithoautotrophic conditions, i.e., H_2 was present but carbon sources excluded (Doyle and Arp, 1987). However, *Alcaligenes latus* is an organism that can live on hydrogen as an energy source.

The effect of different carbon sources in the growth medium on the expression of uptake hydrogenase activity of *Frankia* cultures was studied by Sellstedt et al. (1994). Static cultures of *Frankia* were grown with addition of different carbon sources. It was shown that carbon sources stimulating growth repressed uptake hydrogenase activity. However, the conditions in static cultures are different to those in stirring conditions, so care must be exercised in drawing further conclusions. It was shown by Sellstedt and Mattsson (1994) that hydrogenase was localized mainly in the vesicles. A correlation was found by Mattsson and Sellstedt (unpubl. results) between nitrogenase and uptake hydrogenase activities in *Frankia* strain isolated from *C. equisetifolia*, with an earlier onset of nitrogenase activity compared to uptake hydrogenase activity.

VII. ENVIRONMENTAL EFFECTS ON GROWTH, NITROGEN FIXATION AND BIOMASS PRODUCTION

Factors affecting nodulation, nitrogen fixation and symbiotic efficiency most mandatorily be studied. Factors of relevance under field conditions are salt and phosphorous concentration, as well as temperature. Many *Casuarina* species grow in saline environments. A way to handle this high salinity is very much needed for the survival of the tree. Reddell et al. (1986) showed that treating *C. obesa* root nodules with high salt concentration resulted in high concentration of salt in the non-infected cells but not in the infected cells. This suggests that there is a semi-permeable layer in the wall of the infected cell. Trials performed by El-Lakany and Luard (1982) showed that *C. obesa*, *C. glauca* and *C. equisetifolia* var *incana* were the most salt-tolerant species of Casuarinaceae, surviving salt concentrations up to 500 mM NaCl.

However, the effects of nitrogen fixation were not assessed in that study. Girgis et al. (1992) showed that increasing NaCl concentration up to 250 mM significantly decreased *Frankia* growth and also growth and nitrogen fixation *C. glauca*. Also, one *Frankia* strain was shown to be more salt tolerant than the other studied. Ng (1987) stressed the importance of inoculation of saline-tolerant *Frankia* strains when introducing *C. equisetifolia* into saline sites. The choice of combination of *Frankia* and *Casuarina* sp. is also very important.

A recent study by Yang (1995) investigated the effect of phosphorus on nodule formation and function in the *Casuarina-Frankia* symbiosis. Supplies of P to the plant nutrient solution increased shoot and nodule dry weights, nodule number per pouch and average nodule size per cent of total dry weight in nodules. It was indicated that P deficiency limited the growth of host plants more severely than the nitrogen fixation process, and that P deficiency operated indirectly on nodulation and nitrogen fixation via reducing host plant growth.

The optimum soil temperature for the growth of *C. cunninghamiana* plants dependent on symbiotic nitrogen fixation was 25°C (Reddell et al., 1985). Lowering the soil temperature below 25°C significantly decreased plant growth.

VIII. PLANT SELECTION AND IMPROVEMENTS FOR INCREASED SYMBIOTIC PERFORMANCE

Species of *Casuarina* are widely planted outside their native range, due to their fast growth and excellent wood quality, suggesting a potential to overcome fuelwood shortage. Torrey (1982) stated that species of *Casuarina* are the most important trees for the tropics. The importance of actinorhizal plants in forestry was reviewed by Dawson (1986). In addition, the nitrogen fixation capacity can be improved through clonal selection (Sougoufara et al., 1992). Significantly different symbiotic performances were obtained with different *Casuarina* clone-*Frankia* combinations (Sougoufara et al., 1992). These authors suggested a two-step selection for increased *Casuarina-Frankia* symbiotic systems, starting with the host plant and commencing with the microbe. Conventional vegetative propagation through cuttings have been reported on *Casuarina* species by Lundquist and Torrey (1984). Young softwood cuttings taken from the growing parts of the mature tree responded to auxin treatment with callus and/or root initiation (Lundquist and Torrey, 1984). Clonal selection is one way to improve performance of *Casuarina*, while micropropagation is another convenient approach to obtaining efficient symbiotic systems. *In-vitro* initiation from explants (e.g. initial piece of plant cultivated *in vitro*, such as stem tips and immature female inflorescences) from mature *Casuarina* spp. was reported without rejuvenation treatment (Duhoux et al., 1986; Abo-El-Nil, 1987). The conditions for vegetative propagation through cuttings and micropropagation of *Casuarina* have been reviewed by Séguin and Lalonde (1990) and will not be detailed here. Importance of choice of *Frankia* was shown by Rosbrook and Bowen (1987), for example, who chose three isolates of *Frankia*, all of which were shown to be equally efficient in nodulation and nitrogen fixation. However, this finding is contradicted in reports by Sellstedt et al. (1991a, b) and Reddell et al. (1985), where *Frankia-Casuarina* combinations showed very different symbiotic performances.

It is evident that *Casuarina* plays a significant role in the N economy of the particular site in which it is growing. Hannon (1956) reported that analyses of *C. littoralis* plants revealed 10,000–12,900 ppm N on a dry weight basis. Litter fall and dry matter increment represented 29.0 tons ha^{-1} year^{-1}. An estimate of 1% N per g dry weight would represent an accretion of 290 kgN ha^{-1} year^{-1}, N that mostly came from nitrogen fixation. By measuring soil and plant N, Dommergues (1963) was able to measure the yearly increment of N in soils by *Casuarina equisetifolia* in Cape Verde Islands. He found a yearly increment of 58 kgN ha^{-1}, mostly derived from nitrogen fixation. Silvester (1977) reviewed and confirmed the evidence for the importance of *Casuarina equisetifolia* in the revegetation of Krakatau Island, Indonesia, following volcanic destruction of the vegetation. Parotta et al. (1995) estimated cumulative dinitrogen fixation to be 73 kg N ha^{-1} yr^{-1} in Costa Rica.

IX. CURRENT USES OF *CASUARINACEAE*

This section by no means claims to fully cover this important field, but is rather a brief introduction into how members of Casuarinaceae can be used. Some economically interesting species are listed in Table 5.1.

A. Rehabilitation of Degraded Lands

Acceleration of salinity in soils around the world has increased the need for saline-tolerant species. It is also very high on the priority list to restore potential land suitable for forestry, since there is a global shortage of fuelwood. Some studies have been performed on salt tolerance of *Casuarina*, as mentioned in section VII above. In addition, Malik and Sheikh (1983) tested species for salt tolerance in Pakistan. They found *C. equisetifolia* to be the most salt-tolerant of the ten species tested.

B. *Casuarina* as a Source of Fuelwood

Fuelwood production is greatly needed as firewood supply has decreased. The average inhabitant of a third-world country uses almost one ton of firewood in a year. Species of *Casuarina* are excellent as the wood has a high caloric value; e.g. *C. equisetifolia* has a value of 5000 kcal per kg. In addition, the rotation time is quite low at approximately 6–7 years.

C. Casuarina as an Intercrop in Agroforestry Systems

C. ologodon, Gymnostoma papuanum and *C. equisetifolia* were intercropped with food crops or cash crops (coffee) or used as shade trees in rotation with crops in Papua New Guinea (Thiagalingam, 1983). *C. equisetifolia* is associated with crops such as peanut, sesame and various grain legumes (Kondas, 1983).

X. CONCLUSIONS AND FUTURE PROSPECTS

Casuarina-Frankia symbioses are very interesting actinorhizal systems. The lack of vesicles in the symbioses of *Casuarina-Frankia* has inspired a large number of scientists to perform very important research on these symbioses. In addition to being most interesting for basic research *Casuarina-Frankia* symbioses are also of great interest for applied research. This is mainly due to the fact that nitrogen-fixing species play an important role in the nitrogen status of tropical soils. Interest has mainly focused on legume symbioses but is now on the increase for the actinorhizal species of Casuarinaceae.

LITERATURE CITED

Abo-El-Nil, M.M. 1987. Micropropagation of *Casuarina*. In: *Cell and Tissue Culture in Forestry* Vol. 3 (J.M. Bonga and D.J. Durzan, eds.) Nijhoff, The Hague, The Netherlands, pp. 400–410.
Allen, S.E. 1974. *Chemical Analysis of Ecological Materials*. Blackwell Scientific Publications, Oxford, pp. 25–40.
An, C.S., Riggsby, W.S. and Mullin, B.C. 1985a. Relationship of *Frankia* isolates based on deoxyribonucleic acid homology studies. *Int. J. Syst. Bacteriol.* 35: 140–146.
An, C.S., Riggsby, W.S. and Mullin, B.C. 1985b. Restriction pattern analysis of genomic DNA of *Frankia* isolates. *Plant soil* 87: 43–48.
Appleby, C.A. 1984. Leghaemoglobin and *Rhizobium* respiration. *Annu. Rev. Plant Physiol.* 35: 443–478.
Baker, D.D. 1987. Relationships among pure cultured strains of *Frankia* based on host specificity. *Physiol. Plant* 70: 245–248.
Becking, J.H. 1970. Frankiaceae fam. nov. (Actinomycetales) with one new combination and six new species of the genus *Frankia* Brunchorst 1886, 174. *Int. J. Syst. Bacteriol.* 20: 201–220.
Benoist, P. and Schwencke, J. 1990. Native agarose-polyacrylamide gel electrophoresis allowing the detection of aminopeptidase, dehydrogenase, and esterase activities at the nanogram level: enzymatic patterns in some *Frankia* strains. *Anal. Biochem.* 187: 337–344.
Benoist, P., Müller, A. and Diem, H.G. 1993. Coordinate increase in activity of proteinase subunits of the 1300-kDa megaproteinase of *Frankia* strain BR: role of carbon source depletion and extracellular metabolites. *Can. J. Microbiol.* 39: 32–39.

Benoist, P., Müller, A., Diem, H.G. and Schwencke, J. 1992. High-molecular-mass multicatalytic proteinase complexes produced by the nitrogen-fixing actinomycete *Frankia* strain BR. *J. Bacteriol.* 174: 1495–1504.
Benson, D.R. and Schultz, N.A. 1990. Physiology and biochemistry of *Frankia* in culture. In: *The Biology of Frankia and Actinorhizal Plants* (C.R. Schwintzer and J.D. Tjepkema eds.). Academic Press, San Diego, CA, pp. 107–127.
Benson, D.R. and Silvester, W.B. 1993. Biology of *Frankia* strains, actinomycete symbionts of actinorhizal plants. *Microbiol. Rev.* 57: 293–319.
Berg, R.H. 1983. Preliminary evidence for the involvement of suberization in infection of *Casuarina*. *Can. J. Bot.* 61: 2910–2918.
Berg, R.H. and McDowell, L. 1987. Endophyte differentiation in *Casuarina* actinorhizae. *Protoplasma* 136: 104–117.
Berg, R.H. and McDowell, L. 1988. Cytochemistry of the wall of infected cells in *Casuarina* actinorhizae. *Can. J. Bot.* 66: 2038–2047.
Berry, A.M. and Sunell, L.A. 1990. The infection process and nodule development. In: *The Biology of Frankia and Actinorhizal Plants* (C.R. Schwintzer and J.D. Tjepkema, eds.). Academic Press, San Diego, CA, pp. 61–81.
Bollard, E.G. 1957. Translocation of organic nitrogen in the xylem. *Austr. J. Biol. Sci.* 10: 292–301.
Bond, G. 1957. The development and significance of the root nodules of *Casuarina*. *Ann. Bot.* 83: 373–381.
Bond, G. 1974. Root-nodule symbioses with actinomycete-like organisms. In: *The Biology of Nitrogen Fixation* (A. Quispel, ed.). Elsevier/North Holland Publ., Amsterdam, pp. 342–378.
Bond, G. 1983. Taxonomy and distribution of non-legume nitrogen-fixing systems. In: *Biological Nitrogen Fixation in Forest Ecosystems* (J.C. Gordon and C.T. Wheeler, eds.). Martinus Nijhoff, The Hague, The Netherlands, pp. 342–356.
Bowen, G.D. and Reddell, P. 1986. Nitrogen fixation in Casuarinaceae. Proc. 18th IUFRO World Congress, Ljubljana, Yugoslavia, Sept. 1986, 12 pp.
Burleigh, S.H. and Torrey, J.G. 1990. Effectiveness of different *Frankia* cell types as inocula for the actinorhizal plant *Casuarina*. *Appl. Environ. Microbiol.* 56: 2565–2567.
Burleigh, S.H. and Dawson, J.O. 1991. Effects of sodium chloride and melbiose on the *in vitro* sporulation of *Frankia* strain HFPCcI3 isolated from *Casuarina cunninghamiana*. *Austr. J. Ecol.* 16: 531–535.
Burleigh, S.H., Baker, D. and Torrey, J.G. 1988. Mass inoculum production protocol for *Casuarina*. In: *Abstracts from the 7th Int. Meet. Frankia Actinorhizal. Plants.*
Callaham, D., Del Tredici, P. and Torrey, J.G. 1978. Isolation and cultivation *in vitro* of the actinomycete causing root nodules in *Comptonia*. *Science* 199: 899–902.
Callaham, D., Newcomb, W., Torrey, J.G. and Peterson, R.L. 1979. Root hair infection in actinomycete-induced root nodule initiation in *Casuarina, Myrica* and *Comptonia*. *Bot. Gaz.* 140 (Suppl.): S1–S9.
Christensen, T., Dennis, E.S., Peacock, J.W., Landsmann, J. and Marcker, K.A. 1991. Hemoglobin genes in non-legumes: cloning and characterization of a *Casuarina glauca* hemoglobin gene. *Plant Mol. Biol.* 16: 339–344.
Davenport, H.E. 1960. Haemoglobin in the root nodules of *Casuarina cunninghamiana*. *Nature (London)*. 186: 653–654.
Dawson, J.O. 1986. Actinorhizal plants: their use in forestry and agriculture. *Outlook Agr.* 15: 202–208.
Dawson, J.O. and Gibson, A.H. 1987. Sensitivity of selected *Frankia* isolates from *Casuarina, Allocasuarina* and North American host plants to sodium chloride. *Physiol. Plant.* 70: 272–278.
Diem, H.G. and Dommergues, Y.R. 1982. Isolation, characterization and cultivation of *Frankia*. In: *Biological Nitrogen Fixation. Recent Developments* (N.S.S. Rao, ed.). Oxford

& IBH Publishing Co. Pvt. Ltd., New Delhi, pp. 227-254.
Diem, H.G. and Dommergues, Y. 1983. The isolation of *Frankia* from nodules of *Casuarina*. *Can. J. Bot.* 61: 2822-2825.
Diem, H.G. and Dommergues, Y.R. 1990. Current and potential uses and management of Casuarinaceae in the tropics of subtropics. In: *The Biology of Frankia and Actinorhizal Plants* (C.R. Schwintzer and J.D. Tjepkema, eds.). Academic Press Inc., San Diego, CA, pp. 317-342.
Diem, H.G., Gauthier, D. and Dommergues, Y.R. 1983. An effective strain of *Frankia* from *Casuarina* sp. *Can. J. Bot.* 61: 2815-2821.
Diem, H.G., Duhoux, D., Simonet, P. and Dommergues, Y.R. 1988. Actinorhizal symbiosis biotechnology: The present and future. *Proc. 8th Int. Symp. Biotechnol.* 2: 984-995.
Dixon, R.O.D. 1972. Hydrogenase in legume root nodule bacteriods: occurrence and properties. *Arch. Microbiol.* 85: 193-201.
Dobritsa, S.V. 1985. Restriction analysis of the *Frankia* spp. genome. *FEMS Microbiol. Lett.* 29: 123-128.
Dommergues, Y.R. 1963. Evaluation de du taux de fixátion de lazote dans un sol dunaire reboisé en filao (*Casuarina equisetifolia*). *Agrochimica* 186: 335-340.
Doran, J.C. and Hall, M. 1983. Notes on fifteen Australian *Casuarina* species. In: *Casuarina Ecology, Management and Utilization* (S.J. Midgley, J.W. Turnbull and R.D. Johnston, eds.). CSIRO, Melbourne, Australia, pp. 19-52.
Doyle, C.M. and Arp, D.J. 1987. Regulation of H_2 oxidation and hydrogenase protein levels by H_2, O_2 and oxidation of substrates in *Alcaligenes latus*. *J. Bacteriol.* 169: 4463-4468.
Duhoux, E., Sougoufara, D. and Dommergues, Y.R. 1986. Propagation of *Casuarina equisetifolia* through axillary buds of immature female inflorescenes cultured *in vitro*. *Plant Cell Rep.* 3: 161-164.
El-Lakany, M.H. and Luard, E.J. 1982. Comparative salt tolerance of selected *Casuarina* species. *Aust. For. Res.* 13: 11-20.
Fernandez, M.P., Meugnier, H., Grimont, P.A.D. and Bardin, R. 1989. Deoxyribonucleic acid relatedness among members of the genus *Frankia*. *Int. J. Syst. Bacteriol.* 39: 424-429.
Fleming, A.I., Wittenberg, J.B., Wittenberg, B.A., Dudman, W.F. and Appleby, C.A. 1987. The purification, characterization and ligand-binding kinetics of hemoglobins from root nodules of the non-leguminous *Casuarina glauca-Frankia* symbiosis. *Biochim. Biophys. Acta* 911: 209-220.
Flores, E.M. 1978. The shoot apex of *Casuarina* (Casuarinaceae). *Rev. Biol. Trop.* 26: 247-260.
Frioni, L., Le Roux, C., Dommergues, Y.R. and Diem, H.G. 1994. Inoculant made of encapsulated *Frankia*: assessment of *Frankia* growth within alginate beads. *World J. Microbiol. Biotechnol.* 10: 118-121.
Gardner, I., C. 1986. Mycorrhizae of actinorhizal plants. *MIRCEN J.* 2: 147-160.
Gauthier, D., Doe, H.G. and Dommergues, Y. 1981. *In vitro* nitrogen fixation by two actinomycete strains isolated from *Casuarina* nodules. *Appl. Environ. Microbiol.* 41: 306-308.
Gauthier, D., Diem, H.G., Dommergues, Y. and Ganry, F. 1985. Assessment of N_2 fixation by *Casuarina equisetifolia* inoculated with *Frankia* ORS021001 using ^{15}N methods. *Soil Biol. Biochem.* 17: 375-379.
Girgis, M.G.Z. and Schwencke, J. 1993. Differentiation of *Frankia* strains by their electrophoretic patterns of intracellular esterases and aminopeptidases. *J. Gen. Microbiol.* 139: 2225-2232.
Girgis, M.G.Z., Ishac, Y.Z., Diem, H.G. and Dommergues, Y.R. 1992. Salt tolerance of *Casuarina glauca* and *Frankia*. *Acta Oecolog.* 13: 443-451.
Hannon, N.J. 1956. The status of nitrogen in Hawkesbury sand stone soils and their plant communities in the Sydney district I. The significance and level of nitrogen. *Proc.*

Linn. Soc. N.S.W. 81: 119–143.
Hardy, R.F.W., Holsten, R.D., Jackson, E.K. and Burns, R.C. 1968. The acetylene-ethylene assay for nitrogen fixation: laboratory and field evaluation. Plant Physiol. 43: 1185–1207.
Huang, J.B., Zhao, Z.Y., Chen, G.X. and Liu, H.C. 1985. Host range of Frankia endophytes. Plant Soil 87: 61–65.
Jacobsen-Lyon, K., Ostergaard Jensen, E., Jorgensen, J.-E., Marcker, K.A., Peacock, W.J. and Dennis, E.S. 1995. Symbiotic and nonsymbiotic hemoglobin genes of Casuarina glauca. Plant Cell. 7: 213–223.
Jiabin, H., Zheying, Z., Guanxiong, C. and Huichang, L. 1985. Host range of Frankia endophytes. Plant Soil 87: 61–65.
Johnson, L.A.S. 1982. Notes on Casuarinaceae. II. J. Adelaide Bot. Gard. 6: 73–87.
Johnson, L.A.S. 1988. Notes on Casuarinaceae. III. The new genus Ceuthostoma. Telopae 3: 133–137.
Kondas, S. 1983. C. equisetifolia—a multipurpose tree crop in India. In: Casuarina Ecology, Management and Utilization (S.J. Midgley, J.W. Turnbull and R.D. Johnston, eds.). Commonw. Sci. Ind. Res. Org. Melbourne, Australia, pp. 66–76.
Kortt, A.A., Inglis, A.S., Fleming, A.I. and Appleby, C.A. 1988. Amino acid sequence of hemoglobin I from root nodules of the non-leguminous Casuarina glauca-Frankia symbiosis. FEBS Lett. 231: 341–346.
Koske, K.E., and Halvorson, W.L. 1981. Ecological studies of vesicular-arbuscular mycorrhizae in a barrier dune. Can. J. Bot. 59: 1413–1422.
Leaf, G., Gardner, I.C. and Bond, G. 1958. Observations on the composition and metabolism of the nitrogen-fixing root nodules of Alnus. J. Exp. Bot. 9: 320–331.
Lechevalier, M.P. and Lechevalier, H.A. 1990. Systematics, isolation and culture of Frankia. In: The Biology of Frankia and Actinorhizal Plants (C.R. Schwintzer and J.D. Tjepkema, eds.). Academic Press Inc., San Diego, CA, pp. 35–60.
Lundquist, R. and Torrey, J.G. 1984. The propagation of Casuarina species from rooted stem cutting. Bot. Gaz. 145: 378–384.
Maggia, L., Prin, Y., Picard, B. and Goullet, P. 1993. Esterase diversity among 46 Frankia strains isolated from Casuarina equisetifolia in West Africa, Can. J. Microbiol. 39: 709–714.
Malik, M. and Sheikh, M.I. 1983. Planting of trees in waterlogged and saline areas. Part I. Plant testing at Azakhel. Pak. J. For. 33: 1–17.
Mansour, S. 1994. Production of growth hormones in Casuarina cunninghamiana root nodules by Frankia strain $HFPC_9I_4$. Protoplasma 183: 126–130.
Meesters, T.M., Genesen, S.T.H. van and Akkermans, A.D.L. 1985. Growth, acetylene reduction activity and localization of nitrogenase in relation to vesicle formation in Frankia strains cc 1.17 and cp 1.2. Arch. Microbiol. 143: 137–142.
Miettinen, J.K. and Virtanen, A.I. 1952. The free amino acids in the leaves, roots and root nodules of the Alder (Alnus). Physiol. Plant. 5: 540–557.
Minchin, F.R., Witty, J.F., Sheehy, J.E. and Muller, M. 1983. A major error in the acetylene reduction assay: decrease in nodular nitrogenase activity under assay conditions. J. Exp. Bot. 34: 641–649.
Moore, A.W. 1964. Note on non-leguminous plants in Alberta. Can. J. Bot. 42: 952–955.
Müller, A., Benoist, P., Diem, H.G. and Schwencke, J. 1991. Age-dependent changes in extracellular proteins, aminopeptides and proteinase activities in Frankia isolate BR. J. Gen. Microbiol. 137: 2787–2796.
Murray, M.A., Zhang, Z. and Torrey, J.G. 1985. Effect of O_2 on vesicle formation, acetylene reduction and, O_2-uptake kinetics in Frankia sp. HFPCcI3 isolated from Casuarina cunninghamiana. Can. J. Microbiol. 31: 804–809.
Nazaret, S., Simonet, P., Normand, P. and Bardin, R. 1989. Genetic diversity among Frankia isolated from Casuarina nodules. Plant Soil. 118: 240–247.

Newcomb, W. and Wood, S. 1987. Morphogenesis and fine structure of *Frankia* (Actinomycetales): The microsymbiont of nitrogen-fixing actinorhizal root nodules. *Int. Rev. Cyt.* 109: 1–88.
Newcomb, W., Jackson, S., Racette, S. and Torrey, J.G. 1990. Ultrastructure of infected cells in the actinorhizal root nodules of *Gymnostoma papuanum* (Casuarinaceae) prepared by high pressure and chemical fixation. *Protoplasma* 157: 172–181.
Ng, B.H. 1987. The effects of salinity on growth, nodulation and nitrogen fixation of *Casuarina equisetifolia*. *Plant Soil* 103: 123–125.
Niner, B.M., Brandt, J.P., Reyes, A., Marshall, C.R., Hirsch, A.M. and Valdés, M. 1996 (in press). Analysis of partial rDNA sequences of a novel group of actinomycetes isolated from nodules of *Casuarina equisetifolia* growing in México. *Ann. Microbiol.* 62: 3034–3036.
Normand, P., Simonet, P. and Bardin, R. 1988. Conservation of *nif* sequences in *Frankia*. *Mol. Gen. Genet.* 213: 238–246.
Normand, P., Orso, S., Curnoyer, B., Jeannin, P., Chapelon, C., Dawson, J., Evtushenko, L. and Misra, A.K. 1996. Molecular phylogeny of the genus *Frankia* and related genera and emendation of the family Frankiaceae. *Int. J. Syst. Bacteriol.* 46: 1–9.
Parotta, J.A., Baker, D.D. and Fried, M. 1995. Changes in dinitrogen fixation in maturing stands of *Casuarina equisetifolia* and *Leucaena leucocephala*. *Can. J. For. Res.* 26: 1684–1691.
Parsons, R., Silvester, W.B., Harris, S., Gruitjers, W.T.M. and Bullivant, S. 1987. *Frankia* vesicles provide inducible and absolute oxygen protection for nitrogenase. *Plant Physiol.* 83: 728–731.
Prin, Y., Duhoux, E., Diem, H.G., Roederer, Y. and Dommergues, Y.R. 1991. Aerial nodules in *Casuarina cunninghamiana*. *Appl. Environ. Microbiol.* 57: 871–874.
Puppo, A., Dimitrijevic, L., Diem, H.G. and Dommergues, Y.R. 1985. Homogeneity of superoxide dismutase patterns in *Frankia* strains from Casuarinaceae. *FEMS Microbiol. Lett.* 30: 43–46.
Quispel, A. 1954. Symbiotic nitrogen fixation in non-leguminous plants. II. The influence of the inoculation density and external factors on the nodulation of *Alnus glutinosa* and its importance for our understanding of the mechanism of the infection. *Acta Bot. Neerlandica* 3: 512–532.
Reddell, P. and Bowen, G.D. 1985a. Do single nodules of Casuarinaceae contain more than one *Frankia* strain. *Plant and Soil* 88: 275–279.
Reddell, P. and Bowen, G.D. 1985b. *Frankia* source affects growth, nodulation and nitrogen fixation in *Casuarina* species. *New Phytol.* 100: 115–122.
Reddell, P. and Bowen, G.D. 1986. Host-*Frankia* specificity within the Casuarinaceae. *Plant and Soil* 93: 293–298.
Reddell, P., Bowen, G.D. and Robson, A.D. 1985. The effects of soil temperature on plant growth, nodulation and nitrogen fixation in *Casuarina cunninghamiana* Miq. *New Phytol.* 101: 441–450.
Reddell, P., Foster, R.C. and Bowen, G.D. 1986. The effects of sodium chloride on growth and nitrogen fixation in *Casuarina obesa* Miq. *New Phytol.* 102: 397–408.
Reddell, P., Rosbrook, P., Bowen, G.D. and Gwaze, D. 1985. Growth responses in *Casuarina cunninghamiana* plantings to inoculations with *Frankia*. *Plant Soil* 108: 78–86.
Rosbrook, P.A. and Bowen, G.D. 1987. The abilities of three *Frankia* isolates to nodulate and fix nitrogen with four species of *Casuarina*. *Physiol. Plant.* 70: 373–377.
Rose, S.L. 1980. Mycorrhizal associations of some actinomycete nodulated nitrogen-fixing plants. *Can. J. Bot.* 58: 1449–1454.
Rouvier, C., Prin, Y., Reddell, P., Normand, P. and Simonet, P. 1996. Genetic diversity among *Frankia* strains nodulating members of the family casuarinaceae in Australia revealed by PCR and restriction fragment length polymorphism analysis with crushed nodules. *Appl. Environ. Microbiol.* 62: 979–985.

Safo-Sampah, S. and Torrey, J.G. 1988. Polysaccharide-hydrolyzing enzymes of *Frankia* (Actinomycetales). *Plant Soil* 112: 89–97.
Sasakawa, H., Hiyoshi, T. and Sugiyama, T. 1988. Immunogold localization of nitrogenase in root nodules of *Elaeagnus pungens* Thunb. *Plant Cell Physiol.* 29: 1147–1152.
Sayed, W.F., Wheeler, C.T., Zahran, H.H. and Shoreit, A.A.M. 1997. Effect of temperature and soil moisture on the survival and symbiotic effectiveness of *Frankia* spp. *Cont. Biol. Fert. Soil* 25: 349–353.
Schwencke, J. 1991. Rapid, exponential growth and increased biomass yield of some *Frankia* strains in buffered and stirred mineral medium (BAP) with phosphatidyl choline. *Plant Soil* 137: 37–41.
Séguin, A. and Lalonde, M. 1989. Detection of pectolytic activity and *pel* homologous sequences in *Frankia*. *Plant Soil* 118: 221–229.
Séguin, A. and Lalonde, M. 1990. Micropropagation, tissue culture, and genetic transformation of actinorhizal plants and *Betula*. In: *The Biology of Frankia and Actinorhizal Plants* (C.R. Schwintzer and J.D. Tjepkema, eds.) Academic Press Inc., San Diego, CA, pp. 215–238.
Selim, S. and Schwencke, J. 1994. 1,2-dipalmitoyl phosphatidylcholine, 1,2-dipalmitoyl phosphatidic acid or 1,2-dipalmitoyl-*sn*-glycerol inhibit sporangia formation and promote exponential growth of various *Frankia* isolates from the Casuarinaceae family. *Soil. Biol. Biochem.* 26: 569–575.
Sellstedt, A. 1989. Occurrence and activity of hydrogenase in symbiotic *Frankia* from field collected *Alnus incana*. *Physiol. Plant.* 75: 304–308.
Sellstedt, A. 1995. Specificity and effectivity in nodulation by *Frankia* on southern hemisphere actinorhiza. *FEMS Microbiol. Lett.* 125: 231–236.
Sellstedt, A. and Winship, L.J. 1987. Hydrogen metabolism of *Casuarina* root nodules: A comparison of two inoculum. *Physiol. Plant.* 70: 367–372.
Sellstedt, A. and Smith, G.D. 1990. Nickel is essential for active hydrogenase in free-living *Frankia* isolated from *Casuarina*. *FEMS Microbiol. Lett.* 70: 137–140.
Sellstedt, A. and Winship, L.J. 1990. Acetylene, not ethylene inactivates hydrogenase of actinorhizal nodules during acetylene reduction assays. *Plant Physiol.* 94: 91–94.
Sellstedt, A. and Atkins, C.A. 1991. Composition of amino compounds transported in xylem of *Casuarina* sp. *J. Expt. Bot.* 42: 1493–1497.
Sellstedt, A. and Mattsson, U. 1994. Hydrogen metabolism in *Casuarina-Frankia*: Immunolocalization of nitrogenase and hydrogenase. *Soil Biol. Biochem.* 26: 583–592.
Sellstedt, A., Högberg, P. and Jonsson, K. 1989. Diurnal variation in acetylene reduction and net hydrogen evolution in five tropical and subtropical nitrogen-fixing tree symbioses. *J. Exp. Bot.* 40: 1163–1168.
Sellstedt, A., Reddell, P. and Rosbrook, P. 1991a. The occurrence of haemoglobin and hydrogenase in nodules of twelve *Casuarina-Frankia* symbiotic associations. *Physiol. Plant.* 82: 458–464.
Sellstedt, A., Reddell, P., Rosbrook, P. and Zierhl, A. 1991b. The relations of haemoglobin and lignin-like compounds to acetylene reduction in symbiotic *Casuarina*. *J. Exp. Bot.* 243: 1331–1337.
Sellstedt, A., Wullings, B., Nyström, U. and Gustafsson, P. 1992. Identification of *Casuarina-Frankia* strains by use of polymerase chain reaction (PCR) with arbitrary primers. *FEMS Microbiol. Lett.* 93: 1–6.
Sellstedt, A., Rosbrook, P.A., Kang, L. and Reddell, P. 1994. Effect of carbon source on growth, nitrogenase and uptake hydrogenase activities of *Frankia* isolates from *Casuarima* sp. *Plant Soil.* 158: 63–68.
Sellstedt, A., Ståhl, L., Mattsson, U., Jonsson, K. and Högberg, P. 1993. Can the ^{15}N dilution technique be used to study N_2 fixation in tropical tree symbioses as affected by water deficit? -*J. Exp. Bot.* 44: 1749–1755.

Sempavalan, J., Wheeler, C.T. and Hooker, J.E. 1995. Lack of competition between *Frankia* and *Glomus* for infection and colonization of roots of *Casuarina equisetifolia* (L.). *New. Phytol.* 130: 429–436.

Sempavalan, J., Wheeler, C.T. and Narayanan, R. 1998. The isolation and characterization of *Frankia* from nodules of *Casuarina equisetifolia* L. from Tamil Nadu. *Indian J. Microbiol.* (In Press).

Shipton, W.A. and Burggraaf, A.J.P. 1982. Aspects of the cultural behaviour of *Frankia* and possible ecological implications. *Can. J. Bot.* 61: 2783–2792.

Shipton, W.A. and Burggraaf, A.J.P. 1983. A comparison of the requirements for various carbon and nitrogen sources and vitamins in some *Frankia* strains. *Plant Soil* 69: 149–161.

Silvester, W.B. 1977. Dinitrogen fixation by plant association excluding legumes. In: *A Treatise on Nitrogen Fixation*, sec. IV (R.W.F. Hardy and A.H. Gibson, eds.). John Wiley Inc., New York, pp. 141–189.

Silvester, W.B. and Winship, L.J. 1990. Transient responses of nitrogenase to acetylene and oxygen in actinorhizal nodules and cultured *Frankia*. *Plant Physiol.* 92: 480–486.

Smith, J.D. 1949. The concentration and distribution of haemoglobin in the root nodules of leguminous plants. *Biochem. J.* 44: 585–598.

Sougoufara, B., Danso, S.K.A., Diem, H.G. and Dommergues, Y.R. 1990. Estimating N_2 fixation and N derived from soil by *Casuarina equisetifolia* using labelled ^{15}N fertilizer: Some problems and solutions. *Soil Biol. Biochem.* 22: 695–701.

Sougoufara, B., Maggia, L., Duhoux, E. and Dommergues, Y.R. 1992. Nodulation and N_2 fixation in nine *Casuarina* clone-*Frankia* combinations. *Acta Oecolog.* 13: 497–503.

Tavares, F. and Sellstedt, A. 1997. DNase activities of the extracellular cell wall associated, and cytoplasmic protein fraction of *Frankia* strain R43. *Appl. Environ. Microbiol.* 63: 4597–4599.

Tavares, F., Sellstedt, A., Parente, A. and Salema, R. 1994. Protein excretion from *Frankia* strains HFPCcI3 and R43. *Biol. Plant.* 36: S340.

Thiagalingam, K. 1983. Role of *Casuarina* in agroforestry. In: *Casuarina Ecology, Management and Utilization* (S.J. Midgley, J.W. Turnbull and R.D. Johnston, eds.). Commonw. Sci. Ind. Res. Org. Melbourne, Australia, pp. 175–179.

Tjepkema, J.D. 1978. The role of diffusion from the shoots and the nodule roots in nitrogen fixation by *Myrica gale*. *Can. J. Bot.* 56: 1365–1372.

Tjepkema, J.D. 1983. Hemoglobins in the nitrogen-fixing root nodules of actinorhizal plants. *Can. J. Bot.* 61: 2924–2929.

Tjepkema, J.D. and Asa, D.J. 1987. Total and CO-ractive heme content of actinorhizal nodules and the roots of some non-nodulated plants. *Plant Soil* 100: 225–236.

Tjepkema, J.D. and Schwintzer, C.R. 1992. Factors affecting the acetylene-induced decline during nitrogenase assay in root nodules of *Myrica gale* L. *Plant Physiol.* 98: 1451–1459.

Tjepkema, J.D., Schwintzer, C.R. and Benson, D.R. 1986. Physiology of actinorhizal nodules. *Ann. Rev. Plant Physiol.* 37: 209–232.

Tjepkema, J.D., Schwintzer, C.R. and Monz, C.A. 1988. Time course of acetylene reduction in nodules of five actinorhizal genera. *Plant Physiol.* 86: 581–583.

Torrey, J.G. 1976. Initiation and development of root nodules of *Casuarina* (Casuarinaceae). *Amer. J. Bot.* 63: 335–344.

Torrey, J.G. 1978. Nitrogen fixation by actinomycete-nodulated angiosperms. *Bio-Science* 28: 586–592.

Torrey, J.G. 1981. Root development and root nodulation in *Casuarina*. In: *Casuarina Ecology, Management and Utilization* (S.J. Midgley, J.W. Turnbull and R.D. Johnston, eds.). CSIRO, Melbourne, Australia, pp. 180–192.

Torrey, J.G. 1982. *Casuarina*: Actinorhizal nitrogen-fixing tree of the tropics. In: *Biological Nitrogen Fixation Technology for Tropical Agriculture* (P.H. Graham and S.C. Harris, eds.). CIAT, Cali, Colombia. pp. 427–439.

Torrey, J.G. and Callaham, D. 1982. Structural features of the vesicle of *Frankia* sp. CpI1 in culture. *Can. J. Microbiol.* 28: 749–757.
Torrey, J.G. and Callaham, D. 1979. Early nodule development in *Myrica gale*. *Bot. Gaz.* 140: S10–S14.
Torrey, J.G. and Racette, S. 1989. Specificity among the Casuarinaceae in root nodulation by *Frankia*. *Plant and Soil* 118: 157–164.
Tzean, S.S. and Torrey, J.G. 1989. Spore germination and the life cycle of *Frankia in vitro*. *Can. J. Microbiol.* 35: 801–806.
Walsh, K.B., Ng, B.H. and Chandler, G.E. 1984. Effects of nitrogen nutrition in xylem sap composition of Casuarinaceae. *Plant Soil* 81: 291–293.
Wang, R., Healey, F.P. and Myers, J. 1971. Amperometric measurements of hydrogen evolution in *Chlamydomonas*. *Plant Physiol.* 48: 109–110.
Wheeler, C.T. and Bond, G. 1970. The amino acids of non-legume root nodules. *Phytochemistry* 9: 705–708.
Wheeler, C.T., Henson, I.E. and McLaughlin, M.E. 1979. Hormones in plants bearing actinomycete nodules. *Bot. Gaz.* 140: 52–57.
Winship, L.J., and Tjepkema, J.D. 1990. Techniques for measuring nitrogenase activity in *Frankia* and actinorhizal plants. In: *The Biology of Frankia and Actinorhizal Plants* (C.R. Schwintzer and J.D. Tjepkema, eds.). Academic Press Inc., San Diego, CA, pp. 263–280.
Winship, L.J., Martin, K.J. and Sellstedt, A. 1987. The acetylene reduction assay inactivates root nodule uptake hydrogenase in some actinorhizal plants. *Physiol. Plant.* 70: 361–366.
Witty, J.F., Minchin, F.R., Skot, L. and Sheehy, J.E. 1986. Nitrogen fixation and oxygen in legume root nodules. *Oxfords Surv. Plant Mol. & Cell Biol.* 3: 275–315.
Yang, Y. 1995. The effect of phosphorus on nodule formation and function in the Casuarina-Frankia symbiosis. *Plant Soil* 176: 161–169.
Zeng, S., Tjepkema, J.D. and Berg, R.H. 1989. Gas diffusion pathway in nodules of *Casuarina cunninghamiana*. *Plant Soil* 118: 119–123.
Zhang, Z. and Torrey, J.G. 1985. Studies of an effective strain of *Frankia* from *Allocasuarina lehmanniana* of the Casuarinaceae. *Plant Soil* 87: 1–16.
Zhang, Z., Lopez, M.F. and Torrey, J.G. 1984. A comparison of cultural characteristics and infectivity of *Frankia* isolates from root nodules of *Casuarina* species. *Plant Soil* 78: 79–90.
Zhang, Z., Murry, M.A. and Torrey, J.G. 1986. Culture conditions influencing growth and nitrogen fixation in *Frankia* sp. HFPCcI3. *Plant Soil* 91: 3–15.

Chapter 6

ROLE OF NITROGEN IN PLANT MORPHOGENESIS *IN VITRO*

Rana P. Singh, Susan J. Murch and Praveen K. Saxena

I. Introduction

II. *In vitro* Cell Division and Cell Growth
   A. Inorganic Nitrogen
   B. Organic Nitrogen

III. Somatic Embryogenesis and Organogenesis
   A. Inorganic Nitrogen
   B. Organic Nitrogen

IV. Regeneration via Organogenesis

V. Establishment of Plantlets and Growth of Plants

VI. Conclusions and Future Prospects

Literature Cited

Key Words: Amino acids, amides, cell culture, inorganic nitrogen, morphogenesis, organogenesis, polyamines, somatic embryogenesis.

I. INTRODUCTION

Nitrogen is an essential component of many important biomolecules. However, the metabolic aspects and physiological and molecular regulation of nitrogen metabolism during cell differentiation and morphogenesis are not well understood. A complete study of these aspects should be made to address some basic questions of cell differentiation and morphogenic responses. The aim of this chapter is to examine the available literature and to assess various aspects of the role of nitrogen in regulation of morphogenesis including: 1) a comparison of the uptake and primary assimilation of inorganic nitrogen in plant morphogenesis *in vitro* and *in vivo*, 2) the specific role of certain metabolites, such as amino acids and polyamines etc., 3) Involvement of nitrogenous metabolites in the modulation of plant growth regulators and 4) Evaluation of the current status of the extensively studied tissue culture systems, *Daucus carota*, *Nicotiana tabacum* and *Pinus radiata* as model systems for understanding the common principles of nitrogen metabolism.

Nitrate and ammonia are the most common inorganic nitrogen compounds used as nutrient salts in tissue culture media for *in vitro* plant cultures (Murashige and Skoog, 1962; Wetherell and Dougall, 1976; Mantell and Hugo, 1989; Mordhorst and Lörz, 1993; Niedz, 1994). The pathways for the assimilation and utilization of nitrogen from the environment are well defined for *in vivo* growth of plants. However, very limited information is available for plant tissues grown in culture. The most common inorganic nitrogen form acquired by plants grown *in vivo* is nitrate. Intact plants are usually equipped to acquire optimal quantities of nitrate when the exogenously available nitrate concentrations vary from 10 µM to 100 mM (Crawford, 1995; Glass and Siddiqui, 1995). In the plant cells, nitrate can be either converted to nitrite and subsequently to ammonia by the action of the enzymes nitrate reductase (NR, E.C.1.6.6.1) (Crawford, 1995; Srivastava, 1995) and nitrite reductase (NiR, E.C. 1.6.6.4) (Sawhney, 1995) respectively, or the nitrate may be stored in the vacuoles. Ammonia, produced by these reactions, is further utilized in combination with carbon skeletons to produce glutamine and glutamic acid. This reaction is catalyzed by glutamine synthetase (GS, E.C. 6.3.1.2) and glutamate synthase (GOGAT, E.C.1.4.1.14) (Lam et al., 1995; Singh, 1995). A small fraction may be utilized via glutamate dehydrogenase (GDH, E.C.1.4.1.2-4) (Bhadula and Shargool, 1995; Singh, 1995). Once it has entered into the organic cycle nitrogen can be incorporated into other amino acids, amides, proteins, nucleic acids, chlorophylls, alkaloids, polyamines, vitamins and plant growth regulators. Recent studies also suggest that nitrate may function as a signal molecule of plant growth via increased gene expression for enzymes responsible for the uptake and utilization of nitrate (i.e., NR, NiR, GS, and GOGAT) (Crawford and Arst, 1993; Hoff et al., 1994; Crawford, 1995).

Though nitrate and ammonium salts have been commonly used as nutrients in tissue culture media, many additional reports indicate that reduced nitrogen forms, especially amides and amino acids, e.g. glutamine, glutamic acid, proline and alanine, etc., can enhance cell proliferation as well as regeneration in specific genotypes (Jullien et al., 1979; Stuart and Strickland, 1984a, b; Olsen, 1987; Shetty and Asano, 1991a, b; Shetty et al., 1992a, b; Gill et al., 1993; Thorpe, 1993; Murthy et al., 1996a, b). The role of these compounds in the induction and expression of morphogenesis is not well understood, however.

II. *IN-VITRO* CELL DIVISION AND CELL GROWTH

A. Inorganic Nitrogen

The need for nutrient salts including nitrate, ammonium and some forms of reduced organic nitrogen in plant cell culture media has long been realized. Many of the early tissue culture media contained only nitrate (Gautheret, 1937; White, 1939) and the importance of reduced nitrogen forms for better growth and regeneration emerged later (Murashige and Skoog, 1962; White, 1963; Gamborg et al., 1968). More recently, Preece (1995) emphasized the necessity of optimizing both nutrient requirements and plant growth regulators for morphogenic responses. Various forms of nitrogen in the culture media modulate the endogenous levels of cell metabolites including proteins, nucleic acids, plant growth regulators (PGRs) and other specific regulatory molecules (Preece, 1995). This nitrogen requirement may or may not be met by either inorganic or organic forms of nitrogen depending on the species and culture conditions.

The initial events of cell division, cell size increase and entry into S-phase of DNA synthesis were increased and the rate of cell mortality decreased when *Asparagus officinalis* mesophyll cells were cultured on MS media with 30 mM l-glutamine as the sole nitrogen source (Jullien et al., 1979). In this culture system an organic form of nitrogen is most appropriate since asparagus cells lack NR activity and thus are unable to use nitrate ions (Jullien et al., 1979). Seelye et al. (1995) have recently shown that exogenously supplied ammonium enhanced GS activity, ammonium content and growth of asparagus callus as compared to those with no ammonium supplementation. However, high ammonium supplementation reduced GS activity and growth of the calli. Also a study of nitrate use by suspension cultured tobacco cells during a culture cycle indicated that the patterns of utilization of nitrate ions for cell growth and the expression of nitrate uptake proteins and reduction enzymes were similar to those of plant seedlings (Heimer and Filner, 1971; Behrend and Mateles, 1975; Deane-Drummond, 1990; Zhang and Mackown, 1992; Glass and Siddiqui, 1995).

Another factor which may affect nitrogen modulated cell division and growth is the pH of the growth medium. Ammonium uptake in suspension cultures of *Ipomoea* cells (Martin and Rose, 1976), tomato roots (Sheat et al., 1959), maize roots and leaves (Singh et al., 1984) and soybean plants (Tolley-Henry and Raper, 1986) was dependent on the pH of the media. Ammonium uptake rates in ammonium limited semicontinuous cultures of carrot were 25% higher when the pH of the media was 4.5 compared to a pH of 5.5 or 6.5 (Steiner and Dougall, 1995). It is suggested that this change in culture medium pH may have caused the cells to aggregate or disaggregate, which in turn increased the rate of ammonium uptake from the medium.

B. Organic Nitrogen

The effect of reduced nitrogen form in cell cultures on cell division and cell growth is not always positive. Filner (1966) observed that L-forms of alanine, asparagine, aspartic acid, glutamic acid, proline, valine, histidine, and leucine inhibited cell growth and repressed NR activity during short-term cultivation of tobacco cells in culture media containing one of these amino acids. As products of nitrate assimilation, exogenous and/or endogenous ammonium and amino acids generally inhibit NR activity in plants; however, an ammonium-induced NR activity has been shown in some plants including tissue culture and cell suspension cultures (Srivastava, 1992, 1995). Therefore, amino acids can regulate the nitrogen utilization of *in-vitro* cultures by regulating primary nitrogen assimilation. Simultaneously many amino acids can be readily transformed into other amino acids and get incorporated into proteins during the cell culture (Dougall, 1965, 1966; Thorpe, 1993).

III. SOMATIC EMBRYOGENESIS AND ORGANOGENESIS

A. Inorganic Nitrogen

Cellular totipotency extends to the somatic cells and flexibility of differentiation programmes makes possible differentiation of somatic cells to the specific types of regenerants, i.e., shoots, roots, microtuber or somatic embryos, under defined culture conditions. A balance of nitrogenous compounds as well as phytohormones, particularly auxin and cytokinins, in the media and subsequently in the tissues or cell masses is required for this morphogenesis to occur (Olsen, 1987; Thorpe, 1980, 1988a, 1993). It has been suggested that the level and form of the supplemented nitrogen in the culture media influences the growth and metabolic activity of the cultured cells as well as their morphology and regenerative potentials. Reprogramming of the entire gene expression pattern with specific signals for genes is required to initiate the regeneration process. Dudits et al. (1995) have recently reviewed the current molecular understanding of the events involved in induction of regeneration and indicated a central role of hormone or stress-induced activation of signal transduction systems which may alter chromatin structure, transcription or in other ways induce the series of events that lead to the formation of either dedifferentiated callus tissues or somatic embryos. Nitrogenous compounds may be involved either as carriers, catalysts, transporters or other regulatory molecules in this process.

It has been clearly demonstrated that adequate nitrogen in the medium is important for somatic embryogenesis as no embryos are produced on explants cultured with very low levels of nitrogen or inadequate forms of nitrogen in the medium (Reinert et al., 1967; Tazawa and Reinert, 1969; Reinert and Tazawa, 1969; Wetherell and Dougall, 1976; Nomura and Komamine, 1995). Many studies have shown that inorganic nitrogen in the form of ammonium is required for the initiation of embryogenesis in carrot cell cultures since nitrate alone is insufficient but supplementation of the culture medium with ammonium chloride induced embryo formation (Halperin and Wetherell, 1965; Wetherell and Dougall, 1976). Using ^{15}N NMR analysis, Thorpe and coworkers (1993) demonstrated that ammonium is the preferred source of nitrogen in carrot and white spruce embryogenic tissues as it is taken up from the medium earlier and utilized faster than nitrate. Dougall and Verma (1978) reported that carrot suspension cultures could grow and produce somatic embryos in the presence of ammonium as the sole nitrogen source, if the pH of the medium is controlled by continuous titration. According to Kamada and Harada (1984 a, b) the induction phase of somatic embryogenesis in carrot requires no nitrogenous compounds, if the appropriate level of 2,4-D is available. Reduced nitrogen is required only for further embryo development. Low levels of ammonium (1–5 mM) as sole nitrogen source and low pH have been shown to induce somatic embryogenesis in carrot, even in the absence of the auxin (Smith and Krikorian, 1990; Merkle et al., 1995).

It appears that once reprogramming of the somatic cells is induced by the induction factor, the repetitive cycles of embryogenesis follow the same programme. Tazawa and Reinert (1969) reported that the presence of ammonium in the medium is not essential for embryo formation *in vitro*, but a certain level of intracellular ammonium is a prerequisite for this process. A threshold level of tissue ammonium and its correlation to the embryogenic response of the cells in all the cultures, however, was not apparent. Mordhorst and Lörz (1993) reported that during the embryogenesis and development of isolated barley microspores the level of total nitrogen content in the media, the nitrate: ammonium ratio, and the ratio of inorganic:organic nitrogen were not related to the frequency of initial divisions, had only moderate effects on plating efficiency but had significant effects on embryogenesis and plant regeneration.

When either potassium nitrate (at the level equivalent to MS nitrate), or ammonium sulphate (at the level equivalent to MS ammonium) or glutamine (10 mM) was included in the culture media as the sole nitrogen source, somatic embryos formed at the cotyledonary notch of peanut seed cultures *in vitro*, though they were significantly fewer in number compared with cultures raised in full MS nitrogen (Singh, R.P., Murthy, B.N.S. and Saxena, P.K. unpublished results). However, glutamic acid (10 mM) as a sole nitrogen source in the media was found to satisfy the requirement for nitrogen. Khanna and Raina (1997) have recently shown that the nitrogen content of the callusing medium and the composition of NO_3^- and NH_4^+ nitrogen significantly affect the shoot regeneration from the calli in Basmati rice cultivars.

The differences observed in many studies may be related to genotypic variation, variation in the source tissues, and interaction of nitrogen with other components of the media and to the endogenous status of various metabolites and PGRs. Several other aspects may be involved in nitrogen-mediated regeneration including recycling and metabolism of nitrogen compounds within the plant cells during the culture phase. The reasons why nitrate is not the preferential source of nitrogen for cell growth and regeneration in *in-vitro* cultures remains unresolved, particularly in view of its utilization *in vivo* (Crawford, 1995). It is possible that *in-vitro* cultured cells do not have adequate physiological resources to induce nitrate uptake, transport and assimilation during the initial culture phase. When an intact seedling system was used to test the effect of nitrate on regeneration, the response differed from that observed for the explant culture systems (Wetherell and Dougall, 1976; Singh, R.P., Murthy, B.N.S. and Saxena, P.K. unpublished results). This suggests that the loss of the cellular ability to absorb and assimilate the nitrate may be linked to the physical isolation of the explants and alterated growth of the root system in the presence of high levels of cytokinins. In both carrot and white spruce embryogenic tissues, the incorporation of inorganic nitrogen into glutamine, glutamate and alanine and subsequent embryo development with incorporation into arginine and aliphatic amines was demonstrated using ^{15}N tracer techniques (Thorpe, 1993).

Reduced NR activity in the early stages of carrot somatic embryogenesis has been linked with poor embryogenic potential in the absence of reduced nitrogen (Kamada and Harada, 1984a). The activity of the ammonia assimilating enzyme glutamine synthetase also declined during somatic embryogenesis in carrot following an initial activation (Higashi et al., 1996). Therefore, it is likely that exogenously supplied amino nitrogen in the form of glutamine and/or other amino acids may be essential to provide the adequate nitrogen for the synthesis of metabolites for embryogenesis. It is apparent that during the embryo development, the flow of nitrogen occurred via the GS-GOGAT cycle of ammonia assimilation, leading to its incorporation into ornithine and eventually polyamines. However, biochemical and molecular studies related to uptake and assimilation of nitrate and ammonium using different systems and control forms and levels of nitrogen in the media and tissue are required to provide a definite answer.

B. Organic Nitrogen

There is growing evidence for the usefulness of the supplemented amino compounds in the culture medium in the presence or absence of nutrient nitrogen salts (Table 6.1). The literature summarized in Table 6.1 indicates that many amino acids are effective in improving somatic embryogenesis at various stages of development and conversion. Young embryos apparently do not have the active enzyme systems to assimilate nitrate and ammonium and the addition of certain amino compounds including glutamine, glutamic acid and alanine, improve production, development and conversion frequency of somatic embryos. Glutamine is one of the most commonly effective amino compounds for somatic embryogenesis in many plant species (Table 6.1 and references therein). Asparagine, has been found to be less effective than glutamine for enhancing somatic embryogenesis *in vitro* (Kamada and Harada, 1979; Olsen, 1987). Recently, Higashi et al. (1996) demonstrated that the addition of glutamine, alanine and glutamic acid strongly stimulated somatic embryo formation in carrot cell cultures and were more effective than ammonium. Glutamine supplementation of the culture media during the first 7 days of culture resulted in higher numbers of both globular and heart-shaped embryos of carrot while the development of torpedo-stage embryos was improved by glutamine supplementation in the last 7 days of culture.

Alanine supplementation, either alone or in combination with serine, has been found to increase somatic embryogenesis in some plant

Table 6.1. Stimulation of *in-vitro* somatic embryogenesis in some plants by supplementation of amino nitrogen in the culture media

Plant Species	Source Material	Induction Factor	Form and Level of Nitrogen	Morphogenic Responses	Reference
Daucus carota	cell suspension	2,4-D	alanine (5–10 mM)	Accelerated cell division during the early stages and increased number of embryos/cell cluster	Kamada and Harada, 1979
Daucus carota	cell suspension	2,4-D	glutamine asparagine glutamate aspartate proline (0.1–10 mM)	Increased somatic embryogenesis	Kamada and Harada, 1979
Daucus carota	seedling hypocotyls	2,4-D	proline, serine (5–100 mM)	Enhanced mitotic division and regenerated somatic embryos	Nuti-Ronchi et al., 1984
			glutamine alanine glutamate (10 mM)	Increase in number and rate of development of somatic embryos	Higashi et al. 1996,
Nicotiana tabacum	pollen culture	2,4-D	glutamine (5.5 mM) serine (1.0 mM)	Improved embryo development	Nitsch, 1974
Gyssypium klotzschiamum	cell suspension	2iP & 2,4-D	glutamine (10–15 mM) asparagine (10–15 mM)	Increased somatic embryo differentiation	Price and Smith, 1979

Species	Explant	Hormones	Amino acids/other	Effect	Reference
Oryza sativa	callus cultures	2,4-D, 2,4-5-T, BA, Kinetin	tryptophan (100 mg/L)	Increased formation of embryogenic callus in some cultivars	Siriwardana and Nabors, 1983
Medicago sativa	ovary or petiole	2,4-D & Kinetin	alanine (30 mM) proline (30 mM)	5–10 fold increase in somatic embryogenesis	Skokut et al., 1985
Medicago sativa	petiole	NAA & Kinetin, 2,4-D & Kinetin	proline (50–300 mM) glutamine (1–30 mM) alanine (30–100 mM) arginine (1–30 mM) lysine (1–3 mM) serine (1 mM)	Improved number, structural quality & conversion frequency of somatic embryos	Stuart and Strickland, 1984a, b
Hordeum vulgare	anther culture	BAP	glutamine (5.1 mM) asparagine (5.0 mM)	Induction of microspore embryogenesis	Olsen, 1987
Agrostis alba	callus culture initiated from mature seeds	2,4-D	proline (1–20 mM), glutamine (1–5 mM)	Higher % of embryogenic callus	Shetty and Asano, 1991a, b
Quercus acutissima	callus culture	BAP + IBA	glutamine (1 g/l), proline (5 mM)	High frequency somatic embryogenesis and plant regeneration	Kim et al., 1994
Mallus × domestica	cotyledons of immature zygotic	NAA	adenine (100 mg/l) +	Enhanced frequency of somatic embryogenesis	Paul et al., 1994

(contd.)

Table 6.1 (contd.)

Plant Species	Source Material	Induction Factor	Form and Level of Nitrogen	Morphogenic Responses	Reference
Pelargonium × hortorum	embryo		glutamine (100 mg/l) + glycine (200 mg/l)	Improved numbers of somatic embryos	Gill et al., 1993
Pelargonium × hortorum	hypocotyl sections	TDZ	glutamine (10–100 mM)	Improvement in number and development of somatic embryos	Murthy et al., 1996a Singh et al., 1996
Pelargonium × hortorum	young cotyledon from mature seeds	TDZ	proline (5 mM) glutamine (5 mM) metionine (1 mM)		
Cicer arietinum	seed culture	TDZ	proline (5 mM)	Shift in morphogenic pathway from organogenesis to somatic embryogenesis	Murthy et al., 1996b
Arachis hypogaea	seed culture	TDZ	glutamine (10 mM) + proline (10–50 µM); as sole N source	Improved number and development of somatic embryos	Singh, R.P., Murthy, B.N.S. and Saxena, P.K. (unpublished results)
Dactylis glomerata	cell suspension cultures	dicamba	proline (12.5 mM) serine (12.5 mM)	Stimulation of embryogenic callus	Trigiano and Conger, 1987
Zea mays	callus culture		proline	Stimulation of embryogenic callus	Armstrong and Green, 1985

species. It has been shown that glutamine and alanine stimulate cell division at the initial stage of culture which may substantially increase the embryogenic potential of the tissues (Jullien et al. 1979; Kamada and Harada, 1979; Salonen, 1980; Higashi et al., 1996). Kamada and Harada (1984 b, c) found that alanine was metabolized to glutamic acid via alanine aminotransferase and further utilized as a nitrogen source. Similarly, glutamine is also metabolized to glutamic acid via glutamate synthase in a primary route for nitrogen flow during inorganic nitrogen assimilation (Lam et al., 1995; Singh, 1995). In this case, glutamic acid should be more readily available and thus a suitable amino acid for induction of a higher frequency of *in-vitro* somatic embryogenesis than glutamine or alanine. TDZ-induced somatic embryogenesis in peanut seed cultures was found to improve when glutamic acid was the sole nitrogen source (Singh, R.P., Murthy, B.N.S. and Saxena, P.K. unpublished results).

Experiments with proline and thioproline (thiazolidine-4-carboxylic acid), a proline analogue, supplementation of culture media have led to another area of recent interest. Proline has been shown to enhance somatic embryogenesis in many plant species (Table 6.1) and also to regulate plant regeneration via organogenesis, especially shoot formation (Shetty et al., 1992a, b; Rajasubramaniam and Saradhi, 1994; O'Neill et al., 1996; Singh, R.P. and Saxena, P.K. unpublished results). It has recently been demonstrated that proline induces a shift in morphogenic pathways from organogenesis to somatic embryogenesis in chick-pea seedling cultures (Figure 6.1; Murthy et al., 1996b). Proline is known to stimulate embryogenic callus in maize (Armstrong and Green, 1985), carrot (Nuti-Ronchi et al., 1984), orchard grass (Trigiano and Conger, 1987), alfalfa (Stuart and Strickland, 1984a, b) and turfgrass (Shetty et al., 1991a, b). However, the mechanism of action of proline-induced stimulation of *in-vitro* regeneration is not known.

The stimulatory effect of proline may be dependent on many other factors in the culture media, e.g. the level and forms of other nitrogen compounds, PGRs and pH, etc. (Nuti-Ronchi et al., 1984, Stuart and Strickland, 1984a, b; 1985; Trigiano and Conger, 1987; Shetty et al., 1991a, b). A significantly high level of endogenous proline, glutamine, glutamic acid, aminobutyric acid (GABA) and other amino acids in the embryonic tissues have been reported, even if these substances have not been supplied exogenously through the culture media (Shetty and Asano, 1991b; Kamada and Harada, 1984c; Murch, S.J. and Saxena, P.K., unpublished results). Proline accumulation may be merely a symptom of hormone-induced alterations in cell metabolism during *in-vitro* regeneration, which may have a role as osmolyte, a nitrogen storage compound, or an energy source through generation of $NADP^+$ during proline

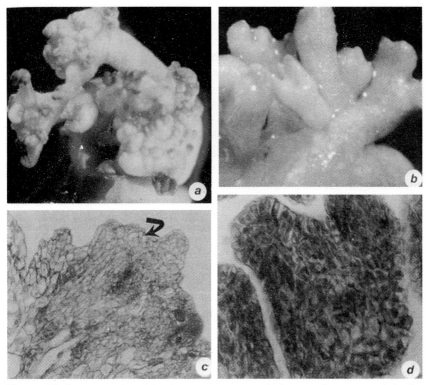

Fig. 6.1. Thidiazuron-induced *in vitro* regeneration in chickpea. a. Organogenesis in the presence of TDZ alone; b. Proline supplementation resulting in somatic embryogenesis; c. Histological profile of organogenesis from TDZ-treated seedlings (a); d. Histological examination of a somatic embryo from seedlings grown on media containing TDZ and proline (b).

synthesis (Shetty and Asano 1991a; Chiang and Daandekar, 1995). An osmotic component of importance for shoot formation in tobacco callus cultures had been reported (Thorpe, 1982). Shetty and Asano (1991b) reported that cell lines of *Agrostis alba* which have the capacity to overproduce proline were embryogenic and tolerant to the proline analogue thioproline (1.0 mM) which effectively inhibited the growth of non-overproducing cell lines. They suggest the possible selection of embryogenic cell lines in other species using thioproline as a selection marker. Proline accumulated at 35 d of culture in tissues of somatic embryo producing seedlings of peanut; however, this did not correlate with embryogenic efficiency when different nitrogen forms were included in the media (Singh, R.P., Murthy, B.N.S., and Saxena, P.K., unpublished results). James et al. (1993) have shown that proline could replace another osmoprotectant, betaine phosphate, if supplied with a signal molecule, acetosyringone to increase the transformation efficiency of *Agrobacterium tumefaciens*-mediated transformation of apple. Proline and other osmolytes, e.g. GABA, glycine betaine, etc. have been reported to accumulate under many environmental stresses, however, their specific role in stress protection is yet not clear (Flasinski and Ragozinski, 1985; LaRosa et al., 1991; Emons et al., 1993; Ober and Sharp, 1994).

Though a specific nutritional effect of proline and serine cannot be excluded, the similarity of the response with the action of fusicoccin during carrot somatic embryogenesis (Nuti-Ronchi et al., 1984) and shift in morphogenetic pathway in chickpea (Murthy et al., 1996b) suggests a specific auxin-mediated effect of proline and serine on the regeneration process. Nitsch (1974) has also suggested a specific stimulatory role of proline and serine on embryo formation during androgenesis. A change in rooting pattern in *in-vitro* seed cultures of grain legumes in the presence of exogenously supplemented proline has also been found (Singh, R.P. and Saxena, P.K., unpublished results). Further, proline (as hydroxyproline) and serine are the two major constituents of the cell wall glycoprotein extensine which may have a morphoregulatory role (Nuti-Ronchi et al., 1984). In alfalfa, however, the increase in somatic embryogenesis in response to alanine and proline could not be correlated with the utilization of nitrogen for the synthesis of protein in the embryogenic tissues as alanine, proline and two other non-stimulating amino acids, glutamate and glycine, were utilized for protein synthesis in a similar manner (Skokut et al., 1985). Overall, the metabolism of these amino acids in regenerating and non-regenerating genotypes was similar except for the accumulation of high levels of free amino acids in non-regenerating cell lines or during suppression of regeneration in the regenerating lines.

The endogenous levels of free amino acids have been examined during somatic embryogenesis in several embryogenic systems (Hirsh, 1975; Wetherhead and Henshaw, 1979; Thorpe, 1983; Kamada and Harada, 1984c; Skokut et al., 1985; Shetty et al., 1991b; Claporols et al., 1993; Magnaval et al., 1995; Murch, S.J. and Saxena, P.K., unpublished results). The total amount of amino acids in embryogenic carrot tissues was found to increase rapidly during cell proliferation and globular embryo formation (Kamada and Harada, 1984c). This accumulation phase of endogenous free amino acids in carrot was dependent, however, on the nitrogen supplemented media (Krikorian and Steward, 1969; Sangwan, 1978; Kamada and Harada, 1984c, Murch, S.J., Singh, R.P., Murthy, B.N.S., and Saxena, P.K., unpublished results). GABA was also found to be a major component of tissues during somatic embryogenesis (Steward et al., 1958; Vickremasinghe et al., 1963; Hirsh, 1975; Ojima and Ohira, 1978; Sangwan, 1978; Kamada and Harada, 1984c;

and our own results), however its precise role has not yet been determined. GABA is formed through decarboxylation of glutamic acid with transamination of succinic semi-aldehyde and is catabolized through rapid oxidization to organic acids in green leaves of tobacco (Steward and Shantz, 1956; Mizusaki et al., 1964). A significant role for GABA as a nitrogen storage compound in the cell, especially during stress, has been proposed (Bown and Shelp, 1989; Satyanarayan and Nair, 1990). GABA and other amino acids, e.g. asparagine, alanine, and serine, were found to decrease in the later phases of carrot embryogenesis, possibly either as a result of metabolism to the other amino compounds or incorporation into the rapidly synthesized proteins (Kamada and Harada, 1984b). In peanut seed cultures, embryogenic seedlings accumulated high levels of proline, GABA, glutamine and glutamic acid during the initial phases (6–10d) only (Murch, S.J. and Saxena, P.K., unpublished results); accumulation of these amino acids was affected by altering the available nitrogen forms in the media (Murch, S.J., Singh, R.P., Murthy, B.N.S. and Saxena, P.K., unpublished results). A rapid turnover of amino acids and active protein synthesis during cell proliferation and the formation of globular embryos has been reported (see Nomura and Komamine, 1995). To maintain this cycle, the cells require a regular supply of amino acids and/or a rapid assimilation of inorganic nitrogen. Considerable amounts of glutamine, alanine and glutamic acid have been shown to accumulate and be metabolized during the maturation and germination of zygotic embryos of *Gossypium* (Capdevila and Dure, 1977).

Magnaval et al. (1995) examined a phase-wise accumulation of endogenous amino acids in coconut callus cultures and found an increase in proline, valine and leucine in the embryogenic calli concomitant with the formation of storage proteins observed histologically during embryogenesis initiation. In this case, non-embryogenic calli had a different pattern of amino acid accumulation. In alfalfa cell cultures no significant difference in amino acid utilization in embryogenic and non-embryogenic cell lines could be identified but higher accumulation of free amino acids was found in non-embryogenic and embryogenically suppressed cell lines (Skokut et al., 1985).

Polyamines may be involved in somatic embryogenesis in carrot, as the endogenous level of polyamines was found to be higher in embryogenic cell lines (Montague et al., 1978; Fienberg et al., 1984; Thorpe, 1993; Nomura and Komamine, 1995). The enzymes of polyamine metabolism in plants, arginine decarboxylase and ornithine carbamoyltransferase, were present at higher levels in embryogenic cell clusters than in non-embryogenic cells (Montague et al., 1979; Baker et al., 1983; Dannin et al., 1993; Bastola and Minocha, 1995). A possible explanation for polyamine profiles is found in the suppressed activities of arginine carboxylase and S-adenosyl methionine decarboxylase by auxins, thereby reducing the polyamine level in the tissue and raising the cellular levels of l-aminocyclopropane-1-carboxylic acid, a precursor of ethylene (Feinberg et al., 1984; Thorpe, 1993). This indicates a competition between ethylene and polyamine biosynthesis which may affect organized development and induce a shift in metabolism during the embryogenesis. Polyamines are thought to be involved in the regulation of transcription and translation in animal cells and may have various functions including: membrane stabilization, free radical scavenging, effects on DNA, RNA and protein synthesis during plant development (see Nomura and Komamine, 1995). The role of polyamine metabolism during somatic embryogenesis can be partially explained by studies with α-difluoromethyl arginine, an inhibitor of arginine decarboxylase (Feirer et al., 1984) and methylglyoxal-bis(guanylhydrasone) (MGBG), an inhibitor of s-adenosylmethionine (SAM) decarboxylase. These inhibitors resulted in significant reductions in cellular levels of spermidine and sperimine and allowed high putrescine accumulation, thereby completely inhibiting somatic embryogenesis in carrot (Minocha et al., 1991). It appears, therefore, that the accumulation of spermidine is more important for embryogenesis than either putrescine or spermine in carrot. An elevated accumulation of putrescine and spermidine was found in carrot cells during somatic embryogenesis although the spermine level was still low (Montague et al., 1978). Recently, Bastola and Minocha (1995) demonstrated an increase in putrescine biosynthesis and a subsequent promotion of somatic embryogenesis in the carrot cells transformed with an ornithine decarboxylase overexpression gene in an *Agrobacterium tumefaciens*-mediated transformation. The incorporated gene was expressed in the plant cell and transgenic cells resulting in a significant increase in the cellular putrescine and an acquired tolerance to α-difluoromethyl arginine.

Recently, the effects of inhibitors of polyamine biosynthesis on the development of embryogenic cell cultures of celery (*Abium graveolux*) were further elucidated (Danin et al., 1993). Cyclohexylamine and particularly methylglyoxal-bis(guanylhydrasone) were found to inhibit both cell division and the organization of polar embryos from globular embryos. Difluoromethylornithine promoted embryo development and especially cell division. The onset of embryogenesis may be characterized by a higher content of putrescine and cytokinins while a decrease in putrescine synthesis and cytokinin content and an increase in spermidine and spermine content may be involved in further embryo development and plantlet formation. A marked decrease in the polyamine titer, especially putrescine, in the presence of higher K^+

supplementation was reported in wheat seedlings (Reggiani et al., 1993). Polyamine metabolism and *in-vitro* cell multiplication and differentiation in leaf explants of *Chrysanthemum morifolium* have also been reported in relation to hormone treatments (Aribaud et al., 1994). Arginine decarboxylase, ornithine decarboxylase and diamine oxidase activities increased during rapid cell multiplication and sharply declined during cell differentiation. Diamine oxidase activity increased rapidly in proliferating and growing organs and decreased during maturity. This variation in DAO occurred concomitant with that of the other two enzymes and the level of polyamines. Bonneau et al. (1995) have recently reported stimulation of rooting activity in non-embryogenic calli in the presence of α-difluromethyl ornithine (DFMO), a specific irreversible inhibitor of ornithine decarboxylase. DFMO partially inhibited putrescine accumulation in embryogenic calli but had no effect on callus growth and significantly reduced the time of emergence of roots as well as stimulating somatic embryogenesis (Bonneau et al., 1995).

The maturation phase is considered the period of embryo development following cell division and histodifferentiation, in which cell expansion and reserve deposition occur (Verhagen and Wann, 1989). There are reports of both similarities and differences in the accumulation of nutrient reserves in these developing embryos in comparison to zygotic embryos (Merkle et al., 1995). Whether the observed differences in the two processes can be attributed to real differences in metabolism or are due to non-optimal conditions of the somatic embryo development is not clear. As discussed earlier, a variety of different media containing various forms and levels of nitrogen and combinations of inorganic and organic nitrogen have been shown to influence embryo development and maturation. It may be concluded, therefore, that nitrogen has a role in the development and maturation of somatic embryos and in addition to an adequate amount of the inorganic nitrogen forms nitrate and ammonium, and certain amino acids, especially glutamine, proline, alanine and serine, etc., can enhance embryo maturation in a defined time phase and thus a better embryo conversion rate can be obtained.

IV. REGENERATION VIA ORGANOGENESIS

Like somatic embryogenesis, *in-vitro* organogenesis also involves a switch in development. At the cellular level, this is reflected by the level of macromolecular synthesis and accumulation mediated by increased enzyme activities as are required for the production of multiple shoots (Thompson and Thorpe, 1990; Thorpe, 1993). Shoot formation *in-vitro* is a highly energy-consuming process with rapid metabolic rates involving increased breakdown of starch and free sugars accompanied with increased respiration rates in shoot-forming calli in comparison to non-shoot-forming ones (Thorpe and Laishley, 1973). This energy generation and the metabolic conversions involve increased activities of the glycolytic and pentose phosphate pathways and increase in the glucose oxidation result increased rates of production of reducing power NAD(P)H and energy (ATP) which could be utilized for shoot formation (Brown and Thorpe, 1980, 1982; Biondi and Thorpe, 1982; Patel and Thorpe, 1984). A high availability of the reducing power and C-skeleton favours the assimilation of inorganic nitrogen and subsequently entry of nitrogen into the organic cycle. It has been reported that shoot forming calli had higher activities of the enzymes of the shikimate pathway when compared to non-shoot-forming calli and increased activities of tyrosine ammonia lyase catylase, leading to more rapid conversion of tyrosine to p-coumaric acid (Beaudoin-Eagen and Thorpe, 1983, 1984, 1985). Higher levels of total nitrogen, protein nitrogen, nitrate and ammonium accompanied by higher levels of amino acids especially proline, threonine and serine were found in shoot-forming tobacco calli than in non-shoot-forming tissues (Thorpe, 1983; Hardy and Thorpe, 1990).

Polyamine accumulation has also been shown to be related to *in-vitro* shoot production. In radiata pine cotyledons spermidine is the major polyamine but primarily ^{14}C-putrescine metabolized to the amino acids and $^{14}CO_2$ during short-term incubations and to spermidine and spermine in longer incubations (Thorpe, 1993). Recently, shoot regeneration from cotyledon explant cultures of *Brassica compestris* in the presence of 1–20 mM putrescine, 0.1–2.5 mM spermidine or 0.1–1 mM spermine was elevated after three weeks of culture (Chi et al., 1994). This study confirms a polyamine requirement for shoot regeneration from the cotyledons of *B. compestris in-vitro* and indicates that the promotive effect of PAs on regeneration may not be related to an inhibition of ethylene biosynthesis.

Exogenously supplied nitrogen, its form and the ammonium:nitrate ratio can affect *in-vitro* organogenesis. The concentration of ammonium is important for shoot formation in *Solanum tuberosum* and it has been reported that a 2:1 ratio for nitrate:ammonium improved shoot formation (see Preece, 1995). In *Rhododendron*, however, the ratio of nitrate to ammonium did not improve the regenerative response but keeping the same nitrogen level, with reduced potassium nitrate and increased ammonium nitrate did improve shoot growth (Anderson, 1984). The number of shoots per culture was found to be related to the concentration of ammonium nitrate in this species (Anderson, 1984). Preece (1995) recently documented the relation of nutrient salts and PGRs to the formation and development of multiple shoots *in-vitro*. Like inorganic nitrogen forms, supplementation of amino nitrogen has also been shown to be beneficial for increased shoot production in many plants (Stuart and Strickland, 1984a, b, 1985; Shetty et al., 1992a, b; O'Neill et al., 1996; Singh, R.P. and Saxena, P.K., unpublished results). The main amino compounds which have been shown to influence *in-vitro* shoot regeneration are allantoin, amides and amino acids proline and proline analogue, thioproline. These reports indicate that organized development involves a shift in metabolism that leads to a change in the content and spectrum of both structural and enzymatic proteins (Thorpe, 1993). However, the signal transduction system involved in the induction of gene expression for these kinds of shift is not characterized. Further, it is not clear whether other plant systems and culture procedures follow similar patterns during shoot formation.

V. ESTABLISHMENT OF PLANTLETS AND GROWTH OF PLANTS

Once conversion of the regenerants be the somatic embryos, multiple shoots or other kinds of regenerants is achieved, the plantlets are grown and acclimatized for field conditions and cultivation of a healthy plant. It is apparent that plantlet development from the regeneration system is a process similar to *in-vivo* seedling development and, therefore, most of the metabolic processes leading to the growth and development of plants must be similar. However, in initial plantlet establishment and acclimatization there may be some requirement for supplemental organic nitrogen, especially when photosynthetic activities are not fully achieved in the closed vessels of laboratory cultures. The inadequate supply of CO_2 in such systems during plantlet development is followed by reduced rates of photosynthesis (Niedz, 1994). A pronounced reduction in NO_3^- and NH_4^+ uptake at 10 and 28 days of culture was observed which was correlated to low CO_2 concentrations in the culture tubes during plantlet growth (Hdider et al., 1994).

VI. CONCLUSIONS AND FUTURE PROSPECTS

In-vitro studies to determine the effects of different nitrogen sources in the culture medium have been conducted on primary explants, callus and developing embryos focussing on carrot, tobacco and radiata pine as most highly characterized systems. It appears that many nutritional, hormonal and environmental effects are specific to the different plant species, cultivars and the culturing procedure. Nitrogen, its forms, and the ammonium to nitrate ratio can affect cell division, differentiation, growth and development of somatic embryos or multiple shoots *in-vitro*. Dividing and differentiating cells, tissues, or explants and even young immature somatic embryos do not contain an adequate active system for uptake, transport and assimilation of inorganic nitrogen, unlike *in-vivo* plant seedlings. And thus, supplemented organic nitrogen forms especially glutamine, proline, alanine, serine and glutamic acid, etc. enhance the differentiation and development of the regenerants. Proline and some other amino acids may enhance the *in-vitro* regeneration process through fulfilling nutritional requirements, availability and incorporation into regeneration related proteins, through stored nitrogen forms for energy, through regulating auxin metabolism and/or through some unknown specific regulatory function. Polyamines are another group of nitrogen compounds that may regulate *in-vitro* regeneration in many species. Elevated levels of endogenous polyamines or exogenously supplied polyamines enhance somatic embryogenesis and organogenesis. However, the signal transduction systems during *in-vitro* plant morphogenesis, gene expression for induction of various shifts in the pathways and regulation of gene(s) and proteins involved in cell division, differentiation and *in-vitro* regeneration are poorly understood.

LITERATURE CITED

Anderson, W.C. 1984. A revised tissue culture medium for shoot multiplication of rhododendron. *J. Amer. Soc. Hort. Sci.* 109: 343–347.

Aribaud, M., Carre, M. and Martin-Tanguy, J. 1994. Polyamine metabolism and *in-vitro* cell multiplications and differentiations in leaf explants of *Chrysanthemum morifolium* Ramat. *Plant Growth Regul.* 15: 143–155.

Armstrong, C.L. and Green, C.E. 1985. Establishment and maintenance of friable, embryogenic maize callus and the involvement of L-proline. *Planta* 164: 207–214.

Baker, S.R., Jones, L.H. and Yon, R.J. 1983. Ornithine carbamoyl transferase activity and embryogenesis in a carrot cell suspension culture. *Phytochemistry* 22: 2167–2169.

Bastola, D.R. and Minocha, S.C. 1995. Increased putrescine biosynthesis through transfer of mouse ornithine decarboxylase cDNA in carrot promotes somatic embryogenesis. *Plant Physiol.* 109: 63–71.

Beaudoin-Eagan, L.D. and Thorpe, T.A. (1983). Shikimate pathway activity during shoot initiation in tobacco callus culture. *Plant Physiol.* 73: 228–232.

Beaudoin-Eagan, L.D. and Thorpe, T.A. 1984. Turnover of shikimate pathway metabolites during shoot initiation in tobacco callus cultures. *Plant Cell Physiol.* 25: 913–921.

Beaudoin-Eagan, L.D. and Thorpe, T.A. 1985. Tyrosine and phenylalanine ammonia lyase activities during short initiation in tobacco callus cultures. *Plant Physiol.* 78: 438–441.

Behrend, J. and Mateles, R.I. 1975. Nitrogen metabolism in plant cell suspension cultures: Effect of amino acids on growth. *Plant Physiol.* 56: 584–589.

Bhadula, S.K. and Shargool, P.D. 1995. Glutamate dehydrogenase: purification, properties and regulation. In: *Nitrogen Nutrition in Higher Plants* (H.S. Srivastava and R.P. Singh, eds.). Assoc. Publ. Co., New Delhi, pp. 205–217.

Biondi, S. and Thorpe, T.A. 1982. Growth regulator effects, metabolic changes and respiration during shoot initiation in cultured cotyledon explants of *Pinus radiata*. *Bot. Gaz.* 143: 20–25.

Bonneau, L., Beranger-Novat, N., Monin, J. and Martin-Tanguy, J. 1995. Stimulation of root and somatic embryo production in *Euonymus europaeus* L. by an inhibitor of polyamine biosynthesis. *Plant Growth Regul.* 16: 5–10.

Bown, A. and Shelp, B.J. 1989. The metabolism and physiological roles of 4-aminobutyric acid. *Biochem.* (Life Sci. Adv.) 8: 21–25.

Brown, D.C.W. and Thorpe, T.A. 1980. Changes in water potential and its components during shoot formation in tobacco callus. *Physiol. Plant.* 49: 83–87.

Brown, D.C.W. and Thorpe, T.A. 1982. Mitochondrial activity during shoot formation and growth in tobacco callus. *Physiol. Plant.* 54: 125–130.

Capdevila, A.M. and Dure, L. 1977. Developmental biochemistry of cotton seed embryogenesis and germination. VIII. Free amino acid pool composition during cotyledon development. *Plant Physiol.* 59: 268–273.

Chi, G.L., Lin, W.S., Lee, J.E.E. and Pua, E.C. 1994. Role of polyamines in *de novo* shoot morphogenesis from cotyledons of *Brassica campestris* ssp. *pekinensis* (Lour) Olsson *in vitro*. *Plant Cell Rep.* 13: 323–329.

Chiang, H.H. and Daandekar, A.M. 1995. Regulation of proline accumulation in *Arabidopsis thaliana* (L.) Heynh during development and in response to desiccation. *Plant Cell Env.* 18: 1280–1290.

Claporols, I., Santos, M.A. and Thorne, J.M. 1993. Influence of some exogenous amino acids on the production of maize embryogenic callus and on endogenous amino acid content. *Plant Cell Tissue Org. Cult.* 34: 1–11.

Crawford, N.M. 1995. Nitrate: nutrient and signal for plant growth. *Plant Cell* 7: 859-868.

Crawford, N.M. and Arst, H.N.J. 1993. The molecular genetics of nitrate assimilation in fungi and plants. *Ann. Rev. Genet.* 27: 115–146.

Danin, M., Upfold, S.J., Levin, N., Nadel, B.L., Altman, A. and van Staden, J. 1993. Polyamines and cytokinins in celery embryogenic cell cultures. *Plant Growth Regul.* 12: 245–254.

Deane-Drummond, C.E. 1990. Biochemical and biophysical aspects of nitrate uptake and its regulation. In: *Nitrogen in Higher Plants* (Y.P. Abrol, ed.). John Wiley & Sons, Inc. New York, pp. 1–37.

Dougall, D.K. 1965. The biosynthesis of protein amino acids in plant tissue culture. I. Isotope competition experiments using glucose $u^{-14}c^-$ and the protein amino acids. *Plant Physiol.* 40: 891–897.

Dougall, D.K. 1966. The biosynthesis of protein amino acids in plant tissue culture. II. Further isotope competition experiments using protein amino acids. *Plant Physiol.* 41: 1411–1415.

Dougall, D.K. and Verma, D. 1978. Growth and embryo formation in wild carrot suspension cultures with ammonium ion as a sole nitrogen source. *In Vitro* 14: 180–188.

Dudits, D., Gyorgyey, J., Bogre, L. and Bako, L. 1995. Molecular biology of somatic embryogenesis. In: *In vitro Embryogenesis in Plants* (T.A. Thorpe, ed.) Kluwer Acad. Publ, Dordrecht, The Netherlands, pp. 267–308.

Emons, A.M.C., Samallo-Droppers, A. and Van der Toorn, C. 1993. The influence of sucrose, mannitol, 1-proline, abscisic acid and gibberellic acid on the maturation of somatic embryos of *Zea mays* L. from suspension cultures. *J. Plant Physiol.* 142: 597–604.

Feirer, R.P. Mignon, G. and Litvay, J.D. 1984. Arginine decarboxylase and polyamines required for embryogenesis in the wild carrot. *Science* 223: 1433–1439.

Fienberg, A.A., Choi, J.H., Lubich, W.P. and Sung, Z.R. 1984. Developmental regulation of polyamine metabolism in growth and differentiation of carrot culture. *Planta* 162: 532–539.

Filner, P. 1966. Regulation of nitrate reductase in cultured tobacco cells. *Biochem. Biophys. Acta* 118: 299–310.
Flasinski, S. and Ragozinski, J. 1985. Effect of water deficit on proline accumulation, protein and chlorophyll content during flowering and seed formation in winter rape. *Acta Agrobotanica* 38: 11–21.
Gamborg, O.L., Miller, R.A. and Ojima, K. 1968. Nutrient requirements of suspension cultures of soybean root cells. *Exp. Cell Res.* 50: 151–158.
Gautheret, R.J. 1937. Nouvelles recherches sur la culture de tissu cambial. *C.R. Acad. Sci.* 20: 572–577.
Gill, R., Gerrath, J.M. and Saxena, P.K. 1993. High-frequency direct somatic embryogenesis in thin layer cultures of hybrid seed geranium (*Pelargonium* × *hortorum*). *Can. J. Bot.* 71: 408–413.
Glass, A.D.M. and Siddiqui, M.Y. 1995. Nitrogen absorption by plants. In: *Nitrogen Nutrition in Higher Plants* (H.S. Srivastava and R.P. Singh, eds.), Assoc. Publ. Co., New Delhi, pp. 21–56.
Halperin, W. and Wetherell, D.F. 1965. Ammonium requirement for embryogenesis *in vitro*. *Nature* 205: 519–520.
Hardy, E.L. and Thorpe, T.A. 1990. Nitrate assimilation in shoot-forming tobacco callus cultures. *In vitro Cell Develop. Biol.* 26: 525–530.
Hdider, C., Vezina, L.P. and Desjardins, Y. 1994. Short-term studies of $^{15}NO_3^-$ and $^{15}NH_4^+$ uptake by micropropagated strawberry shoots cultured with or without CO_2 enrichment. *Plant Cell Tissue Org. Cult.* 37: 185–191.
Heimer, Y.M. and Filner, P. 1971. Regulation of the nitrate assimilation pathway of cultured tobacco cells. III. The nitrate uptake system. *Biochem. Biophys. Acta* 230: 362–372.
Higashi, K., Kamada, H. and Harada, H. 1996. The effects of reduced nitrogenous compounds suggests that glutamine synthetase activity is involved in the development of somatic embryos in carrot. *Plant Cell Tissue Org. Cult.* 45: 109–114.
Hirsh, A.M. 1975. Sur la culture de tissus de fruits *d'Actinidia chinensis* et la metabolisme des amines libres de fragments de tiges cultives *in vitro*. *C.R. Acad. Sci. Paris*, Serie D, 280: 1369–1372.
Hoff, T., Truon, H.M. and Caboche, M. 1994. The use of mutants and transgenic plants to study nitrate assimilation. *Plant Cell Environ.* 17: 489–506.
James, D.J., Uratsu, S., Cheng, J., Negri, P., Viss, P. and Dandekar, A.M. 1993. Acetosyringone and osmoprotectants like betaine or proline synergistically enhance *Agrobacterium* mediated transformation of apple. *Plant Cell Reports* 12: 559–563.
Jullien, M., Rossini, L. and Guern, J. 1979. Some aspects of the induction of the first division and growth in cultures of mesophyll cells obtained from *Asparagus officinalis* L. In: *Proc. 5th Int. Asparagus Symp.* (G. Reuther, ed.). Eucarpia section: vegetables. Geisenheim Forschungsansdt, Germany, pp. 103–124.
Kamada, H. and Harada, H. 1979. Studies on the organogenesis in carrot tissue culture. II. Effects of amino acids and inorganic nitrogenous compounds on somatic embryogenesis. *Z. Pflanzenphysiol.* 91: 453–463.
Kamada, H. and Harada, H. 1984a. Changes in nitrate reductase activity during embryogenesis in carrot. *Biochem. Physiol. Pflanzen.* 179: 403–410.
Kamada, H. and Harada, H. 1984b. Studies on nitrogen metabolism during somatic embryogenesis in carrot. I. Utilization of α-alanine as a nitrogen source. *Plant Sci. Lett.* 33: 7–13.
Kamada, H. and Harada, H. 1984c. Changes in endogenous amino acid compositions during somatic embryogenesis in *Daucus carota* L. *Plant Cell Physiol.* 25: 27–38.
Kim, Y.W., Lee, B.C., Lee, K.K. and Jang, S.S. 1994. Somatic embryogenesis and plant regeneration in *Quercus acutissima*. *Plant Cell Rep.* 13: 315–318.
Khanna, H. and Raina, S.K. 1997. Enhanced *in vitro* plantlet regeneration from mature embryo-derived primary callus of a Basmati rice cultivar through modification of

nitrate nitrogen and ammonium-nitrogen concentrations. *J. Plant Biochem. Biotech.* 6: 85–89.

Komamine, A., Kawahara, R., Matsumoto, M., Sunabori, S., Toya, T., Fujiwara, A., Tsukahara, M., Smith, J., Ito, M., Fukuda, H., Nomura, K. and Fujimura, T. 1992. Mechanisms of somatic embryogenesis in cell cultures: physiology, biochemistry and molecular biology. *In vitro Cell Dev. Biol.* 28: 11–14.

Krikorian, A.D. and Steward, F.C. 1969. Biochemical differentiation: The biosynthetic potentialities of growing and quiescent tissue. In: *Plant Physiol, a Treatise*, vol. V (F.C. Steward, ed.). Acad. Press, New York, pp. 227–326.

La Rosa, P.C., Rhodes, D., Rhodes, J.C., Bressan, R.A. and Csonka, L.N. 1991. Elevated accumulation of proline in NaCl adapted tobacco cells is not due to altered Δ^1-pyrroline-5-carboxylate reductase. *Plant Physiol.* 96: 245–250.

Lam, H.M., Coschigano, K., Schultz, C., Melo-Oliveira, R., Tjaden, G., Olveira, I., Nagai, N., Hseih, M.H. and Coruzzi, G. 1995. Use of *Arabidopsis* mutants and genes to study amide amino acid biosynthesis. *Plant Cell* 7: 887–898.

Lo-Schiavo, F., Quesada-Allue, L. and Sung, Z.R. 1985. Synthesis of mannosylated dolichyl derivatives in proliferating cells. In: *Somatic Embryogenesis* (M. Terzi, L. Pitto and Z.R. Sung, eds.). IPRA, Rome, pp. 64–76.

Magnaval, C., Noirot, M., Verdeil, J.L., Blattes, A., Huet, C., Grosdemange, F. and Buffard-Movel, J. 1995. Free amino acid composition of coconut (*Cocos nucifera* L.) calli under somatic embryogenesis induction conditions. *J. Plant Physiol.* 146: 155–161.

Mantell, S.H. and Hugo, S.A. 1989. Effects of photoperiod, mineral medium strength, inorganic ammonium, sucrose and cytokinin on root, shoot and microtuber development in shoot cultures of *Dioscorea alata* L. and *D. bulbifera* L. yams. *Plant Cell Tissue Org. Cult.* 16: 23–37.

Martin, S.M. and Rose, D. 1976. Growth of plant cell (*Ipomoea*) suspension cultures at controlled pH levels. *Can. J. Bot.* 54: 1264–1270.

Merkle, S.A., Parrot, W.A. and Flinn, B.S. 1995. Morphogenic aspects of somatic embryogenesis. In: *In vitro Embryogenesis in Plants* (T.A. Thorpe, ed.). Kluwer Acad. Publ., Dordrecht, The Netherlands, pp. 155–203.

Minocha, S.C., Papa, N.S., Khan, A.J. and Samuelsen, A.I. 1991. Polyamines and somatic embryogenesis in carrot. III. Effects of methylglyoxal bis (guanylhydrazone). *Plant Cell Physiol.* 32: 395–402.

Mizusaki, S., Noguchi, M. and Tamaki, E. 1964. Studies on nitrogen metabolism in tobacco plants. *Arch. Biochem. Biophys.* 105: 599–605.

Montague, M.J., Koppenbrink, J.W. and Jaworski, E.G. 1978. Polyamine metabolism in embryogenic cells of *Daucus carota*. I. Changes in intracellular content and rates of synthesis. *Plant Physiol.* 62: 430–433.

Montague, M.J., Armstrong, T.A. and Jaworski, E.G. 1979. Polyamine metabolism in embryogenic cells of *Daucus carota*. II. Changes in arginine decarboxylase activity. *Plant Physiol.* 63: 341–345.

Mordhorst, A.P. and Lörz, H. 1993. Embryogenesis and development of isolated barley (*Hordeum vulgare*) microspores are influenced by the amount and composition of nitrogen sources in culture media. *J. Plant Physiol.* 142: 485–492.

Murashige, T. and Skoog, F. 1962. A revised medium for rapid growth and bioassays with tobacco cultures. *Physiol. Plant.* 15: 473–497.

Murthy, B.N.S., Singh, R.P. and Saxena, P.K. 1996a. Induction of high-frequency somatic embryogenesis in geranium (*Pelargonium* × *hortorum* Bailey cv. Ringo Rose) cotyledonary cultures. *Plant Cell Rep.* 15: 423–426.

Murthy, B.N.S., Victor, J., Singh, R.P., Fletcher, R.A. and Saxena, P.K. 1996b. *In vitro* regeneration of chickpea (*Cicer arietinum* L.): Stimulation of direct organogenesis and somatic embryogenesis by thidiazuron. *Plant Growth Reg.* 19: 233–240.

Niedz, R.P. 1994. Growth of embryogenic sweet orange callus on media varying in the ratio of nitrate to ammonium nitrogen. *Plant Cell Tissue Organ Cult.* 39: 1–5.

Nitsch, C. 1974. La culture de pollen isole sur milieu synthetigue. *C.R. Acad. Sci. Paris* (Ser. D), 278: 1031–1037.

Nomura, K. and Komamine, A. 1995. Physiological and biochemical aspects of somatic embryogenesis. In: *In vitro Embryogenesis in Plants* (T.A. Thorpe, ed.). Kluwer Acad. Publ., Dordrecht, The Netherlands, pp. 249–265.

Nuti-Ronchi, V., Caligo, M.A., Nozzdini, M. and Luccarini, G. 1984. Stimulation of carrot somatic embryogenesis by proline and serine. *Plant Cell Rep.* 3: 210–215.

Ober, E.S. and Sharp, R.E. 1994. Proline accumulation in maize (*Zee mays* L.) primary roots at low water potentials. I. Requirement for increased levels of abscisic acid. *Plant Physiol.* 105: 981–987.

Ojima, K. and Ohira, K. 1978. Nutritional requirements of callus and cell suspension cultures. In: *Frontiers of Plant Tissue Culture* (T.A. Thorpe, ed.). Kluwer Acad. Publ., Dordrecht, the Netherlands, pp. 265–275.

Olsen, F.L. 1987. Induction of microscope embryogenesis in cultured anthers of *Hordeum vulgare*. The effects of ammonium nitrate, glutamine, and asparagine as nitrogen sources. *Carlsberg Res. Commun.* 52: 393–404.

O'Neill, C.M., Arthur, A.E. and Mathias, R.J. 1996. The effects of proline, thioproline and methylglyoxal-bis-(guanylhydrazone) on shoot regeneration frequencies from stem explants of *B. napus*. *Plant Cell Rep.* 15: 695–698.

Patel, K.R. and Thorpe, T.A. 1984. Histochemical examination of shoot formation in cultured cotyledon explants of radiata pine. *Bot. Gaz.* 145: 312–322.

Paul, H., Belaizi, M. and Sangwan-Norreel, B.S. 1994. Somatic embryogenesis in apple. *J. Plant. Physiol.* 143: 78–86.

Preece, J.E. 1995. Can nutrient salts partially substitute for plant growth regulators? *Plant Tissue Cult. Biotech.* 1: 26–37.

Price, H.J. and Smith, R.H. 1979. Somatic embryogenesis in suspension cultures of *Gossypium klotzchianum* Andress. *Planta* 145: 305–307.

Rajasubramaniam, S. and Saradhi, P.P. 1994. Organic nitrogen stimulates caulogenesis from hypocotyl callus of *Phyllanthus fraternus*. *Plant Cell Rep.* 13: 619–622.

Reggiani, R., Aurisano, N., Mattana, M. and Bertani, A. 1993. Influence of K^+ ions on polyamine level in wheat seedlings. *J. Plant Physiol.* 141: 136–140.

Reinert, J. and Tazawa, M. 1969. Influence of nitrogen compounds and auxin on embryogenesis in tissue culture. *Planta* 87: 239–248.

Reinert, J., Tazawa, M. and Semenoff, S. 1967. Nitrogen compounds as factors of embryogenesis *in vitro*. *Nature* 216: 1215–1216.

Salonen, M.L. 1980. Glutamate and aspartate derived amino acids as nitrogen source for the callus of *Atropa belladonna* L. *Ann. Bot. Fennici* 17: 357–378.

Sangwan, R.S. 1978. Amino acid metabolism in cultured anthers of *Datura metel*. *Biochem. Physiol. Pflanzen.* 173: 355–364.

Satyanarayan, V. and Nair, P.M. 1990. Metabolism, enzymology and possible roles of 4-aminobutyrate in higher plants. *Phytochemistry* 29: 336–375.

Sawhney, S.K. 1995. Nitrite reductase. In: *Nitrogen Nutrition in Higher Plants* (H.S. Srivastava and R.P. Singh, eds.). Assoc. Publ. Co., New Delhi, pp. 165–188.

Seelve, J.F., Borst, W.M., King, G.A., Patricia, J.H. and Maddocks, D. 1995. Glutamine synthetase, ammonium accumulation and growth of callus cultures of *Asparagus officinalis* L. exposed to high ammonium or phosphinothricin. *J. Plant Physiol.* 146: 686–692.

Sheat, D.E.G., Fletcher, B.H. and Street, H.E. 1959. Studies on the growth of excised roots. VIII. The growth of excised tomato roots supplied with various inorganic sources of nitrogen. *New Phytol.* 58: 128–141.

Shetty, K. and Asano, Y. 1991a. The influence of organic nitrogen sources on the induction of embryogenic callus in *Agrostis alba* L. *J. Plant Physiol.* 139: 82–85.

Shetty, K. and Asano, Y. 1991b. Specific selection of embryogenic cell lines in *Agrostis alba* L. using the proline analog thioproline. *Plant Sci.* 79: 259–263.

Shetty, K., Asano, Y. and Oosawa, K. 1992a. Stimulation of *in vitro* shoot organogenesis in *Glycine max* (Merrill) by allantoin and amides. *Plant Sci.* 81: 245–251.

Shetty, K., Shetty, G., Nakazaki, Y., Yoshioka, K., Asano, Y. and Oosawa, K. 1992 b. Stimulation of benzyladenine-induced *in vitro* shoot organogenesis in *Cucumis melo* L. by proline, salicylic acid and aspirin. *Plant Sci.* 84: 193–199.

Singh, R.P. 1995. Ammonia assimilation. In: *Nitrogen Nutrition in Higher Plants* (H.S. Srivastava and R.P. Singh, eds.). Assoc. Publ. Co., New Delhi, pp. 189–203.

Singh, R.P., Mehta, P. and Srivastava, H.S. 1984. Characteristics of ammonium absorption by excised root and leaf tissues of maize. *Physiol. Plant* 60: 119–124.

Singh, R.P., Murthy, B.N.S. and Saxena, P.K. 1996. *In vitro* morphogenetic competence of diploid zonal geranium (*Pelargonium* × *hortorum* Bailey cv. Scarlet Orbit Improved) cotyledonary tissue induced with phenylurea compounds. *Physiol. Mol. Biol. Plants* 2: 53–58.

Siriwardana, S. and Nabors, M.W. 1983. Tryptophan enhancement of somatic embryogenesis in rice. *Plant Physiol.* 73: 142–146.

Skokut, T.A., Manchester, J. and Schaefer, J. 1985. Regeneration in alfalfa tissue culture. Stimulation of somatic embryo production by amino acids and ^{15}N NMR determination of nitrogen utilization. *Plant Physiol.* 79: 577–583.

Smith, D.L. and Krikorian, A.D. 1990. Somatic proembryo production from excised, wounded zygotic carrot embryos on hormone-free medium: evaluation of the effects of pH, ethylene and activated charcoal. *Plant Cell Rep.* 9: 34–37.

Srivastava, H.S. (1992). Multiple functions and forms of nitrate reductase in higher plants. *Phytochemistry* 31: 2941–2947.

Srivastava, H.S. 1995. Nitrate reductase. In: *Nitrogen Nutrition in Higher Plants* (H.S. Srivastava and R.P. Singh, eds.). Assoc. Publ. Co., New Delhi, pp. 145–164.

Steiner, H.Y. and Dougall, D.K. 1995. Ammonium uptake in carrot cell cultures is influenced by pH-dependent cell aggregation. *Physiol. Plant* 95: 415–422.

Steward, F.C. and Shantz, E.M. 1956. The chemical induction of growth in plant tissue cultures. In: *The Chemistry and Mode of Action of Plant Growth Substances* (R.L. Wain and F.C. Wightman, eds.). Academic Press, New York. pp. 165–187.

Steward, F.C., Mapes, M.O. and Mears, K. 1958. Growth and organized development of cultured cells. II. Organization in cultures grown freely from suspended cells. *Amer. J. Bot.* 45: 705–709.

Stuart, D.A. and Strickland, S.G. 1984a. Somatic embryogenesis from cell cultures of *Medicago sativa* L. I. The role of amino acid additions to the regeneration medium. *Plant Sci. Lett.* 34: 165–174.

Stuart, D.A. and Strickland, S.G. 1984b. Somatic embryogenesis from cell cultures of *Medicago sativa* L. II. The interaction of amino acids with ammonium. *Plant Sci. Lett.* 34: 175–181.

Stuart, D.A., Nelson, J., Strickland, S.G. and Nichol, J.W. 1985. Factors affecting developmental processes in alfalfa cell cultures. In: *Tissue Culture in Forestry and Agriculture* (R.R. Henke, K.W. Hughes, M.J. Constantin and A. Hollainder, eds.). Plenum Pub. Corp., New York, pp. 59–74.

Tazawa, M. and Reinert, J. 1969. Extracellular and intracellular chemical environments in relation to embryogenesis *in vitro*. *Protoplasma* 68: 157–163.

Thompson, M.R. and Thorpe, T.A. 1990. Biochemical perspectives in tissue culture for crop improvement. In: *Biochemical Aspects of Crop Improvement* (K.R. Khanna, ed.). CRC Press, Boca Raton, FL, pp. 328–358.

Thorpe, T.A. 1980. Organogenesis *in vitro*: structural, physiological and biochemical aspects. *Int. Rev. Cyto. Suppl.* 11A: 71–111.
Thorpe, T.A. 1982. Callus organization and *de novo* formation of shoots, roots and embryos *in vitro*. In: *Techniques and Applications of Plant Cell and Tissue Culture to Agriculture and Industry* (D. Tomes, B.E. Ellis, P.M. Harney, K.J. Kasha and R.L. Peterson, eds.). University of Guelph, Ontario, pp. 115–138.
Thorpe, T.A. 1983. Morphogenesis and regeneration in tissue culture. In: *Genetic Engineering: Applications in Agriculture*, Beltsville Symp. No. 7, (L.D. Owens, ed.). Rowan and Allanheld, Totowa, N.J., pp. 258–302.
Thorpe, T.A. 1988a. *In vitro* somatic embryogenesis. *ISI Atlas of Science — Animal and Plant Sciences* 1: 81–88.
Thorpe, T.A. 1993. *In vitro* organogenesis and somatic embryogenesis: physiological and biochemical aspects. In: *Morphogenesis in Plants* (K.A. Roubelakis-Angelakis and K. Tran Thanh Van, eds.). Plenum Press, New York, pp. 19–38.
Thorpe, T.A. and Laishley, E.J. 1973. Glucose oxidation during shoot initiation in tobacco callus cultures. *J. Exp. Bot.* 24: 1082–1089.
Tolley-Henry, L. and Raper, C.D. 1986. Utilization of ammonium as a nitrogen source. Effects of ambient acidity on growth and nitrogen accumulation by soybeans. *Plant Physiol.* 82: 54–60.
Trigiano, R.N. and Conger, B.V. 1987. Regulation of growth and somatic embryogenesis by proline and serine in suspension cultures of *Dactylis glomerata*. *J. Plant Physiol.* 130: 49–55.
Verhagen, S.A. and Wann, S.R. 1989. Norway-spruce somatic embryogenesis: high frequency initiation from light cultured mature embryos. *Plant Cell Tissue Org. Cult.* 16: 103–108.
Vickremasinghe, R.L., Swain, T.S. and Goldstein, J.L. 1963. Accumulation of amino acids in plant cell cultures. *Nature* 19: 1302–1303.
Wetherell, D.F. and Dougall, D.K. 1976. Sources of nitrogen supporting growth and embryogenesis in cultured wild carrot tissue. *Physiol. Plant* 37: 97–103.
Wetherhead, M.A. and Henshaw, G.G. 1979. The induction of embroid in free pollen culture of potatoes. *Z. Pflanzenphysiol.* 94: 441–447.
White, P.R. 1939. Controlled differentiation in plant tissue culture. *Bull. Torrey Bot. Club.* 66: 507–513.
White, P.R. 1963. *The Cultivation of Animal and Plant Cells*, 2nd Ed. Ronald Press, New York.
Zhang, N. and Mackown, C.T. 1992. Nitrate use by tobacco cells in response to N-stress and ammonium nutrition. *Plant Cell Rep.* 11: 470–475.

Chapter 7

ROLE OF NITROGEN SOURCE IN CARBON BALANCE

Maria Amélia Martins-Loução and Cristina Cruz

I. Introduction

II. Nitrogen Sources and Growth

    A. Nitrogen Concentrations

    B. Root Temperature

    C. pH

III. Interactions Between C and N Metabolism

    A. Carbon assimilation

    B. C Skeletons and Carbon Flow

    C. Nitrogen Uptake

IV. Ecological Strategies

Literature Cited

Abbreviations: a.a, amino acids; ABA, abcisic acid; cHATS, constitutive high-affinity transport system; Glut, glutamine; GOGAT, glutamate synthase; GS, glutamine synthetase, HATS, high-affinity transport system; iHATS, inductive high affinity transport system; LATS, low affinity transport system; NAP, nitrate reductase activating peptide; Nr, organic nitrogen; NR, nitrate reductase; NR-P-I, inactive nitrate reductase; NUE, nitrogen use efficiency; PEPcase, phosphoenolpyruvate carboxylase; PEPcase-P, phosphorylated phosphoenolpyruvate carboxylase; Pi, inorganic phosphate; R/S, root/shoot dry weight ratio; RNA, ribonucleic acid; RO, organic acids; Rubisco, ribulose-1,5-biphosphate carboxylase/oxygenase; RZT, root zone temperature; SPS, sucrose phosphate synthase; SPS-P, phosphorylated sucrose phosphate synthase; TCA, tricarboxylic acid cycle.

I. INTRODUCTION

The basis of our current knowledge of carbon-nitrogen relationships is only 35 to 50 years old. Leaf photosynthetic capacity usually increases with leaf nitrogen level (Evans, 1989) but these appear to be regulated in relation to species' growth potential. Furthermore, the relation between CO_2 assimilation and nitrogen reduction directly or indirectly linked to photosynthesis, is not constant but varies with environmental conditions and the availability of resources, including nitrogen itself.

Carbon and nitrogen have a central role in plant function, so their acquisition must be balanced. For example, carbon fixation takes place in the leaves, but there is a need for nitrogen supply, and thus for root activity, to make and to maintain key components of the photosynthetic apparatus, so that a tightly integrated system of relatively constant efficiency is formed. Thus, in order to understand growth and productivity it is important to consider the plant's ability to produce carbon and nitrogen assimilates and to ascertain the extent to which the genetic potential can be exploited.

The form in which nitrogen is acquired also has profound implications for the essential core of plant constituents during growth (Raven, 1988). Nitrate ions are generally regarded as the major form of inorganic nitrogen available to higher plants and the broad outlines of nitrogen uptake, assimilation and transport have been established for a range of crop plants fed with nitrate. Where soils have a low capacity for nitrification, for instance acid or waterlogged soils, then ammonium may be found in soil solution (Rice and Pancholy, 1972; Ellenberg, 1977; Smirnoff and Stewart, 1985; Stewart, 1991). In terrestrial communities dominated by perennial plants, there is also a general inhibition of nitrification, ammonium (plus organic N) being the major nitrogen sources. While ammonium is tightly bound to negative charges of soil particles, nitrate, as a mobile anion, is susceptible to leaching (Raven et al., 1992). For these reasons the vertical concentration gradient of nitrate in the soil is more variable than that of ammonium. This gives ammonium a more significant role and a new insight to the study of carbon nitrogen relations in both physiological and ecological approaches.

A recent review has considered the involvement of carbon assimilation and respiration in the assimilation of nitrogen from a biochemical point of view (Huppe and Turpin, 1994). In this paper we emphasize the importance of the nitrogen sources in controlling the feedbacks and constraints of nitrogen and carbon metabolism, particularly in woody plants, some of which were considered in less detail by those authors. The influence of nitrogen sources in carbon partitioning between roots and shoots, on comparative energetic costs of root and shoot nitrogen assimilation, and on carbon-nitrogen interactions are considered, taking into account physiological and ecological aspects.

II. NITROGEN SOURCES AND GROWTH

As in plants, nitrogen in soil occurs both in organic and inorganic forms. In organic matter nitrogen is in a reduced form. Some occurs as amino and amide nitrogen; some as a constituent of large and often resistant molecules with nitrogen in heterocyclic aromatic rings, associated with humic substances such as protein-tannin complexes (Tamm, 1991; Northup et al., 1995). The importance of dissolved organic nitrogen in ecosystem nutrient fluxes and plant nutrition has only recently been appreciated (Qualls et al., 1991; Chapin et al., 1993). In adapting to nitrogen limited ecosystems on strongly acidic and infertile soils with cold climate, plants developed specialized systems to utilize organic soil nitrogen, either by hydrolysis of proteins in ecto- or ericoid mycorrhizae associations (Read, 1993), or by preferential uptake of amino acid, as in the case of the non-mycorrhizal arctic sedge *Eriophorum vaginatum* (Chapin, 1980).

In spite of the presence of organic nitrogen in soil, nitrate and ammonium are the major nitrogen sources available for plant growth. Nitrate is the dominant source for plants in most arable areas while ammonium is the dominant source in most stabilized ecosystems. This difference in the chemical nature of the soil nitrogen pool is believed to be a result of a multifactorial suppression of nitrification in later stages of succession in forest soils, caused by such factors as low soil temperatures, low pH, and allelopathic soil conditions (Verhangen et al., 1995). In agricultural, perturbed or early successional soils, nitrification can proceed at high rates.

Ammonium (NH_4^+) and nitrate (NO_3^-) can have different effects on plant development. Bioenergetic considerations reveal that the assimilation of nitrate nitrogen by plants requires more energy than the assimilation of ammonium (Raven, 1985; Salsac et al., 1987). Thus, ammonium nutrition should be an interesting alternative to that based on nitrate and there are indeed some plants which take this option (Ingestad, 1979; Martins-Loução, 1985; Martins-Loução and Duarte, 1987).

According to Pate (1973), Smirnoff et al. (1984) and Pearson and Stewart (1993), plants can be roughly divided into three groups: i) those that prefer or utilize ammonium; ii) those that preferentially utilize nitrate; iii) those that use both ammonium and nitrate. Generally the type of inorganic nitrogen nutrition and tissue of assimilation reflects the ecological niche that a particular species occupies (Pearson and

Stewart, 1993). However, numerous experiments have revealed that, as a rule, plants perform best on mixtures of nitrate and ammonium (Table 7.1).

Table 7.1 presents the effects of ammonium compared to nitrate on maximum growth of different plant species (for a range of published papers). It is important to emphasize that this table was based on experiments where comparisons with growth conditions and measurements were possible. This type of comparison, taking nitrate as the control, is justified since nitrate is the most used nitrogen source in experiments with crops. The first column of the table lists plants by their Latinic names in alphabetical order. The next two columns refer to the nitrogen source available in the root media: ammonium and nitrate separately. The fourth column refers to the comparison of ammonium-fed plants to nitrate-fed plants. Wherever the simultaneous presence of ammonium and nitrate were investigated, the ratio of ammonium to nitrate as well as the total nitrogen concentration is indicated in the fifth column. Where better growth is attained in the presence of an ammonium and nitrate solution an asterisk is inserted. In order to allow comparison between the experiments with the same plant species, some culture conditions are also included in the table's sixth column. In one of the experiments the effects of ammonium nutrition were compared to those of ammonium nitrate since ammonium nitrate was considered the true control, being closer to the situation in soils (Maghrabi et al., 1985). In a similar fashion, Table 7.2 presents the effects of ammonium, also in comparison to nitrate, on the growth of woody plants.

The data in Table 7.1 show that NH_4^+ impedes growth more than NO_3^- in 55% of the plants studied. At low concentrations, in a nitrate-free medium, or added in small amounts to a nitrate solution medium, growth in the presence of ammonium equals (25%) or exceeds (30%) that under NO_3^- nutrition. Most of the experiments were designed to compare biomass production of plants grown with either nitrate or ammonium at the same concentration. However, the ammonium concentrations needed for maximum growth are usually lower than those of nitrate (Figs. 7.1 and 7.2). The poor growth of many plants supplied with NH_4^+ may result from excessive ammonium concentration in the root medium, which leads to an acidification of the root surface. Hence, the toxicity of ammonium may result from effects of low pH rather than from the presence of ammonium *per se* (Rufty et al., 1983; Tolley-Henry and Raper, 1986; Rideout and Raper, 1994; Mata et al., 1996).

When ammonium and nitrate are simultaneously assimilated the pH stress in the cytoplasm and vacuole of the assimilating cells can be lower than when only a single nitrogen source is present. The manifestations of pH regulation — organic acid accumulation, extracellular pH increase or decrease — on the mixed nitrogen sources are between those found when nitrate and ammonium are supplied separately (Raven and Smith, 1976; Raven, 1985, 1988).

Ammonium uptake and assimilation improved the growth of 53% of the woody species studied while it did not affect the growth of 37% of those species compared to nitrate-grown plants (Table 7.2). This interesting contrast in nitrogen utilization between herbaceous and woody plants shows that indigenous species in later successional forest habitats prefer to utilize ammonium as their inorganic nitrogen source, whereas species indigenous to agricultural, perturbed or early successional forest soils prefer nitrate (Stewart et al., 1990). The assumption that in general trees prefer NH_4^+ to NO_3^- does not always correspond to reality (Stadler et al., 1992). This preference is strongly related to the relative growth rate of the species. According to the ecological explanation developed by Stewart (Smirnoff et al., 1984; Stewart et al., 1988, 1989) and Raven (Raven et al., 1992), ammonium preference tends to be found in perennial, or slow-growing plants which are present in climax or stable communities. Factors affecting the relative response to different nitrogen sources include soil adsorption properties and microbial activities leading to immobilization, nitrification or denitrification.

Considerable genotypical differences must also be taken into consideration when analyzing these adaptations to different nitrogen sources. The ecological relevance of differences between species may be related to the capability of roots to change the rhizosphere pH to favour micronutrient acquisition from soils (Marschner, 1991b). The nitrogen form taken up by the plant determines the excretion of H^+, HCO_3^- or OH^- that in turn influences the pH of the rhizosphere.

A. Nitrogen Concentrations

Generally, the major nutrients in culture solutions are added in concentrations several times higher than those present in soil solutions (Gigon and Rorison, 1972; Ingestad, 1981). The amounts of nitrogen required for improving growth vary with age, pH, temperature, soil water availability, mineral composition of the soil and nitrogen source. It is thus important to evaluate the optimum concentration for each nitrogen source rather than compare growth rates in the presence of equal concentrations of NO_3^- and NH_4^+. Generally, the utilization of equal concentrations of NO_3^- and NH_4^+ has been the base of most of the experiments presented in both Tables 7.1 and 7.2. Only a few studies compared NO_3^--grown to NH_4^+-grown plants at lower NH_4^+ concentrations. The results obtained so far are very controversial and depend on plant species, NH_4^+ concentrations and growth conditions. Plant growth can be unchanged (Allen and Smith, 1986), inhibited (Allen and Smith,

Table 7.1. Effects of different NH_4^+ concentrations on plant growth in comparison with NO_3^- grown plants.

Plant Species	$[NO_3^-]$ (mM)	$[NH_4^+]$ (mM)	Effect of NH_4^+	$NH_4^+:NO_3^-$ [N] (mM)	Observations	References
Arachis hypogea L.	2	2	stimulates	1:1 [2]	SC	Silberbush and Lips, 1988
Arachis hypogea L.	6	6	stimulates	1:1 [6]	SC	Silberbush and Lips, 1988
Beta vulgaris L.	3.75	3.75	inhibits		H	Raab and Terry, 1994
Brassica napus L.	0.01	0.01	equals		H, pH control	MacDuff et al., 1987
Chrysanthemum multiflorum Ramat	0.1	0.05(a)	stimulates		H	Elliot and Nelson, 1983
Chrysanthemum multiflorum Ramat	0.1	0.3(a)	stimulates		H	Elliot and Nelson, 1983
Chrysanthemum multiflorum Ramat	0.1	0.05(a)	inhibits		H, greenhouse	Elliot and Nelson, 1983
Chrysanthemum multiflorum Ramat	0.1	0.3(a)	inhibits		H, greenhouse	Elliot and Nelson, 1983
Deschampsia flexuosa L.	4	4	stimulates		H, $4.2 \leq pH \leq 7.2$	Gigon and Rorison, 1972
Glycine max (L.) Merrill	1	1	equals	1:1* [1]	H, pH control	Rufty et al., 1983
Glycine max (L.) Merrill	3	3	slightly inhibits	1:4 [3]	H, pH control	Imsande, 1986
Glycine max (L.) Merrill	1	1	stimulates		H, $5.1 \leq pH \leq 6.1$	Tolley-Henry and Raper, 1986
Glycine max (L.) Merrill	1	1	inhibits		H, pH 4.1	Tolley-Henry and Raper, 1986
Glycine max (L.) Merrill	0.5	0.5	inhibits	1:1* [1]	H, pH control	Chaillou et al., 1994
Glycine max (L.) Merrill	1	1	equals		H, pH control	Rideout and Raper Jr., 1994
Glycine max (L.) Merrill	10	10	equals		H, pH control	Rideout and Raper Jr., 1994
Hordeum vulgare L.	2	2	equals	1:1* [2]	H, pH control	Lewis and Chadwick, 1983
Hordeum vulgare L.	3	3	inhibits		H, $4.0 \leq pH \leq 6.8$	Lang and Kaiser, 1994
Lolium multiflorum L.	10	10	stimulates		H	Griffith and Streeter, 1994

Species						
Lolium perenne L.	0.072		stimulates		H,	Reisenauer et al., 1982
Lolium perenne L.	1.96	0.036(a)	equals		H, pH control	Jarvis, 1987
Lolium perenne L.	3.57	1.96	inhibits		H, pH control	Jarvis, 1987
Lycopersicum esculentum L.	4	3.57	inhibits		H,	Wilcox et al., 1977
Lycopersicum esculentum L.	3	4	stimulates		H, pH control	Schenk and Wehrmann, 1979
Lycopersicum esculentum L.	3	5(a)	stimulates		H, pH control	Schenk and Wehrmann, 1979
Lycopersicum esculentum L.	10	10(a)	stimulates	1:1* [10]	H, pH control, RZT 8°C	Ganmore-Neuman and Kafkafi, 1980
Lycopersicum esculentum L.	10	10	inhibits	1:1* [10]	H, pH control, 16≤RZT≤34°C	Ganmore-Neuman and Kafkafi, 1980
Lycopersicum esculentum L.	8	10	inhibits		H	Magalhaes and Wilcox, 1982
Lycopersicum esculentum L.		8	inhibits	1:1* [1]	H	Maghrabi et al., 1985
Lycopersicum esculentum L.		1(b)	inhibits	1:1* [3]	H	Maghrabi et al., 1985
Lycopersicum esculentum L.		3(b)	inhibits	1:1* [6]	H	Maghrabi et al., 1985
Lycopersicum esculentum L.	1	6(b)	slightly stimulates		H	Peet et al., 1985
Lycopersicum esculentum L.	0.05	1	inhibits		H, pH control	Smart and Bloom, 1993
Lycopersicum esculentum L.	0.2	0.02	inhibits	1:2.5 [0.07]	H, pH control	Smart and Bloom, 1993
Lycopersicum esculentum L.	10	0.1	stimulates		SC	Ali et al., 1994
Phaseolus vulgaris L.	1	2(a)	inhibits		H	Chaillou et al., 1986
Phaseolus vulgaris L.	1	1	inhibits	1:2*	H	Volk et al., 1992
Raphanus sativus L.	1.7	1	inhibits		H, nitrapyrin	Goyal et al., 1982
Raphanus sativus L.	17	1.7	inhibits	10:1 [18]	H, nitrapyrin	Goyal et al., 1982
Ricinus communis L.	12	17	inhibits		V	Allen et al., 1985
Ricinus communis L.	12	8	stimulates		H	Allen and Smith, 1986
Ricinus communis L.	12	0.5	inhibits		H	Allen and Smith, 1986

(Contd.)

Table 7.1 Contd....

Plant Species	[NO$_3^-$] (mM)	[NH$_4^+$] (mM)	Effect of NH$_4^+$	NH$_4^+$:NO$_3^-$ [N] (mM)	Observations	References
Ricinus communis L.	12	8	inhibits		H	Allen and Smith, 1986
Rumex acetosa L.	4	4	inhibits		H, 4.2≤pH≤7.2	Gigon and Rorison, 1972
Scabiosa columbaria L.	4	4	inhibits		H, 4.2≤pH≤7.2	Gigon and Rorison, 1972
Sinapis alba L.	0.67	5	inhibits		H	Mehrer and Mohr, 1989
Sinapis alba L.	0.67	20	inhibits		H	Mehrer and Mohr, 1989
Sorghum bicolor (L.) Moench.	13.5	0.7[a]	stimulates		H	Bernardo et al., 1984
Sorghum bicolor (L.) Moench.	11.4	2.9[a]	stimulates		H	Bernardo et al., 1984
Triticum aestivum L.	0.05	0.05[a]	stimulates		H	Cox and Reisenauer, 1973
Triticum aestivum L.	2	2	inhibits	1:1* [2]	H, pH control	Lewis et al., 1987
Triticum aestivum L.	4	4	inhibits		H	Lewis et al., 1989
Triticum aestivum L.	4	4	inhibits		H	Leidi et al., 1991
Triticum aestivum L.	7	7	inhibits	1:2.5* [7]	H, greenhouse	Sandoval-Villa et al., 1992, 1995
Triticum aestivum L.	4	4	slightly inhibits	1:1* [4]	H	Botella et al., 1993
Zea mays L.	0.4	0.4	inhibits	1:1 [0.4]	SC	Bennett et al., 1964
Zea mays L.	4	4	inhibits	1:1 [4]	SC	Bennett et al., 1964
Zea mays L.	8	8	inhibits	1:1 [8]	SC	Bennett et al., 1964
Zea mays L.	5	5	inhibits	1.5:3.5* [5]	V, pH control	Handa et al., 1984
Zea mays L.	25	25	inhibits		V, pH control	Handa et al., 1984
Zea mays L.	50	50	inhibits		V, pH control	Handa et al., 1984
Zea mays L.	4	4	stimulates		H	Lewis et al., 1989
Zea mays L.	6.1	2.9[a]	stimulates		T	Alexander et al., 1991
Zea mays L.	1	1	stimulates		H, pH control	Anderson et al., 1991
Zea mays L.	0.2	0.2	equals		H, pH control	Alfoldi and Pinter, 1992

Zea mays L.		2	equals	1:1* [2]	H, pH control	Alfoldi and Pinter, 1992
Zea mays L.	20	20	inhibits		H, pH control	Alfoldi and Pinter, 1992
Zea mays L.	0.4	2.8[(a)]	equals		H, pH control	Barber et al., 1994
Zea mays L.	2.9	3.2[(a)]	equals		H, pH control	Barber et al., 1994
Zea mays L.	7.9	0.7[(a)]	slightly stimulates		H, pH control	Barber et al., 1994

SC – sand culture; H – hydroponic culture; V – vermiculite; T – turface; * – better growth; (a) – additions of NH_4^+ to NO_3^- solution media, (b) – compared to ammonium nitrate fed plants

Table 7.2. Effects of different NH_4^+ concentrations on woody plant growth in comparison with NO_3^- grown plants.

Plant Species	$[NO_3^-]$ (mM)	$[NH_4^+]$ (mM)	Effect of NH_4^+	$NH_4^+:NO_3^-$ [N] (mM)	Observations	References
Ceratonia siliqua L.	3	3	stimulates	1:1 [3]	H	Martins-Loução, 1985
Ceratonia siliqua L.	3	3	stimulates	1:1 [3]	H	Martins-Louçao and Duarte, 1987
Ceratonia siliqua L.	3	3	stimulates	1:1 [3]	H, 10°≤RZT≤38°C	Cruz et al., 1993 a, b
Ceratonia siliqua L.	3	3	inhibits	1:1 [3]	H, RZT≥38°C	Cruz et al., 1993 b
Fraxinus pennsylvanica Marsh.	0.18	0.18	stimulates		S, P, V	Truax et al., 1994
Fraxinus pennsylvanica Marsh.	18	18	stimulates	1:1 [18]	S, P, V	Truax et al., 1994
Picea abies Karst.	6	6	equals	1:1.5 [6]	H, pH control	Van der Burg, 1971
Picea sitchensis (Bong.) Carr.	6	6	stimulates	1:1.5 [6]	H, pH control	Van der Burg, 1971
Pinus contorta Dougl.	2	2	slightly stimulates	1:7 [2]	SC	Krajina et al., 1973
Pinus nigra Arnold	6	6	equals	1:1.5 [6]	H, pH control	Van der Burg, 1971
Pinus sylvestris L.	6	6	stimulates	1:1.5 [6]	H, pH control	Ingestad, 1979
Pinus sylvestris L.	15	7.5	stimulates	1:1 [15]	H, pH control	Flaig and Mohr, 1992
Pseudotsuga menziesii (Mirb.) Franco.	2	2	inhibits	1:7 [2]	SC	Krajina et al., 1973
Quercus rubra L.	0.18	0.18	equals		S, P, V	Truax et al., 1994
Quercus rubra L.	18	18	equals	1:1 [18]	S, P, V	Truax et al., 1994
Quercus suber L.	0.5	0.5	equals		H, pH control	Mata et al., 1995
Tsuga heterophylla (Raf.) Sarg.	2	2	stimulates	1:7 [2]	SC	Krajina et al., 1973
Tsuga plicata Donn. ex. D. Donn.	2	2	inhibits	1:7 [2]	SC	Krajina et al., 1973
Vaccinium vitis idaea L.	6	6	equals	1:1.5 [6]	H, pH control	Ingestad, 1973

SC – sand culture; H – hydroponic culture; S,P,V – sand + peat + vermiculite (1:1:1); P – perlite; * – better growth; RZT – root zone temperature

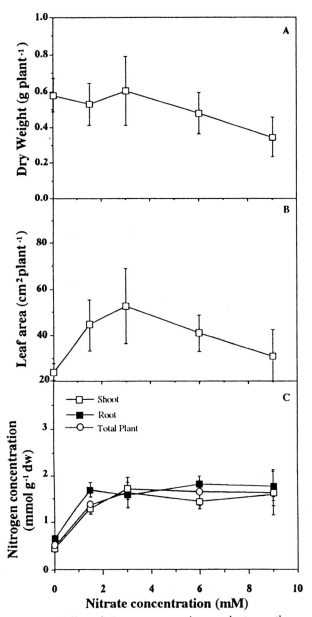

Fig. 7.1. Effect of nitrate concentration on plant growth.
Carob (*Ceratonia siliqua* L.) seedlings were grown hydroponically for 9 weeks in solutions with different nitrate concentrations (0, 1.5, 3, 6 and 9 mM). At the end of the growth period plants were harvested and dry weight, leaf area and nitrogen concentration of the plant material were determined. Values represent the means of three experiments with ten replicates each. Vertical bars represent the standard deviation.

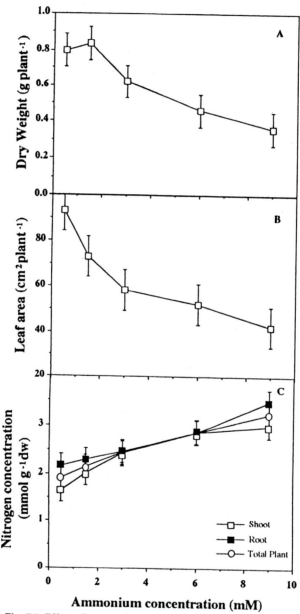

Fig. 7.2. Effect of ammonium concentration on plant growth.
Carob (*Ceratonia siliqua* L.) seedlings were grown hydroponically for 9 weeks in solutions with different ammonium concentrations (0.5, 1.5, 3, 6, and 9mM). At the end of the growth period plants were harvested and dry weight, leaf area and nitrogen concentration of the plant material were determined. Values represent the means of three experiments with ten replicates each. Vertical bars represent the standard deviation.

1986; Smart and Bloom, 1993) or stimulated (Cox and Reisenauer, 1973; Reisenauer, et al., 1982; Elliot and Nelson, 1983; Bernardo et al., 1984; Alexander et al., 1991; Ali et al., 1994; Barber et al., 1994) when ammonium is added in low concentrations to nitrate nutrient solutions. Other studies have shown growth stimulation when ammonium is added in high concentrations to NO_3^- nutrient solution (Schenk and Wehrmann, 1979; Elliot and Nelson, 1983; Barber et al., 1994, Cruz et al., 1993c).

The presence of ammonium, either alone or in combination with nitrate, appears to favour the growth of carob (*Ceratonia siliqua* L.) seedlings (Martins-Loução, 1985; Martins-Loução and Duarte, 1987; Cruz et al., 1993a). This species is a Mediterranean hardwood which belongs to the Leguminosae family but lacks symbiotic fixation of nitrogen (Martins-Loução, 1985; Martins-Loução, 1990). Initially experiments were performed to define the optimum NO_3^- concentration to attain an improved growth. Evidence based on growth parameters, leaf area and nitrogen content indicated 3 mM as the best NO_3^- concentration (Fig. 7.1). From thereon all the experiments were carried out using this concentration of nitrogen (Martins-Loução, 1985; Martins-Loução and Duarte, 1987; Cruz et al., 1993c). A comparison of carob growth with nitrate, ammonium or ammonium nitrate at a 3 mM nitrogen concentration showed that NH_4^+ stimulates carob growth. However, 3 mM of ammonium seems to be a high concentration for carob growth, since maximum growth has been obtained in the presence of 0.5 to 2 mM (Fig 7.2). It is interesting that carob plants present a similar response pattern to increased nitrate and ammonium nutrient solution concentrations, regardless of whether the comparison involves biomass, leaf area or nitrogen content measurements (Figs. 7.1 and 7.2). Ammonium toxicity symptoms have only been observed in the presence of high root temperatures or high pH (Cruz et al., 1993b, 1995).

Toxicity symptoms may be related to the adaptation by plants to different nitrogen levels or availabilities. For example, nitrogen levels which are considered moderate for other plants induce nitrate toxicity in *Boronia megastigma* Nees., a western Australian native woody plant. The phenomenon of nitrate toxicity in *Boronia* is consistent with the hypothesis proposed by Chapin (1980) that plants from nutrient-poor habitats have difficulty in growing rapidly under conditions of high nutrient availability and so accumulate nutrients to toxic levels under those conditions. The low level of nitrate reductase activity in these plants may be one of the factors responsible for this toxicity (Reddy and Menary, 1990).

Ingestad (1973, 1979) developed a technique of hydroponic culture of woody plants in which the nutrients needed for biomass production are added in a way that the relative growth rate is kept constant. This concept led to a nutrient flux model in which high minimum amounts of macronutrients are required (Ingestad, 1982; Agren, 1985; Ingestad and Agren, 1988).

B. Root Temperature

Temperature has a strong impact on the natural geographical distribution of plants. This is because temperature affects the rates of different biochemical processes in different ways and thus induces imbalances between partial processes and metabolic pathways. Root growth is often limited by low or high soil temperature; the optimum temperature varies with species and tends to be lower for root than for shoot growth (Brouwer, 1981). Adjustments to buffer root temperature effects are seen in most plant processes such as growth, development, photosynthesis and ion uptake.

The maximum rate of plant growth is regulated by environmental conditions, particularly temperature and nitrogen supply, in a manner characteristic of plant species (Lawlor et al., 1988). Some workers have shown that the effects of root zone temperature (RZT) are very dependent on nitrogen fertilization (Ganmore-Neumann and Kafkafi, 1980; MacLeod and Onmnod, 1985; Gosselin and Trudel, 1986; Cruz et al., 1993b). NO_3^- fed plants are more sensitive to low RZT whereas NH_4^+ fed plants are more sensitive to high RZT, which is noted when considering the increment of total dry weight, as the ratio between total plant dry weight at different temperatures and total plant dry weight at 10ºC RZT (Fig. 7.3).

The decrease in dry matter production observed in ammonium-fed plants grown at 40ºC root temperature (Fig. 7.3) can be related to the accumulation of ammonium in plant tissue and/or its subsequent transport to shoots (Magalhães and Huber, 1991). Low RZT can limit growth through an assortment of factors including reduced ion uptake, water stress, membrane permeability and changes in hormonal balance and root/shoot ratios (Osmond and Raper, 1981). On the other hand, high temperature can inhibit cell elongation (Pardales et al., 1992) and increase root maintenance respiration (Smakman and Hofstra, 1982). Temperature is thought to affect partitioning mainly through effects on the specific activities of shoot and root, which in turn affect the size of carbon and nitrogen pools (MacDuff et al., 1987; Cruz et al., 1993a, b). Comparative studies of nitrate and ammonium nutrition of plants suggest that nitrate uptake is more severely curtailed by low root temperatures than ammonium uptake (Lycklama 1963; Ganmore-Neumann and Kafkafi 1980; Clarkson et al., 1986; Cruz et al., 1993a), but the increment in growth observed in nitrate-fed plants with a slight increase of RZT is greater than in ammonium-fed plants (Fig. 7.3). Also, the growth

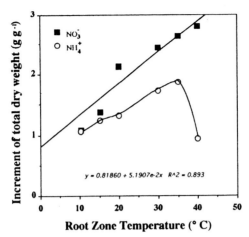

Fig. 7.3. Responses of nitrate- and ammonium-fed plants to root zone temperature. Carob (*Ceratonia siliqua* L.) seedlings were grown hydroponically for 9 weeks in solutions containing 3 mM nitrogen as either nitrate or ammonium. During growth, plants were subjected to the same atmospheric temperature (20–24°C) but to several root-zone temperatures (10, 15, 22, 30, 35 and 40°C). At the end of the growth period plants were harvested and dry weight was determined. Each symbol in the graph represents the ratio between the dry weight of plants grown at x°C and plants grown at 10°C with the same nitrogen sources. Values represent the means of three experiments with ten replicates each.

increase in nitrate-fed plants at enriched CO_2 levels was much more remarkable than that of ammonium-fed ones, though this response is entirely dependent on RZT (Fig. 7.4). There is a decrease in growth increment with an increase in RZT in the presence of nitrate in the nutrient solution. On the contrary, the presence of ammonium seems to buffer the growth response to two abiotic factors: RZT and enriched CO_2 (Fig. 7.4).

High RZT can by itself have detrimental effects on root growth, which are probably related to an insufficient supply of carbohydrate to the root meristems, since high temperatures increase respiration rates (Cumbus and Nye, 1982). This can be of importance in ammonium-fed plants, because ammonium cannot be accumulated and carbohydrates are needed to metabolize it. Ammonium accumulation in NH_4^+-fed plants growing at high RZT may therefore be the reason for the development of ammonium toxicity (Cruz et al., 1993b, 1995).

C. pH

In acid mineral soils, where nitrification is slowed down (Tamm, 1991), plant growth can be limited by a variety of factors (Marschner, 1991b). Plant communities adapted to these soil conditions can sustain

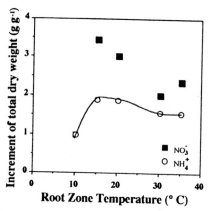

Fig. 7.4. Responses of nitrate- and ammonium-fed plants to CO_2 concentrations at several root-zone temperatures.

Carob (*Ceratonia siliqua* L.) seedlings were grown hydroponically for 9 weeks in solutions containing 3 mM nitrogen either as nitrate or as ammonium. During growth, plants were subjected to the same atmospheric temperature (20–24°C), but to several root zone temperatures (10, 15, 22, and 30°C) and to two CO_2 concentrations (360 and 800 ppm). At the end of the growth period plants were harvested and dry weight was determined. Each symbol in the graph represents the ratio between the dry weight of plants grown at 800 and 360 ppm CO_2 at the same root temperature and nitrogen source. Values represent the means of three experiments with ten replicates each.

productivity despite low nitrogen availabilities and a high potential for nitrogen loss from the ecosystem volatilization of NH_3, by volatilization. Although nutrient studies have tended to focus on mineral forms of available nitrogen released from litter, dissolved organic nitrogen has been found to be the dominant vehicle of nitrogen transport (Qualls et al., 1991). Although dissolved organic nitrogen, contained in protein-tannin complexes, is more important during mineralization, NH_4^+ is almost entirely released rather than NO_3^- (Northup et al., 1995). The impact of plant polyphenols on nitrogen immobilization is interpreted as a plant adaptation to nitrogen limitation as well as a possible maximization of litter nitrogen recovery. However, the role of these polyphenols in acidic and infertile ecosystems has not been entirely explored. Since there is a significant correlation between the release rate of mineral forms of nitrogen and litter polyphenol concentrations (Northup et al., 1995), the availability of ammonium can vary along a gradient of different plant characteristics. Under these conditions the combination of ammonium nutrition and low pH can affect root growth (Marschner et al., 1991c), probably due to ammonium accumulation or Mg^{2+} deficiency rather than impaired carbohydrate or ATP supply (Lang and Kaiser, 1994).

In general, plants alter the pH of their root surface to a value different from that in the bulk medium. This pH modification can be achieved by additional organic acid synthesis in the plant with excretion of H⁺ or organic acid, which increase acidification and reduce or reverse the alkalinization of the medium (Raven, 1990). Independent of the circulating ionic currents present in all the growing and differentiating roots, this pH modulation can be induced during nitrogen assimilation and is very important for nutrient acquisition from soil (Raven, 1987, 1991; Raven et al., 1990; Martins-Loução et al., 1993). Theoretically, the assimilation of ammonium and nitrate involves the generation of an excess of about 1.22 H⁺ and 0.78 OH⁻ per atom of nitrogen assimilated (Raven, 1988). Acid-base regulation within the cells is achieved by the net release of excess H⁺ or OH⁻, or by net production of intracellular organic acids (Kirkby and Mengel, 1967; Ben-Zioni et al., 1971; Kirkby 1974; Raven and Smith, 1976; Kirkby, 1981).

III. INTERACTIONS BETWEEN C AND N METABOLISM

The metabolic pathways of nitrogen and carbon are linked in several ways. Roots from higher plants are completely dependent upon translocation from the shoots of carbohydrates needed for the production of energy and the provision of carbon skeletons. On the other hand, photosynthesis is also completely dependent on the uptake of nutrients provided by the root system. The integration of these two important metabolic processes, nutrient uptake and photosynthesis, must involve extensive co-regulation between the root and the shoot (Bloom et al., 1993). If the activity of either assimilation system is disrupted, adjustment occur in the other.

A. Carbon Assimilation

Gas exchange measurements reflect the effect of nitrogen on carbon flow, since net leaf CO_2 fixation rates are dependent on the amount and source of nitrogen available to the roots. However, carbon-nitrogen relationships are intrinsically complex, since photosynthesis represents the integrated operation of a series of processes which are sensitive to environmental factors as well as leaf physiology and structure (Field and Mooney, 1986). Nutrient uptake and metabolism are similarly dependent on environmental factors, metabolic processes and root morphology and physiology.

The nitrogen investment in many of the compounds involved in photosynthetic metabolism is not precisely known, but the proportion of the total leaf nitrogen allocated to photosynthetic reactions is undoubtedly large. For this reason a number of studies have related light-saturated rates of photosynthesis to the nitrogen content of leaves (Field and

Mooney, 1986; Evans, 1989; Makino and Osmond, 1991; Pons et al., 1994). In all these studies, correlations between photosynthetic capacity and leaf nitrogen concentration were calculated in order to compare variations within single species under different nitrogen supplies or even to characterize slow- and fast-growing species (Makino and Osmond, 1991; Pons et al., 1994).

To provide an overview of the influence of nitrogen source influence on the photosynthesis-nitrogen relationship, we have summarized several studies reporting carob net photosynthesis rate and leaf nitrogen content on the basis of leaf dry weight (Fig. 7.5). The slopes of the relation between nitrogen concentration in the leaf and net photosynthetic rates express the photosynthetic nitrogen use efficiency as a function of leaf nitrogen concentration for each nitrogen source (nitrate, ammonium or different ammonium additions to nitrate solutions). The general comparison provided here shows that NO_3^--grown plants have a high nitrogen use efficiency. Nevertheless, a marked effect on photosynthesis is only observed when leaf nitrogen concentration is above 0.5 mmol g^{-1} dw.

The increment in photosynthesis is also dependent on growth conditions, namely light, temperature and CO_2 concentrations (Stitt, 1991; Cruz et al., 1993a, b; 1995). The net photosynthetic rate was independent of leaf nitrogen content for carob seedlings grown hydroponically at various nitrate to ammonium ratios, as well as for fertilized mature

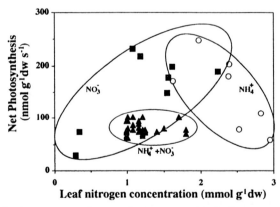

Fig. 7.5. Relation between net photosynthetic rates and leaf nitrogen concentrations. Net photosynthetic rates were measured using a portable infrared gas analyzer, during the period of maximal photosynthetic activity (10:00–14:00h). Rates were determined among the youngest completely developed leaflets. Measurements were taken with mature trees growing in the field, or with carob (*Ceratonia siliqua* L.) seedlings grown hydroponically with nitrate or ammonium as nitrogen source. After measurements the leaflets were collected, dried and their nitrogen concentration was determined. Each point represents the mean of ten measurements.

trees grown in the field. In NH_4^+-grown plants net photosynthetic rate correlated negatively with leaf nitrogen content (Leidi and Lips, 1990; Lips et al., 1990; Botella et al., 1993; Cruz et al., 1993a, b). This implies that a small increment in leaf nitrogen concentration causes a decrease in photosynthesis and consequently a low nitrogen use efficiency.

Several factors can be responsible for changes in nitrogen use efficiency (Losada et al., 1981; Nijs et al., 1995) including differences in the nature of nitrogen compounds present in the leaf and changes in the hormonal balance. Nitrate- and ammonium-fed plants differ in their chemical composition, particularly in the concentration of nitrate and ammonium present in their tissues (Chaillou et al., 1986; Griffith and Streeter 1994). In barley, levels of ammonium in leaves increase during the day, reaching values of 3.5 µmol g^{-1} fresh weight (Lang and Kaiser, 1994), and shoots of carob may achieve concentrations of 7 µmol g^{-1} dry weight when plants are grown with 9mM of ammonium (Fig. 7.6). Ammonium is an uncoupler of photophosphorylation (Izawa and Good, 1972; Zornoza and Carpena, 1992) and interferes with the partition of metabolites between cytoplasm and vacuoles. Thus, lower nitrogen use efficiencies can occur if ammonium ions are transported and accumulated in the leaves. Nevertheless, in carob plants neither leaf ammonium nor nitrate concentrations have a marked effect on net photosynthetic rate (Fig. 7.7).

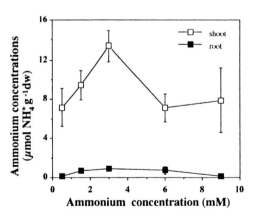

Fig. 7.6. Ammonium in plant tissue as a function of ammonium concentration in the root medium.
Carob (*Ceratonia siliqua* L.) seedlings were grown hydroponically for 9 weeks in solutions with different ammonium concentrations (0.5, 1.5, 3, 6, and 9 mM). At the end of the growth period shoots and roots were harvested separately and their concentrations in ammonium were determined. Values represent the means of three experiments with ten replicates each. Vertical bars represent the standard deviation.

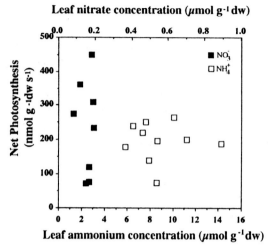

Fig. 7.7. Relation between net photosynthetic rates and leaf nitrate and ammonium concentrations.

Net photosynthetic rates were measured using a portable infra-red gas analyzer during the period of maximal photosynthetic activity (10:00–14:00h). Rates were determined among the youngest completely developed leaflets. Measurements were taken with carob (*Ceratonia siliqua* L.) seedlings grown hydroponically with nitrate or ammonium as nitrogen source. After measurements the leaflets were collected and nitrate and ammonium concentrations were determined. Each point represents the mean of ten measurements.

Using the technique of compartmental analyses with ^{13}N ammonium, concentrations as high as 33–38 mM were found in the cytoplasm when ammonium concentrations of the root medium ranged from 1 to 1.5 mM (Kronzucker et al., 1995c; Wang et al., 1993a). These investigations were performed with rice (*Oryza sativa*) and spruce (*Picea glauca*), species known for their preference for ammonium in relation to nitrate (Wang et al., 1993a, b; Kronzucker et al., 1995b, c), similar to carob. The amount and/or type of organic nitrogen, determined by the nitrogen source available in the root medium, appears to be an important factor in photosynthesis regulation. However, it has been observed on several occasions that ammonium-fed plants are associated with the predominant presence of ABA in the root, while nitrate-fed plants seem to be richer in cytokinins (Wang et al., 1994; Lips et al., 1995). It is likely that changes in the phytohormone level, and in cytokinin in particular, are involved in the regulation of the balance between carbon and nitrogen, since it is known that exogenous cytokinin applications increase the translocation of nitrogen to the shoot (Dale, 1979; Simpson et al., 1982; Marschner, 1986; Wang and Below, 1996).

Generally nitrate-fed plants present higher rates of net photosynthesis per leaf area than ammonium-fed plants (Lewis et al., 1989; Leidi and Lips, 1990; Cruz et al., 1993c). These differences can be related to differences imposed on the metabolism by the two ions. If low net photosynthetic rates can explain the poor growth exhibited by some ammonium-fed plants (Raab and Terry, 1994), they cannot explain the stimulation of growth presented by some species under ammonium nutrition (Tables 7.1 and 7.2). This may be related to higher photosynthetic rates of ammonium-fed plants on a plant basis, due to a larger leaf area per plant (Cruz et al., 1993b).

The synthetic phase of photosynthesis generates ATP and NADPH which are used, in part, for carbon reduction through the Calvin cycle. CO_2 is fixed in the leaf by the enzyme ribulose 1,5-biphosphate carboxylase/oxygenase, known as Rubisco. The decrease in Rubisco activity in the presence of ammonium may explain the lower net photosynthesis rates. In algae, this is the prime cause of photosynthesis limitation (Huppe and Turpin, 1994). In the case of carob, there is no information about Rubisco activity, but photosynthesis is more limited by CO_2 concentrations in NO_3^--than in NH_4^+-grown plants (Cruz et al., 1993a).

The increase in Rubisco concentration may not be related to an increase in activity, since some enzyme binding sites may not be active. For example, in leaves of sugar-beet plants fed with ammonium, the soluble leaf protein levels were 330% higher but Rubisco activity was only 38% higher than in nitrate-fed plants (Raab and Terry, 1994). This explains the above-mentioned lower net photosynthesis rate observed in ammonium-fed sugar-beet plants compared to nitrate-fed ones. Activation of the enzyme involves the formation of a complex with CO_2 and the subsequent addition of a divalent ion (Mg^{2+}) to form the activated ternary complex (Miziorko and Lorimer, 1983). The equilibrium of this reaction is sensitive to pH and low pH in the stroma may lead to its deactivation, which can be counteracted by an influx of Mg^{2+} and Ca^{2+} (MacDonald and Buchanan, 1992). With respect to these cations, NH_4^+-grown carob plants present no statistical differences when compared to those of nitrate ones (Cruz et al., 1993d) so the referred decrease of those cations associated with ammonium nutrition (Kirkby and Mengel, 1967; Lang and Kaiser, 1994; Raab and Terry, 1995) cannot be used to explain the impairment of photosynthesis due to a possible Rubisco deactivation in carob leaves.

NH_4^+-grown plants may contain higher amounts of enzyme per unit leaf area, since their leaves have a higher nitrogen content compared with those of nitrate-grown plants (Cruz et al., 1993c; Martins-Loução 1985). Nevertheless, an increased ratio of Rubisco activity to electron transport activity, ATPase, or chlorophyll content may not affect the balance between the *in-vivo* activities of Rubisco and electron transport, as has been observed in spinach leaves (Evans and Terashima, 1988). This can be explained by the presence of CO_2-transfer resistance between intercellular air spaces and carboxylation sites. This resistance is known to be dependent on carbonic anhydrase activity, mesophyll surface area per unit leaf area, solubility resistance of CO_2 to liquid phase, and leaf air space ratio (Makino et al., 1992). It has been observed in wheat leaves that an increase in leaf nitrogen was responsible for a decrease in the CO_2-transfer resistance associated with a strong stimulation of carbonic anhydrase activity (Makino et al., 1992). These distinctive responses of carbonic anhydrase to different nitrogen levels suggest its importance and physiological role in photosynthesis regulation processes. It would be useful to evaluate its role in CO_2-transfer resistance of plants grown under different nitrogen sources.

Reduction of CO_2 is not the only process of assimilation by which plants are able to synthesize large amounts of organic compounds. The conversion of visible light energy into redox energy and acid-base energy is also used for the reduction of sulphate to sulphide and of nitrate to ammonium in leaves (Losada et al., 1990). Studies carried out with green and blue-green algae (Romero and Lara, 1987) have shown that while cells are fixing CO_2 nitrate stimulates the evolution of O_2, supporting the argument that nitrate assimilation is a direct photosynthetic process that utilizes electrons derived from water photolysis (Losada et al., 1981; Lara and Guerroro, 1989). Nitrate also stimulates O_2 evolution in plant leaves and its assimilation increases the capacity of the photosynthetic apparatus for non-cyclic electron flow, overcoming the limitation imposed by the rate of CO_2 fixation (Torre et al., 1991). So, light plays an important role in NO_3^--assimilation, especially when nitrate reduction takes place in the shoots, since this process uses photosynthetically generated assimilatory power. Under these conditions a major source of cytosolic NADH for the reduction of NO_3^- is a malate/oxaloacetate shuttle between the choloplast and cytosol, especially when nitrate reduction takes place in the shoots. A malate/oxaloacetate shuttle from mitochondria is an alternative source of NADH for reductions localized in the cytosol (Oaks, 1994).

B. C Skeletons and Carbon Flow

Both reducing power and ATP are required for the synthesis of other amino acids within the chloroplast. Thus, carbon supplied by the Calvin cycle is required for the net synthesis of carbohydrates, the carbon skeletons needed for nitrogen assimilation. This feature has been considered a proof of the energy dependence of nitrogen assimilation upon carbohydrate oxidation.

Whether recent or stored photosynthate provides the carbon for amino acid synthesis depends on the capacity of photosynthesis and the availability of stored carbohydrates. The carbon required to support the formation of amino acids from inorganic nitrogen is partially supplied by either a breakdown of starch or recent photosynthate, depending on the rate of nitrogen assimilation and the carbon status of the cells. It appears that the availability of energy and of carbohydrates always matches the overall rate of nitrogen assimilation.

When plants are supplied with nitrogen there is an increase in the synthesis of amino acids, inducing complex strategies to regulate the partitioning of photosynthetic carbon between sucrose and amino acid synthesis. This involves activation of phosphoenolpyruvate carboxylase (PEPcase) and inhibition of sucrose phosphate synthetase (SPS) (Fig 7.8) (Champigny and Foyer, 1992; Foyer et al., 1995). It has been shown that PEPcase, SPS and nitrate reductase respond rapidly to high NO_3^- feeding and that their synchronous modulation serves to regulate, in the short term, the C demand associated with nitrogen assimilation (Champigny and Foyer, 1992). Under physiological conditions, nitrate assimilation in green leaves is restricted at subambient CO_2 by altering the activation state of NR, CO_2 fixation would control nitrate reduction, thus balancing the assimilation of carbon and nitrogen (de Cires et al., 1993).

Photosynthetic carbon fixation produces triose phosphates (Triose P) which are used in the chloroplast or exported to the cytosol via the phosphate translocator in exchange for Pi (Fig 7.8). Triose P export from the chloroplast would increase plastidic Pi inhibiting starch synthesis, thereby diverting carbon to respiratory metabolism. In the cytosol, Triose P is used to synthesize sucrose through the activity of SPS. Sucrose synthesis is co-ordinated by the rate of photosynthesis and is also adjusted to meet the demands of the sink organs. Two forms of SPS with markedly different properties, involving modifications of protein phosphorylation, have been described (Walker and Huber, 1989; Siegl and Stitt, 1990). The phosphorylated form of SPS is characterized by an increased sensitivity to Pi and appears to be related with inactivation of the enzyme (Fig. 7.8). Studies developed by the Champigny group suggest that the activation of SPS and NR differ significantly in terms of regulation. First, inorganic anions (Pi, sulphate and tungstate) inhibit SPS activation both *in vivo* and *in vitro*, whereas in contrast, these anions do not inhibit NR activation. Thus, while the two substrate proteins are dephosphorylated by protein phosphatases of the same class and in the same subcellular compartment, it is clear that differential regulation is somehow employed (Huber et al., 1994).

Co-regulation of C and N metabolism by the stimulation of protein phosphorylation leads to the inhibition of SPS activity and activation of PEPcase (Champigny and Foyer, 1992). PEPcase is a highly regulated enzyme which plays a key role in regulating the flow of carbon through

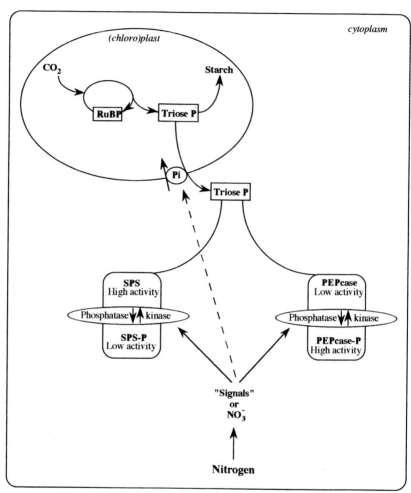

Fig. 7.8. Schematic representation of a possible regulation of non-structural carbohydrate production by nitrogen.

Nitrogen assimilation in the root may influence the activity of cytoplasmatic enzymes such as SPS and PEPcase. Their activities are modulated by phosphorylation/dephosphorylation mechanisms and both enzymes use triose-P as substrate. Pi is responsible for the partition of triose-P between the (chloro)plast and the cytoplasm. If the concentration of Pi in the cytoplasm is low, triose-P will stay in the (chloro)plast, leading to the formation of starch. If the concentration of Pi is higher, triose-P will be exported into the cytoplasm, where it can be used either by SPS to synthesize sucrose, or by PEPcase to give oxaloacetic acid and other products. It is possible that nitrate and ammonium influence these enzymes in different ways as well as the concentration of Pi, leading to different concentrations of carbohydrates in nitrate- and ammonium-fed plants. After Champigny et al., 1992. See the text for more details.

the anapleurotic pathway (Huppe and Turpin, 1994). In all types of plants (C_3, C_4, and CAM), the PEPcase protein is subjected to reversible phosphorylation, and the sensitivity of the enzyme to malate depends on the phosphorylation status of the protein. In C_3 plants, the main function of PEPcase (coupled with malate dehydrogenase) is to replenish the organic acid biosynthesis pathway with malate production when demand of α-ketoglutarate for amino acid biosynthesis is high.

The function of this enzyme has been demonstrated in the N-limited green alga *Scenedesmus minutum*. NH_4^+ addition to these algae causes a 20-40-fold increase in the rate of PEP carboxylation in both light and dark conditions in support of α-ketoglutarate export for amino acid synthesis (Elrifi and Turpin, 1986; Guy et al., 1989). It was also observed that in the presence of mixtures of ammonium and nitrate the activity of this enzyme in the roots increases to meet the demand for carbon skeletons (Sandoval-Villa et al., 1992).

More than a simple function of enabling the cell to cope with pH fluctuations in the cytoplasm, PEPcase is now appointed as the major enzyme involved in replacing the intermediates of the tricarboxylic acid cycle removed for biosynthesis during nitrogen assimilation (Huppe and Turpin, 1994). The gene expression of the nitrogen-dependent PEPcase may be regulated by downstream products of ammonium assimilation. Glutamine is a good candidate for this function since it can work as a signal reflecting nitrogen availability in plant cells for gene expression of PEPcase, a nitrogen consumer, and of NR, a producer of nitrogen availability. However, it is also possible that other products of N assimilation may function as important regulators not only of carbon flow, but also of the gene expression through mechanisms affecting transcription and/or mRNA stability (Sugiharto and Sugiyama, 1992).

Most of the metabolic fluxes induced by NO_3^- are similar but slower than with NH_4^+ triggering similar effectors or inhibitors of PEPcase (Huppe and Turpin, 1994). How these nutrients originate effectors or inhibitors depends on the type of tissue (shoot or root), plant species and its physiological characteristics (Huppe and Turpin, 1994), namely the type of organic nitrogen compounds (Chaillou et al., 1994). In wheat, the activity of PEPcase is modulated by light and nitrate (Van Quy et al., 1991; Champigny and Foyer, 1992). In fact, pH is an important parameter in the regulation of PEPcase activity, the enzyme affinity for PEP increasing when pH is higher than 7 (Iglesias and Andreo, 1984). Nitrate reduction and ammonium assimilation, as well as nitrate accumulation in the vacuole, may change cellular ionic relationships, with simultaneous variations in the cytoplasmatic pH. Although ammonium assimilation acidifies the pH of the cytoplasm, the activity of PEPcase may be increased under ammonium nutrition. Glutamine synthetase/

Glutamate synthase (GS/GOGAT) is the main pathway through which ammonium is incorporated in amino acids, and higher rates of PEPcase are needed to accomplish the supply of carbon skeletons for amino acid synthesis (Huppe and Turpin, 1994). This can explain the decline of soluble sugars observed under ammonium nutrition (Chaillou et al., 1991; Chaillou et al., 1994). It should not be surprising that nitrate- and ammonium-fed plants differ in their content and allocation of soluble sugars (Romero and Lara, 1987; Leidi and Lips, 1990; Botella et al., 1993). Further, besides nitrogen sources, environmental factors also, namely RZT and enriched CO_2, may affect the partitioning of soluble sugars (Cruz et al., 1993a).

Differences in root morphology and root architecture between nitrate- and ammonium-fed plants seem to be a consequence of the concentration of soluble carbohydrates in the plant (Table 7.3), which are known to affect the metabolism of existing meristems, or the initiation of new lateral roots (Williams and Farrar, 1992). The partitioning of photosynthates between starch and soluble carbohydrates is very complex and regulated by the activity of several enzymes (Fig. 7.8) as well as by sink demand.

The demand of carbohydrates can vary, depending on the nitrogen source, because ammonium assimilation always occurs in the root while nitrate assimilation can be divided between roots and shoots. Hence, there is a demand for formed carbohydrates via photosynthetic CO_2 fixation from the site of nitrogen assimilation that can differ between nitrate and ammonium-fed plants (Magalhães and Huber, 1991; Chaillou et al., 1994). The partitioning of photosynthates in the plant can be accessed by the distribution of ^{14}C through the different plant parts. It has been observed in carob plants that 40% of the total CO_2 fixed by nitrate-fed plants was sent to the roots while ammonium-fed plants sent 56% of their newly fixed carbon (Fig. 7.9). This finding is in agreement with the large participation of the root system in nitrate reduction (Cruz et al., 1993a). Reduction in the roots is accompanied by a great translocation of organic nitrogen to the shoots, similar to what happens in

---

Fig. 7.9. Effect of nitrogen source on the partition of newly fixed carbon. Carob (*Ceratonia siliqua* L.) seedlings were grown hydroponically for 9 weeks in solutions with ammonium (0.5, 3 and 6 mM) and nitrate (3 mM) at several root zone temperatures (RZT). At the end of the growth period, 5 plants per treatment were chosen. Each received a pulse of ^{14}C in a young but completely developed leaf (Cruz et al. 1993). Twenty-four hours after the pulse plants were divided into roots and shoots which were analyzed for their content in ^{14}C. Pie segments represent the partition of the ^{14}C that was fixed by the plants of each treatment (100%) between roots and shoots. High ammonium concentrations in the root medium and high RZT seem to increase the percentage of newly fixed carbon that is sent to the roots.

Nitrate-fed plants - 3 mM

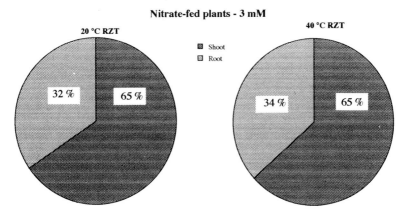

Ammonium-fed plants - 3 mM

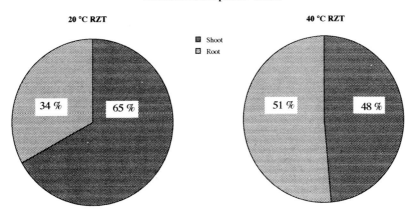

Ammonium-fed plants - 20 °C RZT

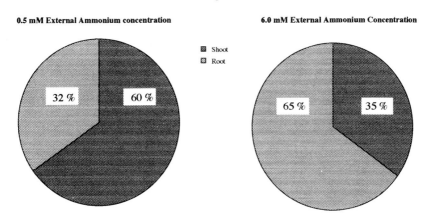

ammonium-fed plants (Cruz et al., 1993d). However, ammonium-fed plants are much more sensitive to stress conditions, namely salt stress (Lewis et al., 1989; Leidi and Lips, 1990; Botella et al., 1993), root zone temperature (Cruz et al., 1993b) or increase in external ammonium concentrations (Fig. 7.8) than nitrate-fed ones. Under these conditions plants can change their carbohydrate partition and allocate more in the roots (Fig. 7.9) in order to cope with the increased NH_4^+ assimilation and, consequently, with further problems of NH_4^+ toxicity.

In nitrate-fed plants the larger CO_2 assimilation and the smaller carbohydrate allocation in the form of starch (Table 7.3) do not contribute to higher growth rates (Cruz et al., 1993a, b). In roots, unlike shoots, photoassimilation cannot occur and so assimilation of NO_3^- may become energetically unfavourable. In barley fed with nitrate, the energy cost of nitrate uptake in the root was much higher than that of those fed with ammonium (Bloom et al., 1992). Under ammonium nutrition 14% of root carbon catabolism was coupled to NH_4^+ uptake and assimilation; under NO_3^- nutrition 23% of root carbon catabolism was associated to NO_3^- uptake and assimilation: 5% coupled to NO_3^- uptake, 15% to NO_3^- assimilation and 3% to NH_4^+ assimilation.

Nitrogen sources also have a remarkable effect on the demand for various carboxylation pathways needed for acid-base homeostasis, altering the extent of non-Rubisco carboxylase contribution to net carbon acquisition (Raven and Farquhar, 1990). Generally, NH_4^+-grown plants make less organic acid per mol carbon assimilated or per unit biomass increase than do plants grown on NO_3^- (Mengel and Kirkby, 1987; Raven

Table 7.3. Concentrations (mg g^{-1} DW) of sucrose and reducing sugars (R.S.) in plant roots and shoots, and of starch in plant shoots. Plants were grown for 9 weeks at different atmospheric CO_2 concentrations (360 and 800 µl l^{-1}). The % of increase in the concentrations of carbohydrates was calculated as [(concentration at 800 µl l^{-1}) * 100/(concentration at 360 µl l^{-1})] − 100. Two nitrogen sources were used, nitrate and ammonium, at a final concentration of 3 mM. Values represent the mean of three experiments with 10 replicates each. Values of the same carbohydrate followed by the same letter are not significantly different at P<0.05 by Duncan's multiple range test

N source	Carbohydrate	Roots			Shoots		
		360	800	Increase (%)	360	800	Increase (%)
NO_3^-	Starch				102^c	140^b	37
	Sucrose	10^c	105^a	950	16^c	108^a	570
	R.S.	22^c	43^a	90	29^b	48^a	65
NH_4^+	Starch				147^b	243^a	65
	Sucrose	62^b	63^b	2	64^b	68^b	6
	R.S.	23^c	30^b	30	24^c	33^b	37

and Farquhar, 1990). However, in most woody plants NO_3^- is assimilated in roots rather than in shoots (Stewart et al., 1988, 1990; Cruz et al., 1993d). Root NO_3^- assimilation requires less organic acid synthesis, since acid-base regulation can be carried out directly through the excretion of either HCO_3^- or OH^- to the root medium (Raven, 1988). In carob, 80% of NO_3^- was reduced in the roots (Cruz et al., 1991) and the site of nitrogen assimilation was thus the same in both nitrate- and ammonium-fed plants. The balance between inorganic cations and anions was therefore not significantly different (Cruz et al., 1993d). It seems that carob seedlings have a cheap way of solving their pH regulation problem by secreting directly to the medium the excess OH^- and H^+ generated during the assimilation of NO_3^- and NH_4^+.

C. Nitrogen Uptake

In assimilation of inorganic nitrogen, uptake from the root medium into root cells is necessarily the first step. For uptake to occur, communication must be established at the interface between the intra- and extracellular space of root membranes. Membrane proteins can function either as signal transducers or as transporters, that is, as permeases or channels able to transfer molecules across cell membranes. The process of nitrogen uptake is self-regulated, and also extensively regulated by the subsequent steps of nitrate and nitrite reduction, and by the assimilation of ammonium to amino acids.

Nitrate uptake from the rhizosphere to the cytoplasm of root cells is thermodynamically disfavoured in terms of both the electric potential gradient (negative inside the cell) and the chemical potential gradient. Nitrate uptake seems to be mediated by specific transport system, which are common to unicellular microalgae or micro-organisms without storage vacuoles, to vacuolated algae or small hypopleurophyte plants, and to roots (attached or detached) of higher terrestrial plants (Ullrich, 1992). The uptake is electrogenic and has been shown to be conducted as an H^+/NO_3^- symport system with a flux stoichiometry of $2H^+/1\ NO_3^-$ (Glass et al., 1992; Mistriki and Ullrich, 1995), producing transient plasmalemma depolarization upon addition of nitrate. This depolarization is counteracted by the plasmalemma H^+-ATPase (Ullrich, 1992) and modulated by induction of the nitrate uptake system and nitrogen status of root cells (Lu and Briskin, 1993; Ruiz-Cristin and Briskin, 1991; Ullrich, 1992). These findings may thus suggest an adaptation of root membranes to exposures of different nitrate concentrations, providing the expression of depolarization responses (Ullrich, 1992).

Most nitrate transport studies to date have used almost exclusively depletion methods and experiments with the stable isotopic tracer ^{15}N.

Recently, however, the short-lived radiotracer ^{13}N was used to provide estimates for nitrate net flux and flux partitioning within the plant as well as cytoplasmatic and apparent free space nitrate concentrations (Kronzucker et al., 1995a, b, c). These experiments enable discussion of the real meaning of the compartment known as apparent free space, in which the aforesaid authors found a high amount of nitrate relative to the external NO_3^- concentrations. The recognition of this space, using depletion methods, is generally done through observation of the increase in net uptake, not affected by root temperature, respiratory inhibitors, anaerobiosis or inhibitors of protein synthesis (Minotti et al., 1969; Goyal and Huffaker, 1986; Cruz et al., 1993e, 1995). After this net uptake the rate decreases. The rapid exchange of nitrogen initially taken up with KCl together with the insensitivity to the inhibitors favours the identification of this space as the apparent free space. The time course of the adsorption varies from minutes to hours, depending on the species studied. Herbaceous crop species require only minutes for the filling of the apparent free space (Lee, 1982) while woody species may take hours or even days (Cruz et al., 1993e, 1995; Kronzucker et al., 1995a). Given the surplus of negative charges present in the apparent free space, which should lead to an exclusion of anions from the free space rather than an enrichment, such increase is surprising and difficult to explain. A possible explanation is that the presumed apparent free space has non-proteinaceous cell wall components (mucilagin or lignin/suberin) which are responsible for the absorption of NO_3^- ions (Cruz et al., 1993e, 1995; Kronzucker et al., 1995a).

Two special nitrate uptake systems can be distinguished: (a) HATS, the high-affinity transport system and (b) LATS, the low-affinity transport system.

Following a lag phase of minutes or hours (depending on species) after first exposure to exogenous nitrate, nitrate uptake rates gradually increase from low to high rates until a plateau is reached. This pattern of uptake is consistent with the existence of two transport systems: a constitutive one, always present in the plasmalemma (cHATS), and an inducible one (iHATS), only present when nitrate is available (Fig. 7.10).

The cHATS, which serves the function of 'NO_3^--sensing' (Behl et al., 1988); is associated with a plasmamembrane-bound nitrate reductase (PM-NR), found in the roots of barley, maize and *Chlorella* cells (Tischner et al., 1993; Tischner et al., 1995). This protein is anchored in cell membranes; its function is the stimulation of nitrate uptake. The products of nitrate reduction, either nitrite, ammonium or glutamine, may serve as promoters for the synthesis of mRNA coding for NO_3^--transport proteins (iHATS) and cytosolic nitrate reductase enzyme (Tischner et al., 1993). In starved carob plants, the cHATS seems to be responsible for

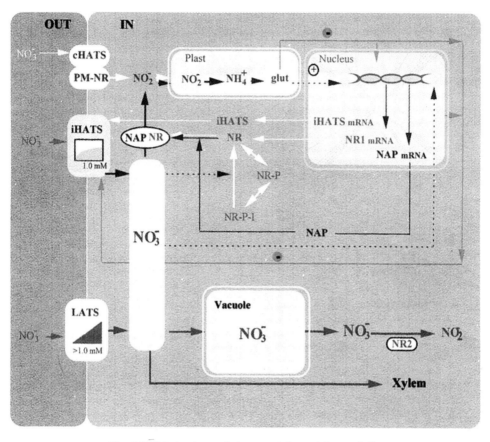

Fig. 7.10. Mechanisms of nitrate uptake and its regulation.
Nitrate can enter the cell through a constitutive high-affinity transport system (cHATS), an inductive high-affinity transport (iHATS) system or a low-affinity transport system (LATS). The uptake system that functions in each circumstance is dependent on the concentration of nitrate in the root medium. cHATS is active at very low nitrate concentrations and works as a sensing system that is associated with a plasmalemma nitrate reductase. The products of this nitrate reduction work as signals for the synthesis of the iHATS. This system is responsible for the uptake of nitrate if the external concentration is below 1 mM. The nitrate taken up will be part of the cytoplasmic nitrate metabolic pool. It can be reduced by the cytoplasmatic nitrate reductase, translocated to the xylem or stored in the vacuole. The LATS becomes functional if nitrate concentrations in the root medium are higher, than 1mM. When this system is working the concentration of nitrate in the medium is high, and consequently the nitrogen content of the tissues is also high, so most of the nitrate is translocated to the xylem or stored in the vacuole. Cytoplasmatic nitrate reductase is very strictly controlled. Its synthesis is induced by the presence of nitrate, but once in the cytoplasm, the activity of the enzyme is modulated by a mechanism of phosphorylation/dephosphorylation as well as by other inhibitors (I) or activators (NAP). For more details see the text.

net transport rates of 0.5 µmol g^{-1} h^{-1} (Cruz et al., 1993). This rate is similar to those observed in crop species (Lee and Drew, 1986; King et al., 1993; Jiao and Lips, 1995), but high in comparison to those of spruce roots (Kronzucker et al., 1995a). These responses may be due to genetic adaptation of the species to particular abiotic conditions, mainly nitrate availability in soil solution, or to differences in growth conditions and experimental procedures.

The iHATS, described by King et al. (1993), is a saturable system activated at low NO_3^- concentrations (<1 mM). Investigations of nitrate transport mutants and respective proteins suggest that this transport mechanism is performed by a carrier protein (MacKown and McCllure, 1988; Tischner et al., 1993). Recently a homologue gene for the high-affinity nitrate transporter was isolated from *Aspergillus nidulans* in a temperate legume. DNA hybridization studies indicated that this gene is both NO_3^--inducible (Trueman et al., 1995).

Activation of cytosolic NR has been studied in depth over the last 20 years (Oaks, 1994). In higher plants, there are a number of examples of NR activation by ferricyanide (de Cires et al., 1993), light (Lillo, 1994), and higher levels of CO_2 (Cruz, 1994), but it seems that redox potential ($NADH/NAD^+$) could be an important regulator of the proportion of NR found in active form (Oaks, 1994). According to Kaiser et al. (1992), there is a light/dark modulation of NR, *via* NR phosphorylation/dephosphorylation reaction, dependent on Mg^{2+} and ATP (Fig. 7.10) and requiring regulatory proteins (Spill and Kaiser, 1994). These proteins are a protein kinase (PK) which phosphorylates NR but without inactivating it, an inhibitor protein (IP) which binds to phospho-NR and inactivates it, and a protein phosphatase (PP) which reactivates NR (Glaab and Kaiser, 1995). Recently, a low molecular weight polypeptide was found in the Lips laboratory, which they called NAP (NR activating peptide) capable of stabilizing the activity of NR or restoring some types of lost activity (Alikulov et al., 1995). They also found that NAP molecules are identical in both higher plants and fungi.

Nitrogen acquisition by plants appears to be regulated by a negative feedback control from cellular nitrogen pools which serve as indicators of plant nitrogen status (Cruz et al., 1993c). Studies carried out in carob and ryegrass suggest that both influx and efflux are strongly influenced by nitrogen status and that cellular nitrate and ammonium may themselves participate in the regulation of nitrogen acquisition (Cruz et al., 1993c). Since the overexpression of NR had little effect on NO_3^- and free amino acid levels in roots, it was hypothesized that phloem transport of amino acids from shoots to roots is responsible for the regulation of NO_3^- uptake in response to the demand at the whole plant level.

The low-affinity system (LATS) (Fig. 7.10) is likely to be a passive transport system representing uniport by a carrier protein or, more likely, by anion channels (Glass and Siddiqui, 1995; Glass, 1988; McClure et al., 1986). It starts to take up nitrate only after the medium concentration of nitrate is above 1.0 mM and seems to be responsible for the transport of nitrate from the apoplast into the symplast of root parenchyma cells. No saturation value is known even when external concentration reaches 40 mM nitrate. This means that once nitrate is present in the root medium at high concentrations the rate of uptake is proportional to its concentration. At rhizosphere concentrations, ranging between 1 and 25 mM, more than one uptake system may be operating and nitrate may accumulate.

Another mechanism proposed for the uptake of nitrate is the K-shuttle, an exchange NO_3^--bicarbonate (Lips et al., 1970; Ben-Zioni et al., 1971; Lips SH et al., 1987). This exchange allows uptake of nitrate from the rhizosphere, directly into the stele of the root through the endodermis. The affinity of this system may be between that of HATS and of LATS. It will be functional at high concentrations of nitrate in plants in which nitrate reduction occurs mainly in the shoots. Since it depends on the rate of nitrate reduction in the shoot and the consequent synthesis of malate, it is never responsible for the accumulation of nitrate in the shoots.

In plants, the most detailed studies on ammonium uptake kinetics and transient electrical changes have been undertaken in rice, *Lemna*, and in woody species (Fried et al., 1965; Ullrich et al., 1984; Cruz et al., 1993e; Wang et al., 1993a, b, 1994; Kronzucker et al., 1995b). Using different methodologies, these studies showed that ammonium uptake presents a biphasic kinetics (Wang et al., 1993a, b; Cruz et al., 1993e) corresponding to two different uptake systems (Fig. 7.11): a high (HATS) and a low-affinity transport system (LATS).

The high-affinity component of the transport system (HATS) is apparently constitutive (Ullrich, 1992), saturable in response to external ammonium concentrations < 1 mM (Cruz et al., 1993; Wang et al., 1993b) and induces rapid depolarization of membrane potential of rice epidermal and cortical cells (Wang et al., 1994). Membrane depolarization reflects that ammonium and not ammonia is the form taken up by HATS, through a proton: NH_4^+ symport (Wang et al., 1993b, 1994). Moreover, the results of inhibitor studies (Cruz et al., 1993e; Wang et al., 1994) and of $^{13}NH_4^+$ influxes (Wang et al., 1994; Kronzucker et al., 1995b) provided evidence for a dependence (either direct or indirect) on a proton motive force. Compartmental analysis of ^{13}N allowed determination of the concentrations of ammonium in the cytoplasm and vacuole and consequently the calculation of the concentrations in the root

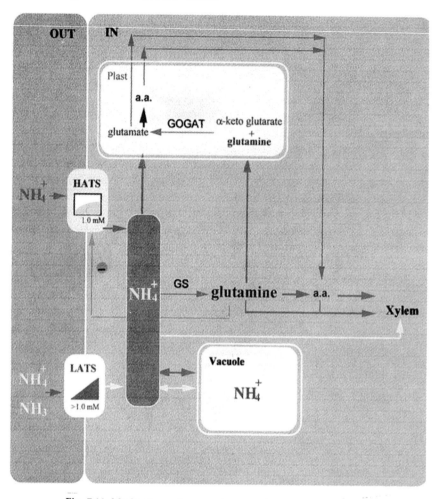

Fig. 7.11. Mechanisms of ammonium uptake and its regulation.
Ammonium can enter the cell through a high affinity transport system (HATS) or a low-affinity transport system (LATS). The uptake system that functions in each circumstance is dependent on the concentration of ammonium in the root medium. HATS is functional at low ammonium concentrations (< 1mM). The ammonium taken up will constitute a matabolic cytoplasmic pool and be metabolized mainly by the GS/GOGAT pathway, leading to the production of amino acids. When ammonium is present at high concentrations in the root medium, the uptake will occur through the LATS. Once in the cytoplasm, ammonium can be metabolized, stored in the vacuole or tranlocated to the xylem.

medium below which the uptake of ammonium is an active process (Kronzucker et al., 1995b). Depending on species, growth conditions and on nitrogen nutritional status of the plant the upper limits of external ammonium concentration for active transport may vary (Cruz et al., 1993e; Wang et al., 1994).

The low-affinity transport system (LATS) is stimulated by high external ammonium concentrations (Fig. 7.11), presents a linear response with increasing external ammonium concentration, is not associated with membrane depolarization (Ullrich et al., 1984; Wang et al., 1994) and is insensitive to metabolic inhibitors (Cruz et al., 1993e; Wang et al., 1993b, 1994). These results support the hypothesis for either a diffusion of uncharged NH_3 across the membrane (Ullrich, 1992; Cruz et al., 1993e) or passive entry of NH_4^+ via an electrogenic uniport (Kleiner, 1981; Ullrich et al., 1984). NH_3 diffuses passively across the membrane and therefore no specific transporters are required (Frommer et al., 1994). However, genetic studies carried out in bacteria (Kleiner, 1981) have shown that mutations can strongly affect ammonium uptake. According to these studies the genes can either encode uptake systems or essential constituents functioning as retrieval systems of ammonia, that is, passively leaking out of the cells. But they favour the hypothesis of NH_4^+/H^+ symport.

This uniport may occur through a specific channel for NH_4^+ or a shared cation channel, namely via the K^+ channel (Ullrich, 1992; Wang et al., 1994). Recently, ammonium transporter genes have been characterized and isolated from yeasts and plasma membranes of plants (Frommer et al., 1994). The transporter mediates high affinity ammonium uptake and is a typical integral membrane protein. Whether it constitutes a typical channel or a transporter remains to be shown. It is important to emphasize that the existence of mutants affected in ammonium transport and the identification of high-affinity transport genes, does not engender a knowledge of putative regulatory functions of those genes. Further studies are needed to enable a better understanding of the role and function of such transporters.

The fact that plants supplied with elevated concentrations of ammonium during growth present low uptake rates indicates that NH_4^+ influx seems to be down-regulated rather than enhanced by the presence of external ammonium concentrations (Wang et al., 1993b; Kronzucker et al., 1995b). At present the mechanism(s) and signals responsible for this down-regulation of uptake is unclear. A negative feedback upon NH_4^+ uptake by cytoplasmic NH_4^+ pool (Glass et al., 1995) or, more likely, by one or several down stream assimilation products produced during nitrogen assimilation (Cruz et al., 1993c; Kronzucker et al., 1995) is also suggested (Fig. 7.11).

The amount of NH_4^+ detected in the xylem of spruce roots was not affected by the down-regulation of NH_4^+ transport (Kronzucker et al., 1995), increasing after prolonged exposure to external ammonium concentration. This trend indicates a possible regulation of ammonium xylem loading, similar to what happens with nitrate (Frommer et al., 1994). In carob, similar quantities of nitrate and ammonium were observed in the xylem, independent of the highest ammonium uptake registered (Cruz et al., 1993d). Since in carob nitrate reduction occurs mainly in roots (Cruz et al., 1993d), it seems that the low nitrate concentrations observed in the xylem are due to a restriction in loading of nitrate into the xylem.

Concerning the intermediary role of ammonium in nitrate assimilation, ammonium appears as a likely candidate for feedback regulation on nitrate uptake and/or reduction. Different reports describe this negative feedback due to a direct effect on nitrate uptake (Lee and Drew, 1989), a low downward carbohydrate flow (Chaillou et al., 1986) or the accumulation of specific negative effectors (Jackson and Volk, 1995). The effect of NO_3^- on net NH_4^+ uptake is less referred to in the literature, but a few reports mention some decrease of nitrate uptake in the presence of NH_4^+ (Warncke and Barber, 1973; Macklon et al., 1990; Raper et al., 1991; Cruz et al., 1993c). Other works, using soybean with split-root system, have shown that the inhibition of NH_4^+ uptake in the presence of exogenous NO_3^- appears to be confined to processes within root tissues, involving a stimulation of efflux rather than an inhibition of influx (Saravitz et al., 1994). This NH_4^+ efflux may be the result of direct cytoplasmic NH_4^+ efflux or endogenous NH_4^+ regeneration from nitrogenous compounds. Work on carob also points out for this possibility, since the amount of NH_4^+ efflux largely exceeds the endogenous NH_4^+ concentrations (Cruz et al., 1993f). It is evident from these studies that regardless of the mechanism of inhibition of NH_4^+ uptake in the presence of NO_3^- or vice versa, the capacity for roots to utilize either NO_3^- or NH_4^+ is dependent on the nitrogen status of the plant (Cruz et al., 1993f), emphasizing the important role of nitrogen assimilation compounds on the down-regulation of both NH_4^+ and NO_3^- uptake, as discussed above.

All these results, some of which are contradictory, reflect the genetic differences which have developed among plants due to years of adaptation to habitats where the rate of nitrification may vary, either due to differences in microbiota, or due to a direct effect of plant root exudation, known to interfere with the nitrogen-mineralization-immobilization turnover in the soil. This supports the hypothesis that plants must have a homeostasis mechanism that controls nitrogen concentration in the plant. Such mechanisms might be related to specific signals, namely carbohydrates, hormones, organic compounds, established between the shoot and the root, to control nitrogen net uptake; either nitrogen influx or efflux. Possibly, the combining structure and physiology of vessels (Vanbel, 1995) may also play a major role in the establishment of these dynamic interactions between root and shoot, contributing to different ecological adaptations.

IV. ECOLOGICAL STRATEGIES

Nitrogen concentrations in soils vary from very low, in most natural ecosystems, to very high concentrations, in agrosystems. Ammonium and nitrate are the main nitrogen sources in the former and latter systems respectively. The existence of strict reciprocal control between the pathways of carbon and nitrogen metabolism relies on the balance between supply and demand in a constantly changing environment. Most of the effects of nitrogen on plant growth are due to changes in carbon partitioning between shoots and roots. In nutrient-rich habitats, where nitrogen availability is high, the ratio between root and shoot dry weights tends to be higher (Fig. 7.12). Growth of the shoot relative to the root implies an increase in the autotrophic parts of the plant in relation to those heterotrophic and consequently results in faster growth (Belanger et al., 1994). However, in nutrient-poor habitats, the ratio between root and shoot dry weights tends to increase. For the reasons explained above such plants tend to grow slowly (Fig. 7.12).

In these habitats the availability of nitrogen is generally less than 1 mM. Nitrogen uptake rates in the roots are regulated by a source-sink effect and depend on growth, reproductive demands, needs for maintenance, storage and defence. When nitrogen is present in limiting concentrations, the mechanisms of feedback regulation allow the development of maximal rates of uptake (Cruz et al., 1993e). These maximal rates can be achieved by increasing the number of transporters per unit of root surface and/or increasing the affinity of the uptake system for the substrate (Ćruz et al., 1993b, e). Hence, plants take up nitrogen, either ammonium or nitrate, through the high-affinity transport systems (HATS) (Fig. 7.12). However, plants can develop efficient down-regulation of uptake with increasing tissue nutrient status (Cruz et al., 1993b, f) avoiding costs with extra nutrient uptake and storage.

In unfertilized soils under a permanent cover of plants (grasslands and forests) the soil generally does not contain nitrate, nitrifying bacteria are low and plants do not contain nitrate reductase (Moore and Waid, 1971; Rice, 1982; Rosswall, 1982; Riha et al., 1986). Since the nitrification process is governed by the C/N ratio in the soil (Woldendorp, 1981; Woldendorp and Laanbroek, 1989), nitrification only takes place when ammonium is present in non-limiting amounts (Verhangen et al., 1995). For these reasons, species indigenous to later successional stages require special adaptations to use and assimilate mostly ammonium (Fig. 7.12). In an evolutionary point of view these plants might have developed special morphological, as well as physiological adaptations and developed much more efficient high-affinity

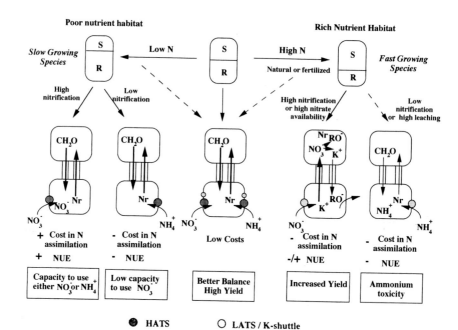

● HATS   ○ LATS / K-shuttle

Fig. 7.12. Plant adaptations to nitrogen availability.

Nitrogen availability (concentration and form) in the soil determines biomass partition between roots and shoots. In poor nutrient habitats nitrogen in the soil may exist in the form of nitrate or ammonium, depending on nitrification rate. Plants tend to grow slowly, presenting high root/shoot ratios. Under low nitrogen concentrations plants develop high-affinity transport systems to either nitrate or ammonium uptake. Presumed energetic costs for nitrogen assimilation are synthesized together with the effect in nitrogen use efficiency (NUE). These assumptions are only taken for the effect of nitrogen source available, and not for other type of limitations that may also induce changes, particularly in NUE. Under high N availability, especially in agrosystems, plants tend to grow quickly presenting low root/shoot ratios. Also, in these systems the amount of nitrate and ammonium may vary depending on the type of added fertilizer and on the agroclimate. In these systems plants have large amounts of nitrogen especially in the form of nitrate. Presumably, high-affinity transport systems are shut off and low-affinity transport systems or/and K-shuttle mechanism prevail. Most of the species in these habitats are shoots nitrate reducers and because of that they have low costs of nitrogen assimilation. Depending on the species and on growth conditions, they can either present high or low nitrogen use efficiencies. In the presence of ammonium these plants cannot cope with ammonium toxicity symptoms, independent of the low costs of assimilation. The presence of both N sources, either in low or high N availability conditions, will promote growth and a balanced root/shoot ratio and carbohydrate partitioning can be achieved, with low cost of assimilation. The different growth rates are a consequence of the mechanisms involved in the nitrogen uptake, the place of nitrogen assimilation and carbon partitioning. For more details see the text S — shoot; R — root; CH_2O — carbohydrates; Nr — organic N; RO — organic acids; NUE — nitrogen use efficiency.

transport systems and, in extreme situations, a low capacity to use nitrate (Reddy and Menary, 1990; Kronzucker et al., 1995a, b, c).

Most of the plants at these stages of succession which can use nitrate are root reducers (Stewart et al., 1990), increasing the energetic costs for nitrogen assimilation. Because nitrogen assimilation is an energy-consuming process it can affect plant carbon balance through an increased carbon allocation for nitrogen acquisition to compensate for intensified nitrogen demand. Although uptake of both ammonium and nitrate may be similar, around 1 ATP, the costs of NO_3^- reduction to NH_4^+ greatly exceed that cost (Bloom et al., 1992). Theoretical and experimental carbon costs of heterotrophic nitrogen assimilation estimate that 2 to 3 mole CO_2 are evolved in the reduction of either NO_3^- or N_2 to NH_4^+. Another 0.02 to 1.1 moles of CO_2 may be used per mole NH_4^+ assimilated in amino acid formation (Amthor, 1994). Without photoassimilation the energetic costs are substantially greater for nitrate than for ammonium (Bloom et al., 1992). Generally, nitrate-fed plants present a high nitrogen use efficiency, either in terms of photosynthetic nitrogen use efficiency (Fig. 7.5) or in terms of the amount of increase biomass per unit of nitrogen assimilated (Cruz, 1994). Comparatively, ammonium-fed plants present a low cost of nitrogen assimilation but less efficiency in nitrogen use (Fig. 7.5; Cruz, 1994).

Changes in root architecture are observed in response to environmental conditions and constitute a strategy to enhance plant growth. These adaptations are very important during the first stages of development in which plants are much more vulnerable to different biotic and abiotic stresses. For instance, plants living in circumstances wherein ammonium is the main nitrogen source may modify their root architecture in order to achieve a better regulation of ammonium uptake, which cannot accumulate in the cells without slowing plant growth (Salsac et al., 1987; Magelhães and Huber, 1991). In carob plants the ammonium uptake mechanism and rate change along the root length is a controlled process in the younger parts of the root, and a passive one in the more mature parts (Cruz et al., 1995). Hence, a root system with a high number of short roots seems to allow a better regulation of ammonium uptake (Martins-Loução, 1985; Cruz et al., 1993e). Nevertheless root elongation can be advantageous for plants in several circumstances: in areas with abundant precipitation, where nitrate can be leached to deeper layers of the soil or in Mediterranean areas, where water is also a limiting factor to plant development (Correia, 1988; Rhizopoulous and Davies, 1991).

Trees and perennial woody shrubs may also develop strategies of either nutrient conservation (Millard, 1993) or nutrient immobilization in the soil in order to avoid nitrogen losses. Nutrient translocation is an important trait since relative allocation patterns are constrained by feedback linkages between storage and growth (Heilmeier et al., 1994). This suggests a tight coupling between growth and filling of storage tissue over a broad range of nitrogen availability. Besides, an increasing translocation efficiency during leaf senescence, allowing nitrogen availability at no cost to growth, is also an adaptive strategy in these systems (Correia, 1996).

Also, variations in phloem architecture and physiology provide different strategies for the distribution of organic matter, which may constitute an important strategy with potential consequences for growth, and carbon and nitrogen partitioning. The possibility to switch from symplasmic to apoplasmic phloem loading is dependent on source-sink effects between shoot and root, which may be associated with climate and/or growth strategy (Vanbel, 1995). These types of adaptations would allow different investments, either in growth or in maintenance and defence, with great adaptive value.

Another type of compensatory mechanism whereby plants can meet higher nutrient demand under nutrient-limited habitats may involve increased symbiotic associations with N_2 fixing and mycorrhizal organisms. But, even when a greater colonization rate is reported, the exact consequences for plant nutrient budget and the physiological interpretation of such effects is often difficult and nuclear (George et al., 1995).

At the other extreme, crop plants, selected on the basis of high-yield productions, are adapted to much larger amounts of available nitrogen through large inputs of fertilization. They are fast-growing species and contrary to what happens in slow-growing species they invest more in photosynthetic biomass and less in roots (Lambers and Poorter, 1992). The nutrient uptake rates per unit root dry weight are higher than those of slow-growing species, but this may well be an effect of rapid growth, rather than its cause (Lambers and Poorter, 1992). Because of high nutrient availability the high-affinity transport systems will be shut off and nitrogen taken up through the low-affinity transport systems (LATS) and K-shuttle mechanism.

Nitrate is commonly considered the preferred nitrogen source for higher plants (Lewis et al., 1989; Pilbeam and Kirkby, 1992). This may reflect the ability of crop plants to photoassimilate NO_3^-, reducing the energetic costs of NO_3^- assimilation, and to divert the excess of energy and reducing power from photosynthesis to NO_3^- reduction. These plants may also have a reduced cost of nitrogen assimilation due to a better use of photosynthetic energy for nitrate reduction (Fig. 7.12). Nevertheless they may present a wide variation in nitrogen use efficiency especially when expressed in terms of photosynthetic nitrogen use efficiency (Pons et al., 1994). Trees and shrubs generally have lower efficiencies while herbaceous species have higher ones. For an ecological and more realistic approach it must be emphasized that all these measurements should be integrated over the whole leaf lifespan. Since nitrogen allocation differs between slow- and fast-growing species (native and crops) the longer leaf life of slow-growing species may give a higher whole-life nitrogen use efficiency when calculated over its whole life (Field, 1988). In terms of biomass increment per unit of nitrogen assimilated fast-growing plants, and particularly crops have a low nitrogen use efficiency. They only present high yield in the presence of high nitrogen concentration. Thus, in this sense, they must have a low nitrogen use efficiency.

However, these plants may develop a greater capacity to adjust to external nitrogen concentration changes, by being able to switch on the high-affinity systems. Since in agrosystems the availability of nitrate is much higher, plants may not cope with equivalent amounts of ammonium. Under high ammonium concentrations ammonium will be taken up through LATS (Fig. 7.11). The amount of carbohydrates will be limited for ammonium assimilation and plants will suffer from both rhizosphere acidification and ammonium toxicity (Fig. 7.12) when the amounts of carbohydrates are limiting. These plants, despite having less costs for nitrogen assimilation, cannot cope with the high root demand on carbohydrates. They reduce their growth and present a low nitrogen use efficiency.

Nevertheless, the presence of ammonium together with nitrate increases nitrogen metabolism (particularly protein synthesis) in roots, thereby increasing the partitioning of photosynthate to the root and whole plant growth (Cruz et al., 1993c). The presence of these two nitrogen sources aids plants in increasing nitrogen assimilation rates, although these responses depend on several interacting environmental factors, namely light, root temperature, differential uptake of nitrate and ammonium and developmental proportions of root and shoot (Lewis et al., 1982; Cruz et al., 1993b, c).

The physiological and ecological costs and benefits of the various inherent adaptations to adverse environments warrant further research since under these conditions plants can develop symbiotic associations from which they can take profit and true benefits. All this information is of great importance for our understanding of whole plant functioning, within the balance of carbon and nitrogen. Sink strength and allocation patterns of carbon and nitrogen and not carbon assimilates *per se* probably determine yield (Martins-Loução, 1985; Cruz et al., 1993b; Cruz, 1994; Correia, 1996). Application of this knowledge to agricultural and environmental issues will help us to better management and ecological approaches both in natural and in exploited crop environments.

LITERATURE CITED

Agren, G.I. 1985. Theory for growth of plants derived from the nitrogen productivity concept. *Physiol. Plant.* 64: 17–28.
Alexander, K.G., Miller, M.H. and Beauchamp, E.G. 1991. The effect of an NH_4^+-enhanced nitrogen source on the growth and yield of hydroponically grown maize (*Zea mays* L.). *J. Plant Nutr.* 14: 31–44.
Alfoldi, Z. and Pinter, L. 1992. Accumulation and partitioning of biomass and soluble carbohydrates in maize seedlings as affected by source of nitrogen, nitrogen concentration and cultivar. *J. Plant Nutr.* 15: 2567–2583.
Ali, I.A., Kafkafi, U. and Inanaga, S. 1994. Response of sand grown tomato supplied with varying ratios of nitrate and ammonium to constant and variable temperatures. *J. Plant Nutr.* 17: 2001–2024.
Alikulov, Z., Savidov, N.A. and Lips, S.H. 1995. A low molecular weight regulator of nitrate reductase in higher plants. In: *Fourth International Symposium on Inorganic Nitrogen Assimilaton*, Seeheim, Southwest Africa, 4 pp.
Allen, S. and Smith, J.A.C. 1986. Ammonium nutrition in *Ricinus comunis:* Its effect on plant growth and the chemical composition of the whole plant, xylem and phloem saps. *J. Exp. Bot.* 37: 1599–1610.
Allen, S., Raven, J.A. and Thomas, G.G. 1985. Ontogenic changes in chemical composition of *Ricinus communis* grown with NO_3^- or NH_4^+ as N source. *J. Exp. Bot.* 36: 413–425.
Amthor, J.S. 1994. Respiration and carbon assimilate use. In: *Physiology and Determination of Crop Yield*, Ameri. Soc. America, Crop Sci. Soci. of America, Soil Sci. Soc. of America, Madison, WI, pp. 221–250.
Anderson, D.S., Teyker, R.H. and Rayburn, A.L. 1991. Nitrogen form effects on early corn root morphological and anatomical development. *J. Plant Nutr.* 14: 1255–1266.
Barber, K.L., Pierzynshi, G.M. and Vanderlip, R.L. 1994. Ammonium/nitrate ratio effects on dry matter partitioning and radiation use efficiency of corn. *J. Plant Nutr.* 17: 869–882.
Behl, R., Tischner, R. and Raschke, K. 1988. Induction of high-capacity nitrate-uptake mechanism in barley roots prompted by nitrate uptake through a constitutive low-capacity mechanism. *Planta* 176: 235–240.
Belanger, G., Gastal, F. and Warembourg, F.R. 1994. Carbon balance of tall fescue (*Festuca arunadinacea* Schreb): Effects of nitrogen fertilization and the growing season. Annals Bot. 74: 653–659.
Ben-Zioni, A., Vaadia, Y. and Lips, S.H. 1971. Nitrate uptake by roots as regulated by nitrate reduction products of the shoot. *Physiol. Plant.* 24: 288–290.
Bennett, W.F., Pesek, J. and Hanway, J.J. 1964. Effect of nitrate and ammonium on growth of corn in nutrient solution sand culture. *Agr. J.*: 342–345.
Bernardo, L.M., Clark, R.B. and Maranville, J.W. 1984. Nitrate/ammonium ratio effects on nutrient solution pH, dry matter yield, and nitrogen uptake of sorghum. *J. Plant Nutr.* 7: 1389–1400.
Bloom, A.J., Sukrapanna, S.S. and Warner, R.L. 1992. Root respiration associated with ammonium and nitrate absorption and assimilation by barley. *Plant Physiol.* 99: 1294–1301.
Bloom, A.J., Jackson, L.E. and Smart, D.R. 1993. Root growth as a function of ammonium and nitrate in the root zone. *Plant Cell and Envir.* 16: 199–206.
Botella, M.C., Cerdá, A.C. and Lips, S.H. 1993. Dry matter production, yield and allocation of ^{14}C assimilates by wheat as affected by nitrogen sources and salinity. *Agronomy J.* 85: 1044–1049.
Botella, M.C., Cruz, C., Martins-Loução, M.A. and Cerdá, A. 1993. Nitrate reductase activity in wheat seedlings as affected by NO_3^-/NH_4^+ ratio and salinity. *J. Plant Physiol.* 142: 531–536.

Brouwer, R. 1981. Coordination of the growth phenomena within a root system of intact maize plants. *Plant and Soil* 63: 65–72.
Chaillou, S., Morot-Gaudry, J.F., Salsac, L., Lesaint, C. and Jolivet, E. 1986. Compared effects of NO_3^- and NH_4^+ on growth and metabolism of french bean. *Physiologie Veg.* 24: 679–687.
Chaillou, S., Rideout, J.W., Raper, D.R. and Morot-Gaudry, J.F. (1994). Responses of soybean to ammonium and nitrate supplied in combination to the whole root system or separately in a split-root system. *Physiol. Plant.* 90: 259–268.
Chaillou, S., Vessey, J.K., Morot-Gaudry, J.F., Raper, C.D., Henry, L.T. and Boutin, J.P. 1991. Expression of characteristics of ammonium nutrition as affected by pH of the root medium. *J. Exp. Bot.* 42: 189–196.
Champigny, M.L. and Foyer, C. 1992. Nitrate activation of cytosolic protein kinases diverts photosynthetic carbon from sucrose to amino acid biosynthesis. *Plant Physiol.* 100: 7–12.
Chapin, F.S.I. 1980. The mineral nutrition of wild plants. *Ann. Rev. Ecol. Syst.* 11: 233–257.
Clarkson, D.T., Hopper, M.J. and Jones, L.H.P. 1986. The effect of root temperature on the uptake of nitrogen and the relative size of the root system in *Lolium perenne*. I. Solutions containing both NH_4^+ and NO_3^-. *Plant Cell Environ.* 9: 535–545.
Correia, O. 1988. *Contribuicão da fenologia e ecofisiologia em estudos da sucessão e dinâmica da vegetação mediterrânica.* Lisboa.
Correia, P.J. 1996. *Efeito da rega e da fertilização com azoto na produtividade da alfarrobeira.* Lisboa.
Cox, W.J. and Reisenauer, H.M. 1973. Growth and ion uptake by wheat supplied nitrogen as nitrate, or ammonium or both. *Plant and Soil* 38: 363–380.
Cruz, C. 1994. *Aspectos do metabolismo do azoto em plântulas de alfarrobeira (Ceratonia siliqua L.).* Lisboa.
Cruz, C., Lips, S.H. and Martins-Loução, M.A. 1993a. The effect of nitrogen source on photosynthesis of carob at high CO_2 concentrations. *Physiol. Plant.* 89: 552–556.
Cruz, C., Lips, S.H. and Martins-Loução, M.A. 1993b. The effect of root temperature on carob growth. Nitrate versus ammonium nutrition. *J. Plant Nutr.* 16: 1517–1530.
Cruz, C., Lips, S.H. and Martins-Loução, M.A. 1993c. Growth and nutrition of carob plants as affected by nitrogen sources. *J. Plant Nutr.* 16: 1–15.
Cruz, C., Lips, S.H. and Martins-Loução, M.A. 1993d. Nitrogen assimilation and transport in carob plants. *Physiol. Plant.* 89: 524–531.
Cruz, C., Lips, S.H. and Martins-Loução, M.A. 1993e. Uptake of ammonium and nitrate by carob (*Ceratonia siliqua*) as affected by root temperature and inhibitors. *Physiol. Plant.* 89: 532–543.
Cruz, C., Lips, S.H. and Martins-Loução, M.A. 1993f. Interactions between nitrate and ammonium during uptake by carob seedlings and the effect of the form of earlier nitrogen nutrition. *Physiol. Plant.* 89: 544–551.
Cruz, C., Lips, S.H. and Martins-Loução, M.A. 1995. Uptake regions of inorganic nitrogen in roots of carob seedlings. *Physiol. Plant.* 95: 167–175.
Cumbus, I.P. and Nye, P.H. 1982. Root zone temperature effects on growth and nitrate absorption in rape (*Brassica napus* cv. Emerald). *J. Exp. Bot.* 33: 1138–1146.
Dale, J.E. 1979. Nitrogen supply and utilization in relation to development of the cereal seedlings. In: *Nitrogen Assimilation of Plants* (E.J. Hewitt and C.V. Cutting, eds.). Academic Press, London, pp. 153–163.
de Cires, A., de la Torre, A. and Lara, C. 1993. Involvement of CO_2 fixation products in the light-dark modulation of nitrate reductase in barley leaves. *Physiol. Plant.* 89: 577–581.
Ellenberg, H. 1977. Stickstoff als Stand ortsfactor, insbesondere fur mitteleuropaische Pflanzengesellschaften. *Oecologia Plant.* 12: 1–22.
Elliot, G.C. and Nelson, P.V. 1983. Relationship among nitrogen accumulation, nitrogen assimilation and plant growth in *Chrysanthemum*. *Physiol. Plant.* 57: 250–259.

Elrifi, I.R. and Turpin, D.H. 1986. Nitrate and ammonium induced photosynthetic suppression in N-limited *Selenastrum minutum*. *Plant Physiol.* 83: 97–104.

Evans, J.R. 1989. Photosynthesis and nitrogen relationships in leaves of C_3 plants. *Oecologia* 78: 9–19.

Evans, J.R. and Terashima, I. 1988. Photosynthetic characteristics of spinach leaves grown with different nitrogen treatments. *Plant Cell Physiol.* 29: 157–165.

Field, C.B. 1988. On the role of photosynthetic responses in constraining the habitat distribution of rainforest plants. *Aust. J. Plant Physiol.* 15: 343–358.

Field, C. and Mooney, H.A. 1986. The photosynthesis-nitrogen relationship in wild plants. In: *On the Economy of Plant Form & Function* (T.J. Givnish, ed.). Cambridge University Press, Cambridge, pp. 25–55.

Flaig, H. and Mohr, H. 1992. Assimilation of nitrate and ammonium by Scots pine (*Pinus sylvestris* L.) seedlings under conditions of high nitrogen supply. *Physiol. Plant.* 84: 568–576.

Foyer, C.H., Valadier, M.H. and Ferrario, S. 1995. Co-regulation of nitrogen and carbon assimilation in leaves. In: *Environment and Plant Metabolism* (N. Smirnoff, ed.). Bios Scientific Publ. Ltd., Oxon, England, pp. 17–33.

Fried, M.F., Zsoldos, F., Vose, P.B. and Shatokin, I.L. 1965. Characterizing the NO_3^- and NH_4^+ uptake process of rice roots by use of ^{15}N labelled NH_4NO_3. *Physiol. Plant.* 18: 313–320.

Frommer, W.B., Kwart, M., Hirner, B., Fischer, W.N., Hummel, S. and Ninnemann, O. 1994. Transporters for nitrogenous compounds in plants. *Plant Molecular Biol.* 26: 1651–1670.

Ganmore-Neumann, R. and Kafkafi, U. 1980. Root temperature and percentage NO_3^-/NH_4^+ effect on tomato plant development. I. Morphology and growth. *Agron. J.* 72: 758–761.

George, E., Marschner, H. and Jakobsen, I. 1995. Role of arbuscular mycorrhizal fungi in uptake of phosphorus and nitrogen from soil. *Crit. Rev. Biotechnol.* 15(3–4): 257–270.

Gigon, A. and Rorison, I.H. 1972. The response of some ecologically distinct plant species to nitrate- and to ammonium-nitrogen. *J. Ecol.* 60: 93–102.

Glaab, J. and Kaiser, W.M. 1995. Regulatory proteins for the modulation of nitrate reductase exist at normal activity in nitrate reductase deficient plants. In: *Fourth International Symposium on Inorganic Nitrogen Metabolism*, Seeheim, Southwest Africa, pp. 44.

Glass, A.D.M. and Siddiqui, M.V. 1995. Nitrogen absorption by plant roots. In: *Nitrogen Nutrition in Higher Plants* (H.S. Srivastava and R.P. Singh eds.). Associated Publ. Co., New Delhi, pp. 21–56.

Glass, A.D.M., Shaff, J.E. and Kochian, L.V. 1992. Studies of the uptake of nitrate in barley. *Plant Physiol.* 99: 456–463.

Glass, A.D.M. 1988. Nitrogen uptake by plant roots. ISI Atlas Animal and Plant Sciences. 1: 151–156.

Gosselin, A. and Trudel, M.J. 1986. Root-zone temperature effects on pepper. *J. Amer. Soc. Hort. Sci.* 111: 220–224.

Goyal, S.S. and Huffaker, R.C. 1986. The uptake of NO_3^-, NO_2^- and NH_4^+ by intact wheat (*Triticum aestivum*) seedlings. *Plant Physiol.* 82: 1051–1036.

Goyal, S.S., Lorenz, O.A. and Huffaker, R.C. 1982. Inhibitory effects of ammoniacal nitrogen on growth of radish plants. I. Characterization of toxic effects of ammonium on growth and its alleviation by nitrate. *J. Amer. Soc. Hort. Sci.* 107(1): 125–129.

Griffith, S.M. and Streeter, D.J. 1994. Nitrate and ammonium nutrition in ryegrass: changes in growth and chemical composition under hydroponic conditions. *J. Plant Nutr.* 17: 71–81.

Guy, R.D., Vanlernerghe, G.C. and Turpin, D.H. 1989. Significance of phosphoenolpyruvate during ammonium assimilation: carbon isotope discrimination in photosynthe-

sis and respiration by the N-limited green alga *Selenastrum minutum*. *Plant Physiol.* 89: 1150–1157.
Handa, S., Warren, H.L., Huber, D.M. and Tsai, C.Y. 1984. Nitrogen nutrition and seedling development of normal and *opaque*-2 maize genotypes. *Can. J. Plant Sci.* 64: 885–894.
Heilmeier, H., Freund, M., Steinlein, T., Schulze, E.D. and Monson, R.K. 1994. The influence of nitrogen availability on carbon and nitrogen storage in the biennial *Cirsium vulgare* (Savi) Ten. I. Storage capacity in relation to resource acquisition, allocation and recycling. *Plant Cell Environ.* 17(10): 1125–1131.
Huber, S.C., Huber, J.L. and Kaiser, W.M. 1994. Differential response of nitrate reductase and sucrose-phosphate synthase-activation to inorganic and organic salts, *in vitro* and *in situ*. *Physiol. Plant.* 92(2): 302–310.
Huppe, H.C. and Turpin, D.H. 1994. Integration of carbon and nitrogen metabolism in plant and algal cells. *Ann. Rev. Plant Physiol. Plant Mol. Biol.* 45: 577–607.
Iglesias, A.A. and Andreo, C.S. 1984. On the molecular mechanism of maize phosphoenolpyruvate carboxylase activation by thiol compounds. *Plant Physiol.* 75: 983–987.
Imsande, J. 1986. Nitrate-ammonium ratio required for pH homeostasis in hydroponically grown soybean. *J. Exp. Bot.* 37: 341–347.
Ingestad, T. 1973. Mineral nutrient requirements of *Vaccinium vitis idaea* and *V. myrtillus*. *Physiol. Plant.* 29: 239–246.
Ingestad, T. 1979. Mineral nutrient requirements of *Pinus sylvestris* and *Picea abies* seedlings. *Physiol. Plant.* 45: 373–380.
Ingestad, T. 1981. Nutrition and growth of birch and grey alder seedlings in low conductivity solutions and at varied relative rates of nutrient addition. *Physiol. Plant.* 52: 454–466.
Ingestad, T. 1982. Relative addition rate and external concentration: driving variables used in plant nutrition research. *Plant Cell and Environ.* 5: 443–453.
Ingestad, T. and Agren, G.I. 1988. Nutrient uptake and allocation at steady-state nutrition. *Physiol. Plant.* 72: 450–459.
Izawa, S. and Good, N.E. 1972. Inhibition of photosynthetic electron transport and photophosphorylation. *Methods Enzymol.* 24: 355–377.
Jackson, W.A. and Volk, R.J. 1995. Attributes of the nitrogen uptake systems of maize (*Zea mays* L.): Maximal suppression by exposure to both nitrate and ammonium. *New Phytol.* 130(3): 327–335.
Jarvis, S.C. 1987. The effects of low, regulated supplies of nitrate and ammonium nitrogen on the growth and composition of perennial ryegrass. *Plant and Soil* 100: 99–112.
Jiao, G. and Lips, S.H. 1995. Characterization of the nitrate uptake capacity developed under a prior feeding of high concentration of nitrate. In: *Fourth International Symposium on Inorganic Nitrogen Assimilation*, Seeheim, Southwest Africa, 61 pp.
Kaiser, W.M., Spill, D. and Brendle-Behnisch, E. 1992. Adenine nucleotides are apparently involved in the light/dark modulation of spinach leaf nitrate reductase. *Planta* 186: 236–240.
Kirkby, E.A. 1974. Recycling of potassium in plants considered in relation to ion uptake and organic acid accumulation. *Proc. 7th Int. Colloq. Plant Anal. Fertilizer Problems.* pp. 557–568.
Kirkby, E.A. 1981. Plant growth in relation to nitrogen supply. *Ecol. Bull.* 33: 249–267.
Kirkby, E.A. and Mengel, K. 1967. Ionic balance in different tissues of the tomato plant in relation to nitrate, urea or ammonium nutrition. *Plant Physiol.* 42: 6–14.
Kleiner, D. 1981. The transport of NH_3^- and NH_4^+ across biological membranes. *Biochem. Biophys. Acta* 639: 41–52.
Krajina, V.J., Madoc-Jones, S. and Mellor, G. 1973. Ammonium and nitrate in the nitrogen economy of some conifers growing in Douglas-fir communities of the Pacific North-West America. *S.H. Biochem.* 5: 143–147.

Kronzucker, H.J., Glass, A.D.M. and Siddiqi, M.Y. 1995a. Nitrate induction in spruce: An approach using compartmental analysis. *Planta* 196(4): 683–690.

Kronzucker, H.J., Siddiqi, M.Y. and Glass, A.D.M. 1995b. Analysis of (NH_4^+) ^{-13}N efflux in spruce roots — a test case for phase identification in compartmental analysis. *Plant Physiol.* 109(2): 481–490.

Kronzucker, H.J., Siddiqi, M.Y. and Glass, A.D.M. 1995c. Compartmentation and flux characteristics of ammonium in spruce. *Planta.* 196(4): 691–698.

Kronzucker, H.J., Siddiqi, M.Y. and Glass, A.D.M. 1995d. Kinetics of NO_3^- influx in apruce. *Plant Physiol.* 109(1): 319–326.

Lambers, H. and Poorter, H. 1992. Inherent variation in growth rate between higher plants: a search for physiological causes and ecological consequences. *Adv. Ecol. Res.* 23: 187–261.

Lang, B. and Kaiser, W.M. 1994. Solute content and energy status of roots of barley plants cultivated at different pH on nitrate- or ammonium-nitrogen. *New Phytol.* 128: 451–459.

Lara, C. and Guerrero, M.G. 1989. The photosynthetic assimilation of nitrate and its interactions with CO_2 fixation. In: *Techniques and New Developments in Photosynthesis Research* (J. Barber and R. Malkin, eds.). Plenum, Publ., New York, pp. 393–411.

Lawlor, D.W., Boyle, F.A., Keys, A.J., Kendall, A.C. and Young, A.T. 1988. Nitrate nutrition and temperature effects on wheat: A synthesis of plant growth and nitrogen uptake in relation to metabolic and physiological processes. *J. Exp. Bot.* 39: 329–343.

Lee, R.B. 1982. Selectivity and kinetics of ion uptake by barley plants following nutrient deficiency. *Ann. Bot.* 50: 429–449.

Lee, R.B. and Drew, M.C. 1986. Nitrogen-13 studies of nitrate fluxes in barley roots. I. Compartmental analysis from measurements of ^{13}N efflux. *J. Exp. Bot.* 37: 1768–1779.

Lee, R.B. and Drew, M.C. 1989. Rapid, reversible inhibition of nitrate influx in barley by ammonium. *J. Exp. Bot.* 40: 741–752.

Leidi, E.O. and Lips, S.H. 1990. Effect of NaCl ion salinity on photosynthesis, ^{14}C translocation and yield of wheat plants irrigated with ammonium or nitrate solutions. *Irrig. Sci.* 11: 155–161.

Leidi, E.O., Silberbush, M. and Lips, S.H. 1991. Wheat growth as affected by nitrogen type, pH and salinity. I. Biomass production and mineral composition. *J. Plant Nutr.* 14(3): 235–246.

Lewis, O.A.M. and Chadwick, S. 1983. An ^{15}N investigation into nitrogen assimilation in hydroponically-grown barley (*Hordeum vulgare* L. cv. Clipper) in response to nitrate, ammonium and mixed nitrate and ammonium nutrition. *New Phytol.* 95: 635–646.

Lewis, O.A.M., James, D.M. and Hewitt, E.J. 1982. Nitrogen assimilation in barley (*Hordeum vulgare* L. cv. Mazurka) in response to nitrate and ammonium nutrition. *Ann. Bot.* 49: 39–49.

Lewis, O.A.M., Fulton B. and Zelewski, A.A.A. 1987. Differential distribution of carbon in response to nitrate, ammonium and nitrate+ammonium nutrition in wheat. In: *Inorganic Nitrogen Metabolism* (P.J. Surett, W.R. Ulrich, P.J. Aparicio and F. Castillo, eds.). Springer-Verlag: Berlin, pp. 240–246.

Lewis, O.A.M., Leidi, E.O. and Lips, S.H. 1989. Effect of nitrogen source on growth response to salinity stress in maize and wheat. *New Phytol.* 111: 155–160.

Lillo, C. 1994. Light regulation of nitrate reductase in green leaves of higher plants. *Plant Physiol.* 90: 616–620.

Lips, S.H., Ben Zioni, A. and Vaadia, Y. 1970. K⁺ recirculation in plants and its importance for adequate nitrate reduction. In: *Recent Advances in Plant Nutrition* (R.M. Samish, Gordon & Breach, eds.). Vol. 1. pp: 207–215.

Lips, S.H., Cramer, M.D. and Sagi, M. 1995. Nitrogen assimilation in plants under stress conditions. In: *Fourth International Symposium on Inorganic Nitrogen Metabolism*, Seeheim, Southwest Africa, 72 pp.

Lips, S.H., Soares, M.I.M., Kaiser, J.J. and Lewis, O.A.M. 1987. K⁺ modulation of nitrogen uptake and assimilation in plants. In: *Inorganic Nitrogen Metabolism* (P.J. Syrett, W.R. Ullrich, P.J. Aparicio and F. Castillo, eds.). Springer-Verlag: Berlin, pp. 233–239.

Lips, S.H., Lidi, E.O., Silberbush, M., Soares, M.I.M. and Lewis, O.A.M. 1990. Physiological aspects of ammonium and nitrate fertilization. *J. Plant. Nutr.* 1: 1271–1289.

Losada, M., Guerrero, M.G. and Vega, J.M. 1981. The assimilatory reduction of nitrate. In: *Biology of Inorganic Nitrogen and Sulfur* (H. Bothe and A., Trebest, eds.). Springer-Verlag, Berlin, pp. 30–68.

Losada, M., Guerrero, M.G., Rosa de la, M.A., Serrano, A. Hervas, M. and Ortega, J.M. 1990. Sequential energy transduction in photosynthesis. In: *Inorganic Nitrogen in Plants and Microorganisms. Uptake and Metabolism* (W.R. Ullrich, C. Rigano, A. Fuggi and P.J. Aparicio, eds.). Springer Verlag, Heidelberg, Berlin, pp. 21–27.

Lu, Q. and Briskin, D. 1993. Modulation of the maize plasma membrane carrier by NO_3^-. *Phytochemistry* 33: 1–8.

Lycklama, J.C. 1963. The absorption of ammonium and nitrate by perennial ryegrass. *Acta Bot. Neerl.* 12: 361–423.

Macdonald, F.D. and Buchanan, B.B. 1992. The reductive pentose phosphate pathway and its regulation. In: *Plant Physiology, Biochemistry and Molecular Biology* (D.T. Dennis and D.H. Turpin, eds.). Longman Sci. Tech. Publ., Essex, England, pp. 239–252.

MacDuff, J.H., Hopper, M.J. and Wild, A. 1987. The effect of root temperature on growth and uptake of ammonium and nitrate by *Brassica napus* L. in flowing solution culture. I. Growth. *J. Exp. Bot.* 38: 53–66.

Macklon, A.E.S., Ron, M.M. and Sim, A. 1990. Cortical cell fluxes of ammonium and nitrate in excised root segments of *Allium cepa* L.; studies using ¹⁵N. *J. Exp. Bot.* 41: 359–370.

MacKown, C.T. and McClure, P.R. 1988. Development of accelerated NO_3^- uptake. *Plant Physiol.* 87: 162–166.

MacLeod, K.C. and Ormrod, D.P. 1985. Responses of white bean to ammonium or nitrate nutrition at three temperatures. *Can. J. Plant Sci.* 65: 201–205.

Magalhães, J.R. and Wilcox, G.E. 1984. Tomato growth and nutrient uptake patterns as influenced by nitrogen form and light intensity. *J. Plant Nutrition* 6: 941–956.

Magalhães, J.R. and Huber, D.M. 1991. Responses of ammonium assimilation enzymes to nitrogen form treatments in different plant species. *J. Plant Nut.* 14(2): 175–185.

Maghrabi, Y.M.S., Younis, A.E. and Abozinah, F.S. 1985. Nitrogen metabolism in tomato seedlings. *Plant and Soil* 85: 403–411.

Makino, A. and Osmond, B. 1991. Effects of nitrogen nutrition on nitrogen partitioning between chloroplasts and mitochondria in pea and wheat. *Plant Physiol.* 96: 355–362.

Makino, A., Sakashita, H., Hidema, J., Mae, T., Ojima, K. and Osmond, B. 1992. Distinctive responses of ribulose-1,5-biphosphate carboxylase and carbonic anhydrase in wheat leaves to nitrogen nutrition and their possible relationships to CO_2-transfer resistance. *Plant Physiol.* 100: 1737–1743.

Marschner, H. 1986. *Mineral Nutrition of Higher Plants.* Academic Press, London, 3rd ed.

Marschner, H. 1991a. Root-induced changes in the availability of micronutrients in the rhizosphere. In: *Plant Roots. The Hidden Half* (Y. Waisel, A. Eshel and U. Kafkafi, eds.). Marcel Dekker, Inc., New York, pp. 503–508.

Marschner, H. 1991b. Mechanisms of adaptation of plants to acid soils. *Plant and Soil* 134: 1–20.

Marschner, H., Haussling, M. and George, E. 1991c. Ammonium and nitrate uptake rates and rhizosphere pH in non-mycorrhizal roots of Norway spruce (*Picea abies* (L.) Karst.). *Trees* 5: 14–21.

Martins-Loução, M.A. 1985. *Estudos fisiológicos e microbiológicos da associação da alfarrobeira (Ceratonia siliqua L.) com bactérias de Rhizobiaceae.* Lisboa.

Martins-Loução, M.A. 1990. Carob (*Ceratonia siliqua* L.). In: *Biotechnology in Agriculture and Forestry*, vol. 10 (Y.P.S. Bajaj, ed.). Springer-Verlag, New York, pp. 658–675.

Martins-Loução, M.A. and Duarte, P. 1987. Effect of ammonium and nitrate nutrition on the growth of carob (*Ceratonia siliqua* L.) plants. In: *Inorganic Nitrogen Metabolism*. (P.J. Surett, W.R. Ulrich, P.J. Aparicio and F. Castillo, eds.). Springer-Verlag: Berlin, pp. 250–252.

Martins-Loução, M.A., Wollenweber, B. and Raven, J.A. 1993. Response of *Salvinia* spp. to different nitrogen sources: the acid-base regulation approach. *Oecologia* 93: 524–530.

Mata, C., Scheurwater, I., Martins-Loução, M.A. and Lambers, H. 1996. Root respiration and growth of *Quercus suber* L. seedlings under nitrate and ammonium nutrition. Plant Physiol Biochem. 34: 727–734.

McClure, P.R., Ohmolt, T.E. and Pace, G.M. 1986. Anion uptake in maize roots: interactions between chlorate and Nitrate. *Physiol. Plant.* 68: 107–112.

Mehrer, and Mohr, H. 1989. Ammonium toxicity: description of the syndrome in *Sinapis alba* and the search for its causation. *Physiol. Plant.* 77: 545–554.

Mengel, K. and Kirkby, E.A. 1987. *Principles of Plant Nutrition*. International Potash Institute. Bern. ISBN 3-906535037.

Millard, P. 1993. A review of internal cycling of nitrogen within trees in relation to soil fertility. In: *Optimization of Plant Nutrition* (M.A.C. Fragoso and M.L.V. Beusichem, eds.). Kluwex Acad. Publ., The Netherlands, pp. 623–628.

Minotti, P.L., Williams, D.C. and Jackson, W.A. 1969. Nitrate uptake by wheat as influenced by ammonium and other cations. *Crop Sci.* 9: 9–14.

Mistriki, E.W. and Ullrich, W. 1995. Ammonium uptake in *Lemna gibba* G1; related membrane potential changes. *Physiol. Plant.* 101: 309–316.

Miziorko, H.M. and Lorimer, G.H. 1983. Ribulose-1,5-biphosphate carboxylase-oxygenase. *Ann. Rev. Biochem.* 52: 507–535.

Moore, D.R.E. and Waid, J.S. 1971. The influence of washings of living roots on nitrification. *Soil Biol. Biochem.* 3: 69–83.

Nijs, I., Behaeghe, T. and Impens, I. 1995. Leaf nitrogen content as a predictor of photosynthetic capacity in ambient and global change conditions. *J. Biogeogr.* 22(2-3): 177–183.

Northup, R.R., Yu, Z., Dahlgren, R.A. and Vogt, K. 1995. Polyphenol control of nitrogen release from pine litter. *Nature* 377 (September): 227–229.

Oaks, A. 1994. Efficiency of nitrogen utilization in C-3 and C-4 cereals. *Plant Physiol.* 106(2): 407–414.

Osmond, D.L. and Raper, C.D. 1981. Growth and nitrogen accumulation in tobacco plants as affected by nitrate concentration, root temperature and aerial temperature. *Agron. J.* 75: 491–496.

Pardales, J.R., Kono, Y. and Yamauchi, A. 1992. Epidermal cell elongation in sorghum seminal roots exposed to high root-zone temperature. *Plant Science* 81: 143–146.

Pate, J.S. 1973. Uptake, assimilation and transport of nitrogen compounds by plants. *Soil Biol. Bioch.* 5: 109–119.

Pearson, J. and Stewart, G.R. 1993. The deposition of atmospheric ammonia and its effects on plants. *New Phytol.* 125: 283–305.

Peet, M.M., Raper, C.D. Jr., Tolley, L.C. and Robarge, W.P. 1985. Tomato responses to ammonium and nitrate nutrition under controlled root-zone pH. *J. Plant Nutr.* 8: 787–798.

Pilbeam, D.J. and Kirkby, E.A. 1992. Some aspects of the utilization of nitrate and ammonium by plants. In: *Nitrogen Metabolism of Plants* (K. Mengel and D.J. Pilbeam, eds.). Clarendon Press, Oxford, pp. 55–70.

Pons, T.L., Van der Werf, A. and Lambers, H. 1994. Photosynthetic nitrogen use efficiency of inherently slow- and fast-growing species: possible explanations for observed differences. In: *A Whole Plant Perspective on Carbon-Nitrogen Interactions* (J. Roy and E. Garnier, eds.). SPB, Acad. Publ., The Hague, The Netherlands, pp. 61–78.

Qualls, R., Haines, B. and Swank, W. 1991. Fluxes of dissolved organic nutrients and humic substances in a deciduous forest. *Ecology* 72: 254–266.

Raab, T.K. and Terry, N. 1994. Nitrogen source regulation of growth and photosynthesis in *Beta vulgaris* L. *Plant Physiol.* 105: 1159–1166.

Raab, T.K. and Terry, N. 1995. Carbon, nitrogen and photosynthesis nutrient interactions in *Beta vulgaris* L. as influenced by nitrogen source, NO_3^- versus NH_4^+. *Plant Physiol.* 107: 575–584.

Raper, C.D. Jr. Vessey, J.K., Henry, L.T. and Chaillou, S. 1991. Cyclic variation in nitrogen uptake rate of soybean plants: effects of pH and mixed nitrogen sources. *Plant Physiol. Soil. Biochem.* 29: 205–212.

Raven, J.A. 1985. Regulation of pH and generation of osmolarity in vascular land plants: costs and benefits in relation to efficiency use of water, energy and nitrogen. *New Phytol.* 101: 25–77.

Raven, J.A. 1987. The role of vacuoles. *New Phytol.* 106: 357–422.

Raven, J.A. 1988. Acquisition of nitrogen by the shoots of vascular plants: its occurrence and implications for acid-base regulation. *New Phytol.* 109: 1–20.

Raven, J.A. 1990. Sensing pH? *Plant Cell and Environ.* 13: 721–729.

Raven, J.A. 1991. Terrestrial rhizophytes and H^+ currents circulating over at least a millimetre: an obligate relationship? *New Phytol.* 117: 177–185.

Raven, J.A. and Smith, F.A. 1976. Nitrogen assimilation and transport in vascular plants in relation to intracellular pH regulation. *New Phytol.* 76: 415–431.

Raven, J.A. and Farquhar, G.D. 1990. The influence of N metabolism and organic acid synthesis on the natural abundance of isotopes of carbon in plants. *New Phytol.* 116: 505–529.

Raven, J.A., Wollenweber, B. and Handley, L.L. 1992. Ammonia and ammonium fluxes between photolithotrophs and environment in relation to the global nitrogen cycle. *New Phytol.* 121: 5–18.

Raven, J.A., Franco, A.A., deJesus, L.L. and Jacob-Neto, J. 1990. H^+ extrusion and organic acid synthesis in N_2-fixing symbioses involving vascular plants. *New Phytol.* 114: 369–389.

Read, D.J. 1993. Plant-microbe mutualisms and community structure. In: *Biodiversity and Ecosystem Function* (E.D. Schulze and H.A. Mooney, eds.). Springer-Verlag, Berlin-Heidelberg, pp. 181–209.

Reddy, K.S. and Menary, R.C. 1990. Nitrate reductase and nitrate accumulation in relation to nitrate toxicity in *Boronia megastigma*. *Physiol. Plant.* 78: 430–434.

Reisenauer, H.M., Clement, C.R. and Jones, L.H.P. 1982. Comparative effect of ammonium and nitrate for grasses. *Plant Nutrition.* (Proc. 9th Int. Plant Nutrition Colloquium) pp. 539–545.

Rhizopoulous, S. and Davies, W.J. 1991. Influence of soil drying on root development, water relations and leaf growth of *Ceratonia siliqua* L. *Oecol.* 88: 41–47.

Rice, E.L. 1982. *Allelopathy.* Academic Press, London, 325 pp.

Rice, E.L. and Pancholy, S.K. 1972. Inhibition of nitrification by climax vegetation. *Amer. J. Bot.* 39: 1033–1040.

Rideout, J.W. and Raper, D.C. Jr. 1994. Dry matter and nitrogen accumulation are not affected by superoptimal concentration of ammonium in flowing solution culture with pH control. *J. Plant Nutr.* 17: 219–228.

Riha, S.J., Campbell, G.S. and Wolfe, J. 1986. A model of competition for ammonium among heterotrophs, nitrifiers and roots. *J. Soil Sci. Sco. Amer.* 50: 1463–1466.

Romero, J.M. and Lara, C. 1987. Photosynthetic assimilation of NO_3^- by intact cells of the cyanobacterium *Anacytis nidulans*. Influence of NO_3^- and NH_4^+ assimilation on CO_2 fixation. *Plant Physiol.* 83: 208–212.

Rosswall, T. 1982. Microbiological regulation of the biogeochemical nitrogen cycle. *Plant and Soil.* 67: 15–34.

Rufty, T.W., Raper, C.D. and Jackson, W.A. 1983. Growth and nitrogen assimilation of soybeans in response to ammonium and nitrate nutrition. *Bot. Gaz.* 144(4): 466–470.

Ruiz-Cristin, J. and Briskin, D.P. 1991. Characterization of a H^+/NO_3^- symport associated with plasma membrane vesicles of maize roots using $^{36}ClO_3^-$ as a radiotracer analog. Arch. Biochem. Biophys. 28: 574–582.

Salsac, L., Chaillou, S., Morot-Gaudry, J.F., Lesaint, C. and Jovilet, E. 1987. Nitrate and ammonium nutrition in plants. Plant Physiol. Biochem. 25: 805–812.

Sandoval-Villa, M., Alcantar-Gonzalez, G. and Tirado-Torres, J.L. 1992. Effect of the NH_4^+/NO_3^- ratio on GS and PEPCase activities and on dry matter production in wheat. J. Plant Nutr. 15: 2545–2557.

Sandoval-Villa, M., Alcantar-Gonzalez, G. and Tirado-Torres, J.L. 1995. Use of ammonium in nutrient solutions. J. Plant Nutr. 18: 1449–1457.

Saravitz, C.H., Chaillou, S., Musset, J., Raper, C.D. and Morotgaudry, J.F. 1994. Influence of nitrate on uptake of ammonium by nitrogen-depleted soybean: Is the effect located in roots or shoots? J. Exp. Bot. 45: 1575–1584.

Schenk, M. and Wehrmann, J. 1979. The influence of ammonia in nutrient solution on growth and metabolism of cucumber plants. Plant and Soil. 52: 403–414.

Siegl, G. and Stitt, M. 1990. Partial purification of two forms of spinach leaf sucrose-phosphate synthetase which differ in their kinetic properties. Plant Sci. 66: 205–210.

Silberbush, M. and Lips, S.H. 1988. Nitrogen concentration, ammonium/nitrate ratio and NaCl interaction in vegetative and reproductive growth of peanuts. Physiol. Plant. 74: 493–498.

Simpson, R.J., Lambers, H. and Dalling, M.J. 1982. Kinetin application to roots and its effect on uptake, translocation and distribution of nitrogen in wheat (Triticum aestivum) grown with a split-root system. Physiol. Plant. 56: 430–435.

Smakman, G.S. and Hofstra, R.J.J. 1982. Energy metabolism of Plantago lanceolata, as affected by change in root temperature. Physiol. Plant. 56: 33–37.

Smart, D.R. and Bloom, A.J. 1993. Relationships between the kinetics of ammonium and nitrate absorption and growth in the cultivated tomato (Lycopersicum esculentum Mill. cv. T-5). Plant Cell and Environ. 16: 259–267.

Smirnoff, N. and Stewart, G.R. 1985. Nitrate assimilation and translocation by higher plants: Comparative physiology and ecological consequences. Physiol. Plant. 64: 133–140.

Smirnoff, N., Todd, P. and Stewart, G.R. 1984. The occurrence of nitrate reduction in the leaves of woody plants. Ann. Bot. 54: 364–374.

Spill, D. and Kaiser, W.M. 1994. Partial purification of two proteins (100 kDa and 67 kDa) cooperating in the ATP-dependent inactivation of spinach leaf nitrate reductase. Planta 192: 183–188.

Stader, J. and Gebauer, G. 1992. Nitrate reduction and nitrate content in ash tree (Fraxinus excelsior L.): distribution between compartments, site comparison and seasonal variation. Trees 6: 236–240.

Stewart, G.R. 1991. The comparative ecophysiology of plant nitrogen metabolism. In: Plant Growth: Interactions with Nutrition and Environment (J.R. Porter and D.W. Lawlor, eds.), Cambridge University Press, Cambridge, pp. 81–97.

Stewart, G.R., Hegarty, E.E. and Specht, R.L. 1988. Inorganic nitrogen assimilation in plants of Australian rainforest communities. Physiol. Plant. 74: 26–33.

Stewart, G.R., Pearson, J., Kershaw, J.L. and Clough, E.C.M. 1989. Biochemical aspects of inorganic nitrogen assimilation by woody plants. Ann. Sci. Forest. 46 (suppl.): 648–653.

Stewart, G.R., Gracia, C.A., Hegarty, E.E. and Specht, R.L. 1990. Nitrate reductase activity and chlorophyll content in sun leaves of tropical Australian closed-forest (rainforest) and open-forest communities. Oecologia. 82: 544–551.

Stitt, M. 1991. Rising levels and their potential significance for carbon flow in photosynthetic cells. Plant Cell Environ. 14: 741–762.

Sugiharto, B. and Sugiyama, T. 1992. Effects of nitrate and ammonium on gene expression of phosphoenolpyruvate carboxylase and nitrogen metabolism in maize leaf tissue during recovery from nitrogen stress. *Plant Physiol.* 98: 1403–1408.
Tamm, C.O. 1991. *Nitrogen in Terrestrial Ecosystems*. Ecological Studies, vol. 81. Springer-Verlag, Berlin-Heidelberg.
Tischner, R., Waldeck, B., Goyal, S.S. and Rains, W.D. 1993. Effect of nitrate pulses on the nitrate uptake rate, synthesis of mRNA coding for nitrate and nitrate reductase activity in the roots of barley seedlings. *Planta* 189: 533–537.
Tischner, R., Stöhr, C., Glogau, U. and Mätschke, M. 1995. The characteristics, the structure and the function of the plasmamembrane-bound nitrate reductase in *Chlorella saccharophila*. In: *Fourth International Symposium on Inorganic Nitrogen Metabolism*, Seeheim, Southwest Africa, p. 72.
Tolley-Henry, L. and Raper, C.D. Jr. 1986. Utilization of ammonium as a nitrogen source. Effects of ambient acidity on growth and nitrogen accumulation by soybean. *Plant Physiol.* 82: 54–60.
Torre, A., Delgado, B. and Lara, C. 1991. Nitrate-dependent O_2 evolution in intact leaves. *Plant Physiol.* 96: 898–901.
Trueman L.J., Onyeocha, I. and Forde, B.G. 1995. Cloning and characterization of putative high affinity nitrate transporters from higher plants. In: *Fourth International Symposium on Inorganic Nitrogen Metabolism* (W. Ullrich ed.) Tech Horschtuhl Darmstad Seeheim pp. 127–134.
Truax, B., Lambert, F., Gagnon, D. and Chevier, N. 1994. Nitrate reductase and glutamine synthetase activities in relation to growth and nitrogen assimilation in red oak and red ash seedlings: effects of N-forms, N concentration and light intensity. *Trees* 9: 12–18.
Ullrich, W.R. 1992. Transport of nitrate and ammonium through plant membranes. In: *Nitrogen Metabolism of Plants* (K. Mengel and D.J. Pilbeam, eds.), Clarendon Press, Oxford, pp. 121–137.
Ullrich, W.R., Larsson, M., Oscarson, P., Lesh, S. and Novacky, A. 1984. Ammonium uptake in *Lemna gibba* GI, related membrane potential changes and inhibition of ion uptake. *Physiol. Plant.* 61: 369–376.
Van der Burg, J. 1971. Some experiments on the mineral nutrition of forest tree seedlings (Report of a study at the Royal College of Forestry, Stockholm) — Forest Research Station "De Dorschkamp". Wageningen, Netherlands. Internal Rapport nr. 8: 1–67.
Van Quy, L., Foyer, C.H. and Champigny, M.L. 1991. Effect of light and NO_3^- on leaf phosphoenolpyruvate carboxylase activity. Evidence for covalent modulation of the C3 enzyme. *Plant Physiol.* 97: 1476–1482.
Vanbel, A.J.E. 1995. The low profile directors of carbon and nitrogen economy in plants: parenchyma cells associated with translocation channels. In: *Plant Stems: Physiology and Functional Morphology* (B.L. Gartner, ed.). Academic Press Inc., San Diego, CA, pp. 205–222.
Verhangen, F.J.M., Laanbroek, H.J. and Woldendorp, J.W. 1995. Competition for ammonium between plant roots and nitrifying and heterotrophic bacteria and the effects of protozoan grazing. *Plant and Soil* 170: 241–250.
Volk, R.J., Chaillou, S., Mariotti, A. and Morot-Gaudry, J.F. 1992. Beneficial effects of concurrent ammonium and nitrate nutrition on the growth of *Phaseolus vulgaris*: A ^{15}N study. *Plant Physiol. Biochem.* 30: 487–493.
Walker, J.L. and Huber, S.C. 1989. Regulation of sucrose-phosphate-synthase activity in spinach leaves by protein level and covalent modification. *Planta* 177: 116–120.
Wang, M.Y., Siddiqi, M.Y., Ruth, T.J. and Glass, A.D.M. 1993a. Ammonium uptake by rice roots. I. Fluxes and subcellular distribution of $^{13}NH_4^+$. *Plant Physiol.* 103: 1249–1258.
Wang, M.Y., Siddiqi, M.Y., Ruth, T.J. and Glass, A.D.M. 1993b. Ammonium uptake by rice roots. II. Kinetics of $^{13}NH_4^+$ influx across the plasmalemma. *Plant Physiol.* 103: 1259–1267.

Wang, M.Y., Glass, A.D.M., Shaff, J.E. and Kochian, L.V. 1994. Ammonium uptake by rice roots. III. Electrophysiology. *Plant Physiol.* 104: 899–906.

Wang, X.T. and Below, F.E. 1996. Cytokinins in enhanced growth and tillering of wheat induced by mixed nitrogen source. *Crop Sci.* 36(1): 121–126.

Wang, Y.Z., Cramer, M.D., Schierholt, A. and Lips, S.H. 1994. The influence of gibberellin and cytokinin on growth and metabolism in salinity stressed *Lycopersicon esculentum* L. cv. F-144. *Biologia Plantarum.* 36(supplem.): 42.

Warncke, D.D. and Barber, S.A. 1973. Ammonium and nitrate uptake by corn (*Zea mays* L.) as influenced by nitrogen concentration and NH_4^+/NO_3^- ratio. *Agron. J.* 65: 950–953.

Wilcox, G.E., Mitchell, C.A. and Hoff, J.E. 1977. Influence of N form on exudation rate and ammonium, amide and cation composition of xylem exudate in tomato. *J. Amer. Soc. Hort. Sci.* 102: 192–196.

Williams, J.H.H. and Farrar, J.F. 1992. Sucrose: a novel plant growth regulator. In: *Molecular, Biochemical and Physiological Aspects of Plant Respiration* (H. Lambers and L.H.W. van der Plas, eds.), SPB Academic Publ. The Hague, pp. 463–469.

Woldendorp, J.W. 1981. Nutrients in the rhizosphere: *Proc. 16th Coll. Int. Potash Inst.* Bern. ISBN 3-906535037, pp. 99–125.

Woldendorp, J.W. and Laanbroek, H.J. 1989. Activity of nitrifiers in relation to nitrogen nutrition of plants in natural ecosystems. *Plant and Soil* 115: 217–228.

Zornoza, P. and Carpena, O. 1992. Study on ammonium tolerance of cucumber plants. *J. Plant Nutr.* 15: 2417–2426.

Chapter 8

NITROGEN, STRESS AND PLANT GROWTH REGULATION

S. Herman Lips

I. Introduction

II. Regulation of Biomass Production

III. Salinity Inhibition of Vegetative Growth

IV. Amide/Nitrate Ratio in the Xylem and Partitioning
    A. Xylem Transport of Nitrate and Growth Rate
    B. Xylem Loading of Nitrogenous Compounds
    C. Regulation of Xylem Loading

V. Nitrate, Aldehyde Oxidase and ABA

VI. Conclusions and Future Prospects

Literature Cited

Abbreviations: ABA, abscisic acid; AO, aldehyde oxidase; CK, cytokinins; MDH, malate dehydrogenase; NR, nitrate reductase; OAA, oxaloacetate; PEPc, phosphoenolpyruvate carboxylase; XP, xylem parenchyma.

I. INTRODUCTION

Organ specialization in plants allows a more efficient uptake and assimilation of inorganic elements from the soil and the atmosphere and facilitates adaptation to a wide range of changing environmental conditions. Roots have developed cellular structures and metabolic mechanisms suited for the uptake of mineral nutrients and water and for their subsequent transport to the shoot. Shoots assimilate CO_2 and harvest solar energy to generate organic matter, part of which is allocated to the root to provide it with energy and to support its growth. Roots and shoots co-ordinate swiftly and rigorously their rates of growth and primary assimilation.

Coordinated shoot-root growth is regulated by hormonal signals (Davis and Zhang, 1991). The growing shoot produces auxin which moves basipetally towards the root together with a flow of assimilates, all of which deliver a message encouraging root growth and activity. Similarly, root-produced cytokinins together with the flow of water and nitrogen signal to the shoot the extent of current root supply of nitrogen, water and other mineral nutrients. Root ABA slows down growth when uptake of nitrate is limited either by its availability or due to environmental stress conditions. Root hormones control stomata opening and through it the rate of water loss (transpiration) and CO_2 influx (photosynthesis). The way in which hormones control basic metabolic functions still remains enigmatic, however.

Sustained accelerated growth requires an interchange of hormonal and metabolic signals between shoot and root, accompanied by an adequate logistic capacity of the primary assimilation process to provide the basic metabolites for biomass production. Metabolite transport between plant parts plays a significant role in the complex exchange of signals and supplies.

One of the best examples of coordination is the uptake and assimilation of nitrate. The K-shuttle (Lips et al., 1970; Ben-Zioni et al., 1971) was one of the first examples of ion (NO_3^-) uptake by the root and was shown to be regulated by products of nitrate assimilation in the shoot. The K-shuttle (Fig. 8.1) comprises the massive transport of $K^+NO_3^-$ from the root to the shoot via the xylem and K^+-malate transport from the shoot to the root through the phloem, fitting the energy expenditure on nitrate uptake by the root to current organic nitrogen requirements of the developing shoot. At the onset of senescence and of flowering, nitrate reductase is rapidly inactivated in the shoot of most annual plants, an event linked to a drastic reduction of nitrate uptake by the roots and the internal redeployment of organic nitrogen from leaves to developing seeds. This pattern was shown to change, however, due to

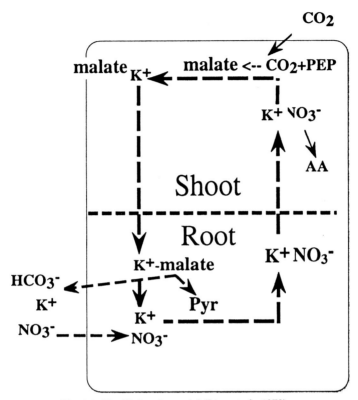

Fig. 8.1. The K-shuttle model (Lips et al., 1970).

exogenous applications of artificial cytokinins to wheat shortly after anthesis, which caused a significant recovery of nitrate uptake, NR activity and nitrogen assimilation (Kaminek et al., 1994). Effects of cytokinins on nitrogen and phosphorus deficiencies were also reported by Horgan and Waring (1980). Since the initial description of the K-shuttle (Lips et al., 1970), other examples of mineral uptake regulated by products transported through the phloem have been described.

Growth regulation of plants has been an intriguing topic studied for its ecological and agricultural implications, since it constitutes an integral part of the mechanisms of adaptation to limited supplies of raw materials (e.g. desert ecosystems) or to a very rich supply of nitrogen and water (agricultural super productivity). In agricultural terms, yield is related in most cases to plant growth, which is boosted by massive fertilization and irrigation, a major environmental pollutant of dwindling water resources around the world. Consequently, an understanding of growth regulation is essential for current efforts to establish new technologies for an environmentally clean or safe production of crops, lowering the level of fertilization required to achieve production goals. Shoot-root coordination is a key element in the determination of the limits of biomass production which seems to bypass genetic limitations of individual components of primary productivity such as photosynthesis and nitrogen assimilation. The implications for plant sciences in the near future and for ecosystem preservation and environmentally safe crop super production are, indeed, immense.

II. REGULATION OF BIOMASS PRODUCTION

One of the most interesting expressions of growth regulation in plants is the relationship between biomass production or yield and the accumulation of excess inorganic cations over anions (Dijkshoorn, 1958) (Fig. 8.2). The relationship described by Dijkshoorn was based on the statistical analysis of a very large number of field experiments with several major crops. Since the life cycle of an annual crop is limited in time, one may conclude that plants accumulating a small excess of cations over anion (poor K-shuttle activity) were slow growers, while plants with a large cation excess (active K-shuttle) were fast growers. These conclusions were based on statistical analysis but the author offered no physiological mechanism to explain the reasons for the correlation observed. In other experiments the excess of cations over anions in the plants was balanced by organic acids, in particular malate (Ullrich, 1941; Burström, 1945; Ben-Zioni et al., 1971). The reduction of nitrate to ammonium, which implies the replacement of an anion by a cation, caused an excess accumulation of inorganic cations, among which K^+ was most prominent. The resultant pH increase triggered synthesis of malate by PEPc (Ting and Dugger, 1967).

Nitrate reduction was coupled to the synthesis of malate by PEPc and MDH as an organic anion compensating the loss of the inorganic NO_3^- anion by its reduction to NH_4^+ (Lips et al., 1970). Leaf K^+ was then recycled to the roots as K-malate, synthesized in response to nitrate reduction in the shoot. Although segregation of malate pools in plants was proven and shown to be linked to different metabolic pathways in maize roots (Lips and Beevers, 1966a, b), it has been far more difficult to explain how the pool of malate resulting from nitrate reduction was kept essentially compartmentalized from other malate pools in the leaf. In the root, K-malate provided by the shoot was oxidized only in the presence of nitrate (Fig. 8.3), releasing HCO_3^- to the root medium in exchange of nitrate (Ben-Zioni et al., 1971).

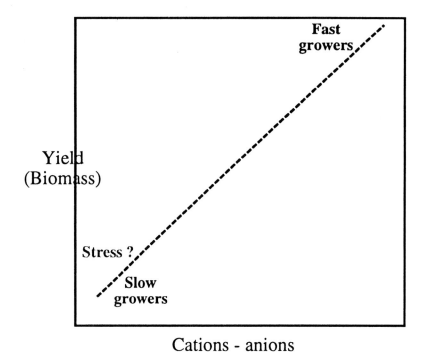

Fig. 8.2. Yield as a function of excess cations over ions (Dijshoorn, 1958).

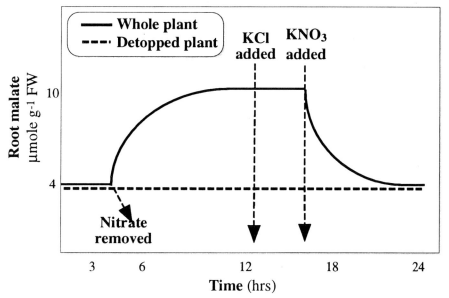

Fig. 8.3. The malate pool in roots as affected by different salts (Ben-Zioni et al., 1970).

Operation of the K-shuttle received additional experimental support from G. Jiao (pers. comm.) who observed, that the $HCO_3^- <\!-\!\!> NO_3^-$ exchange took place under anaerobic conditions in roots of barley seedlings, when other ATP-depending uptakes system (HATS) were totally inactive. The K-shuttle model implied that the extent of nitrate assimilation in the shoots specified the extent of nitrate uptake expected from the roots. When nitrate reduction and the coupled synthesis of malate in the shoot decreased due to aging or flowering (e.g. cereals, sunflower, tobacco, etc.), root nitrate uptake was down-regulated according to the diminished supply of malate by the shoot.

The K-shuttle is a mechanism common to most, if not all plants. Most crops have a very active K-shuttle and exhibit fast growth rates. High-producing crops may have been actually selected as such due to the relative high activity of their K-shuttle, which is linked to rapid growth and productivity (El Bassam, internal report). However, a number of slow-growing species, especially trees, carry out nitrate assimilation "preferentially" in their roots, having a limited K-shuttle activity.

The species selected in our studies to represent slow growth was the carob tree (*Ceratonia siliqua* L.), a leguminous Mediterranean tree which does not fix atmospheric N_2. Carob seedlings reduced about 80% of their nitrate in the roots (Cruz et al., 1991) and only a low level of nitrate reductse was found in the leaves (Table 8.1). The 20% nitrate reduction in the leaves was limited by the extent of nitrate transported from the root. Applications of nitrate directly to the transpiration stream through a cut of the stem cortex above the root elicited a rapid increase of nitrate reduction in the leaves to levels above these observed in the root.

The limited capacity of the carob root to load nitrate into the xylem along the roots (Fig. 8.4) became evident in uptake experiments with root segments and intact roots (Cruz et al., 1993b, c and 1995). It turned out that the 'preference' of the carob tree for ammonium rather than nitrate was due to its very limited capacity to load nitrate into the xylem for its transport to the shoot, apparently because the uptake area (0–2 cm) was in a root section with poorly developed xylem.

Table 8.1. NR distribution in shoots and roots of carob, a Mediterranean leguminous tree (*Cruz et al., 1991), and tomato plants (**Cramer et al., 1995)

Species	NR activity µmole nitrite g^{-1} FW h^{-1}		S/R ratio NR	Growth rate
	Root	Shoot		
Carob*	3.4	0.8	0.24	Slow
Tomato**	0.3	1.8	6.00	Fast

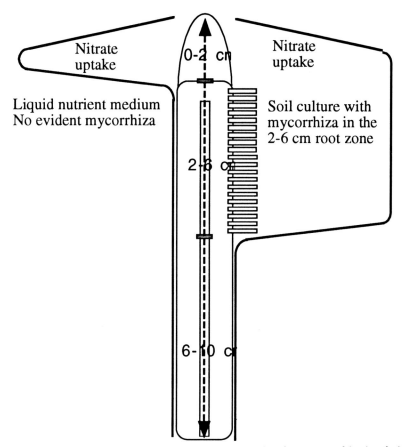

Fig. 8.4. Nitrate uptake regions in the carob root with and without mycorrhiza in relation to xylem development and K-shuttle activity (Cruz et al., 1996).

Without the participation of the shoot in the processes of reduction of nitrate and ammonium assimilation, it became energetically less costly to the root to take up and assimilate ammonium rather than nitrate, saving the energy required for the reduction of nitrate. Recent observations point out a possible gap between the nitrate uptake area located in the youngest 1–2 cm after the root tip and the area of full development of vascular tissue. Furthermore, when mycorrhiza established itself in the 2–6 cm sector of the root, the uptake of nitrate and the growth rate of the carob seedlings were significantly increased (Martins-Louçào et al., 1996).

Consequently, the anatomical characteristics of the carob root may be responsible for the limited K-shuttle activity in the carob seedling.

Similar differences in the development of xylem elements in roots of other plants have been recently described by McCully (1995), who pointed out 'the inadequacy of the classical model of root structure based on studies of seedling roots of a few species', which certainly do not include trees such as the carob. Late maturation of the metaxylem vessel has been observed in soybean (Kevekordes et al., 1988) and other species.

Elucidation of nitrogen uptake and assimilation processes in the slow-growing carob and comparative studies with fast-growing barley tomato and other crops led to the conclusion that the growth rate of these plants was related to the relative distribution of nitrate assimilation between the root and the shoot. The larger the proportion of nitrate reduced in the shoot, supported by adequate root structure and membrane transport mechanisms, the larger the resultants growth rate (Fig. 8.5). Consequently, the rate of activity of the K-shuttle, which determines the distribution of nitrate assimilation between root and shoot, seems to correlate directly with the relative growth rate of the plant.

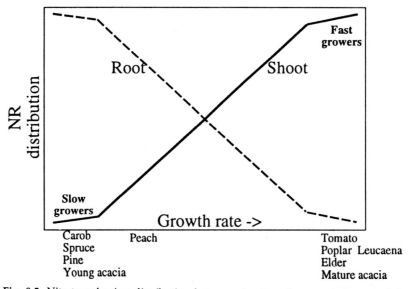

Fig. 8.5. Nitrate reduction distribution between shoots and roots of show- and fast-growing plants.

Not all plants keep up the same growth rate during their life cycle. Some plant species, notably *Acacia saligna*, have a slow initial growth rate during which they reduce nitrate mainly in the root, while developing a long and vigorous tap root, allocating practically no assimilates to the tiny shoot. This pattern is reversed 2–3 months after germination, when the root system starts transporting nitrate to the shoot through new and anatomically different lateral roots while, at the same time, increasing its relative growth rate several fold (Savidova and Lips, unpubl. data).

The most intriguing question arising from these observations is — how does increased nitrate transport to the shoot enhance growth?

III. SALINITY INHIBITION OF VEGETATIVE GROWTH

A fast-growing tomato can be transformed into a slow-growing plant by salinity. Plants grown under moderate saline conditions (6–10 dS m^{-1}) reduced their growth rate even when adapting physiologically to the stress and maintained in most cases normal rates (activity (g^{-1} or m^{-2}) h^{-1}) of ion and water uptake, nitrate assimilation capacity, photosynthesis and transpiration (Leidi et al., 1991). Since growth was limited, the 'normal' levels of these physiological parameters were preserved in plants with a smaller biomass.

Salinity inhibited nitrate transport to the shoot (Table 8.2), apparently due to interference with nitrate uptake and xylem loading, forcing increased reduction of these anions in the root (Cramer et al., 1995). Under these circumstances nitrate reduction in the shoot generally decreased and a concomitant increase of NR activity was observed in the root (Fig. 8.6).

Table 8.2. Concentration of nitrate and amino acids in the xylem sap of tomato plants as affected by NaCl in the nutrient medium (Cramer et al., 1995)

Concentration in xylem	mM NaCl in nutrient medium		
mM	0	100	200
Nitrate	9.8	4.4	4.8
Amino acids	2.9	6.7	6.4

The resultant change in the relative distribution of nitrate reduction sites within the plant was accompanied, as expected from the model relating growth rate to the distribution of reduction sites (Fig. 8.6), by a decrease in growth, leaf area and specific leaf area (Table 8.3).

Salinity increased nitrate reduction in the root through its interference with nitrate transport to the shoot which caused accumulation of nitrate in the root while decreasing drastically the growth rate of tomato plant. Enhancement of root NR under saline conditions has been ob-

Table 8.3. Effect of increasing NaCl concentration in the nutrient medium on growth, leaf area and SLA (specific leaf area) of tomato plants (Cramer et al., 1995)

Growth parameters	mM NaCl in nutrient medium		
	0	100	200
Dry weight (g)	2.11	1.28	0.55
Leaf area (m^2)	0.058	0.031	0.009
SLA (g m^2)	51	46	30

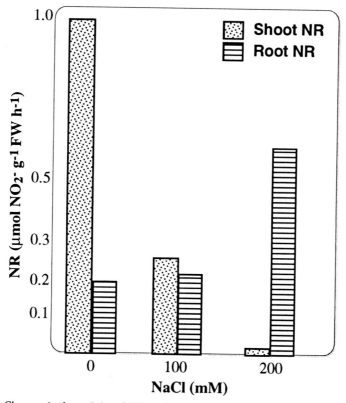

Fig. 8.6. Changes in the activity of NR in shoots and roots of tomato plants induced by increasing concentration of NaCl in the nutrient medium.

served in ryegrass (Sagi et al., unpubl. results) and soybean (Bourgeais-Chaillou et al., 1992).

IV. THE AMIDE/NITRATE RATIO IN THE XYLEM AND PARTITIONING

Plants reducing nitrate in the root enrich the xylem sap with amides, mainly glutamine and asparagine. When growing in saline conditions, nitrate loading onto the xylem is decreased according to the severity of the stress (Cramer et al., 1995; Gao and Lips, 1996). Salinity, consequently, increases very much the amide/nitrate ratio in the xylem (Fig. 8.7), a change accompanied by the enhanced translocation of leaf assimilates to the fruits at the expense of vegetative growth (Gao and Lips, 1996). Consequently, the improved fruit quality of greenhouse tomatoes irrigated with saline water or the earlier flowering and seed formation of plants in harsh ecosystems may be related to a similar increase in the amide/nitrate ratio in the xylem sap reaching the shoots (Gao and Lips, 1996). Bicarbonate fertilization greatly enhances nitrogen assimilation in the root of stressed plants by the anaplerotic supply of OAA which seems to operate as an emergency supply of carbon skeletons (Fig. 8.8) for ammonium assimilation when the provision of oxoglutarate by the respiratory Kreb's cycle is limited by energy demands of the root for growth and osmoregulation (Cramer et al., 1993; Cramer and Lips, 1995).

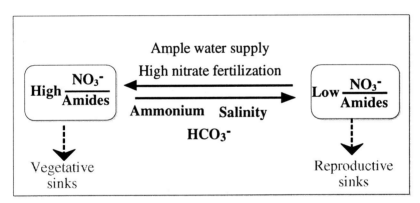

Fig. 8.7. The effect of water and fertilizer management on the ratio of nitrate/amides in the xylem sap on assimilate allocation to vegetative and reproductive sinks.

A higher pod retention or harvest index in actively N_2-fixing soybeans, which generate a high amide/nitrate ratio in the xylem, was accompanied by the repression of vegetative growth (Ishizuka, 1977). The type of nitrogen acquired by the plant had profound effects on its morphology (Hubick, 1990), and physiology (Rabie et al., 1980; Pearen and Huma, 1981; Lips et al., 1990). Bean plants obtaining their nitrogen by N_2-fixation (high amide/nitrate ratio in the xylem) had a higher

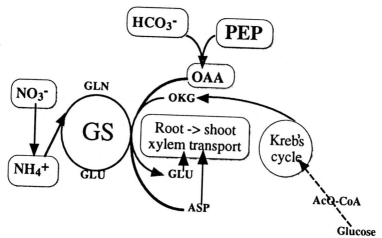

Fig. 8.8. Bicarbonate enhancement of nitrogen assimilation in roots under saline conditions.

harvest index than NO_3^--fed plants (low amide/nitrate ratio in the xylem) (Rodriguez et al., 1995).

A. Xylem Transport of Nitrate and Growth Rate

Several mechanisms may be considered responsible for the enhanced growth rate linked to xylem transport of nitrate to the shoot:

1) *Energetics of Nitrate Assimilation.* Roots receive a given allocation of assimilates (carbohydrate, malate and amino acids) from the shoot, which they use according to strict priorities: (a) osmoregulation; (b) ion uptake; (c) nitrogen uptake and/or assimilation; (d) maintenance and (e) growth. Salinity demands an increased investment of metabolic energy (ATP) for osmoregulation, nitrate reduction in the root and ammonium assimilation, all of which contribute to diminish the allocation of assimilates available for root growth.

The energy cost of nitrogen uptake and assimilation means that when nitrate is transferred to the shoot for its assimilation, a considerable amount of carbohydrate delivered by the shoot to the root could be allocated for root growth instead of generating energy for nitrate reduction and the provision of the carbon skeletons (oxoglutarate) required for ammonium assimilation into amino acids through the GS/GOGAT pathway (Lea, 1985). Under normal growth conditions, carbohydrate limitation does not exist in the shoot, in which the capacity of the chloroplasts to produce reductants by light reactions is generally larger than that required for CO_2 reduction (Torre et al., 1991; Bloom et al., 1989; Robinson, 1988). Consequently, nitrite is reduced in the shoot to ammonium, tapping an energy source which would otherwise be lost as fluorescence, while in the root it requires oxidation of a limited pool of glucose. It follows that for the whole plant energy economy, reducing nitrate to ammonium in the shoot costs less than doing it in the root.

2) *Carbon Paths in the Leaf.* The presence of nitrate in the leaf shifts the pathways of photoassimilated carbon compounds towards the formation of glucose, amino acids and malate (Cao et al., 1993; Foyer et al., 1994). In the absence of nitrate, a large fraction of the photoassimilates is stored in the leaf as starch (Cruz et al., 1993d). Therefore, nitrate presence in the leaf seems to direct traffic of newly formed carbon assimilates out of the chloroplast, increasing the availability of these assimilates for local metabolism and for export to other plant parts as source, supporting growth.

3) *Nitrate Accumulation in Leaves.* Nitrate is readily accumulated in leaf vacuoles, contributing to the turgor pressure required for cell expansion, one of the important elements of growth (Veen and Kleinendorst, 1985).

4) *Nitrate Transport, Cytokinin and ABA.* Xylem parenchyma cells control xylem loading (Fromard et al., 1995). These cells have a very large capacity to accumulate nitrate above its levels in the nutrient medium and discharge nitrate into the xylem vessels. Nitrate fluxes into and out of the XP cells may be regulated by cytokinin and ABA, just as these hormones promote K^+ fluxes into and out of stomata guard cells.

Cytokinin were effective when applied to plants grown under limited nitrogen supply (Trekova et al., 1992), increasing nitrate uptake and stimulating transport of nitrogen and carbon assimilates to the filling grains. Preliminary experiments in our lab have shown that enhanced nitrate transport to the shoot following the increase of nitrate supply to roots was accompanied by a parallel increase of cytokinin transport through the xylem and increased growth rate of tomato and sunflower plants. On the other hand, AO (aldehyde oxidase), the Mo-enzyme responsible for the last step of ABA synthesis, the oxidation of ABAld to ABA, was produced to a larger extent in barley roots of plants growing in ammonium and NaCl, while it was barely evident in plants exposed to nitrate (Fig. 8.9) (Omarov et al., 1998).

The factors described above may be the link between nitrate transport to the shoot and enhanced growth. They are not mutually exclusive and the promotion of growth rate in plants with an active K-shuttle may involve all or part of them.

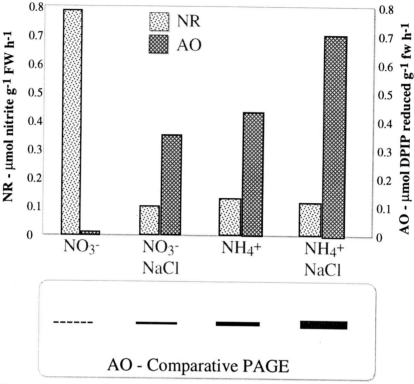

Fig. 8.9. Expression of two MoCo enzymes (NR and AO) in barley roots as affected by nitrogen source (nitrate or ammonium) and NaCl in the nutrient solution.

B. Xylem Loading of Nitrogenous Compounds

The xylem loading of nitrogenous compounds is very selective. Nitrate is readily taken up by the XP cells and discharged into the xylem vessels where its concentration is several fold higher than that in the medium, facilitating water uptake and building up root pressure. Nitrite and ammonium ions are not recognized by the xylem loading system under normal circumstances. Ammonium is generally not detected in the xylem sap. Amides, the main product of ammonium assimilation in the root are loaded by xylem parenchyma cells onto the xylem vessels, through efflux channels which differ from those of nitrate. At least two efflux channels (Fig. 8.10) may exist on the XP membrane in contact with the xylem vessels, an assumption based on the fact that salinity and drought inhibit nitrate but not amide xylem loading (Table 8.2). The uptake specificity of the xylem system determines that, while nitrate taken up by the root can be readily transported to the shoot where it will be reduced and assimilated, ammonium taken up from the medium

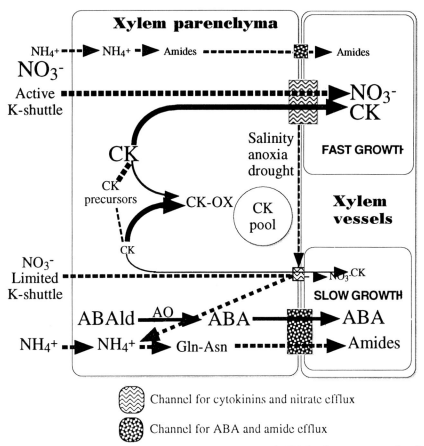

Fig. 8.10. Xylem loading of nitrate, amides, cytokinin and ABA in plant roots as related to slow or fast growth rates.

has to be assimilated in the root prior to its removal to the shoot in the form of amides (glutamine and asparagine) and amino acids (glutamate and aspartate).

The uptake specificity of the xylem determines also that nitrate-fed plants build up a higher water potential than plants grown in ammonium. It has been observed and reported on several occasions that ammonium-fed plants are associated with a predominance of ABA in the root, while nitrate-fed plants seem to be richer in cytokinins. The stress tolerance of nitrate-fed plants is considerably larger than that of plants grown in ammonium (Botella et al., 1993, 1994; Silberbush and Lips, 1991a, b).

Xylem parenchyma cells, responsible for the loading and unloading of the xylem vessels, are presumably the main regulatory stage on the transport of nitrate from the root to the shoot and, consequently, of the plant growth rate. At the same time, increased nitrate transport is associated with a higher cytokinin concentration in the xylem sap (Omarov, Kaminek and Lips, unpubl.) while the activity of AO, the enzyme responsible for the last stage of ABA production, is largest in plants growing in ammonium or under stress conditions (Fig. 8.9).

C. Regulation of Xylem Loading/Unloading

A limitation of xylem loading activity, induced with salinity in crops, takes place under normal conditions in certain plant species such as the carob tree (*Ceratonia siliqua* L.). The natural limitation of xylem loading of nitrate in carob results in the assimilation in the root of about 85% of the nitrate taken up (Cruz et al., 1993a). This situation is coupled to a very slow growth rate. In the case of carob, a natural cause inherent to the pattern of root development does not allow a very active nitrate loading of the xylem. Mycorrhiza will remedy this situation at later stages, by facilitating influx of nitrate through more mature areas of the carob root in which the anion can access the xylem vessels (Fig. 8.4). In both cases, natural or induced, most of the nitrate is reduced and assimilated in the root and the resultant overall growth rate of the plant is slow.

Salinity has been known to inhibit transport of cytokinins through the xylem (Kende, 1965). In view of the recent observations on the effect of this salt on xylem loading of nitrate and other inorganic ions one may consider the possible relationship between both effects. The fact is that NaCl blocks xylem loading of nitrate and cytokinins (Fig. 8.10). Cramer et al. (1995) observed that kinetin added to the nutrient medium of tomato plants affected xylem loading of nitrate in a way similar to NaCl. The key to the regulation of xylem loading must reside in XP cells, protoplasts of which have been isolated from barley (Wegner and Raschke, 1994) and pine trees (Leinhos and Savidge, 1993). These specialized cells, capable of massive influx and efflux of ions, seem to be similar to stomata guard cells. Guard cells are characterized by the involvement of these cells in either rapid influx or efflux according to hormonal signals or the water status of the plant in relation to the atmosphere. The fluxes in XP cells, forming the xylem parenchyma tissue, differ in the sense that they exhibit a rapid influx on the cortex side of the XP tissue but engage in massive efflux on the xylem side. Unfortunately, no details of the metabolism of XP cells *in vivo* are available. Both guard cells and XP cells are sensitive to osmotic changes and to hormones such as CK and ABA which may be involved in the control of the massive and rapid ion fluxes required by the plant. The xylem joint loading of amides and ABA is characteristic of slow plant growth due to stresses such as salinity or insufficient water supply, with the concomitant repression of cytokinin transport (Fig. 8.11). Fast-growing plants show a different activation of efflux channels, which are very active on xylem joint loading of nitrate and cytokinin with a considerable repression of amides and ABA transport (Fig. 8.11).

V. NITRATE, ALDEHYDE OXIDASE AND ABA

Is there any functional link between nitrate uptake by the root and the xylem transport of CK and ABA? Since AO, the enzyme which performs the last step of ABA synthesis is a Mo-enzyme, just as NR, our attention has lately been focussed on the regulation of Mo-enzymes in the root.

The Mo-pterin cofactor (MoCo) is the active component of three distinct enzymes in plants (Fig. 8.12): nitrate reductase (NR), aldehyde oxidase (AO) and xanthine dehydrogenase (XDH) (Rajagopalan and Johnson, 1992).

While expression of MoCo is controlled by a group of 7 cnx genes (Stallmeyer et al., 1995), control of the expression of the three apoproteins is under the control of distinct genes. The allocation of MoCo to these enzymes may determine their relative activities. AO is responsible for both the oxidation of ABAld to ABA as well as of IAld to IAA (Koshiba et al., 1996). The expression of AO of barley seedlings (Omarov et al., 1998) is repressed by nitrate in roots and is enhanced by ammonium and salinity (Fig. 8.9). Drought and salinity enhance expression of root AO, while in well-watered plants and lack of salinity, only traces of the enzyme are observed in the roots. Increased allocation of MoCo to AO takes place under stress or in the absence of nitrate, conditions which have been shown to enhance the concentrations of ABA and amides in the xylem sap, presumably due to change in the activation/repression of different channels in XP cells affecting the selective efflux from these cells into the xylem vessels (Fig. 8.11).

VI. CONCLUSIONS AND FUTURE PROSPECTS

A number of correlations between growth rate, K-shuttle activity and the resultant distribution of nitrate reduction sites in the plant have been observed. Correlations do not describe mechanisms but may be the first indicators of their existence. K-shuttle activity and nitrate reduction and

Fast growth rate

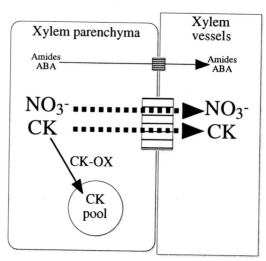

Slow growth rate

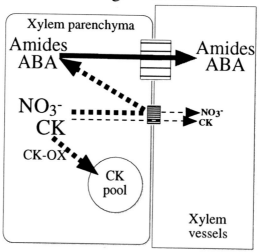

Fig. 8.11. Nitrogen and hormone coupled loading of the xylem sap in fast- and slow-growing plants.

assimilation in the shoot are associated with fast growth rates and have been selected by breeders to produce higher yielding crop cultivars. The mechanisms which determine fast growth in plants reducing nitrate in their shoots may be several: (a) the shoot can provide more and less

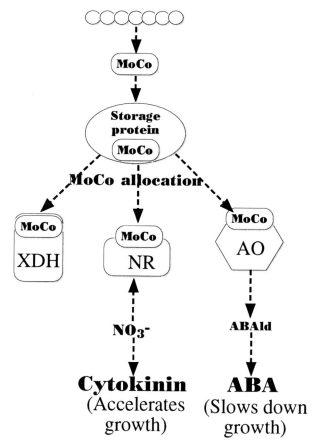

Fig. 8.12. MoCo synthesis and allocation to Mo-enzymes which affect the level of hormones in roots.

costly reductants and carbon acceptors for the reduction and assimilation of nitrate than the root; (b) nitrate readily accumulates in shoot cell vacuoles providing turgor for cell expansion and growth and (c) nitrate xylem loading promotes cytokinin transport to the shoot while depressing ABA levels in the xylem sap. The K-shuttle may promote growth, consequently, by enhancing the supply of growth signals (cytokinin) and the logistic delivery of elements essential for growth and principally nitrate and water to the shoot. Adaptation to stress conditions implies a restraint of the K-shuttle and CK transport to the shoot, while synthesis of root ABA is increased. These changes bring about a severe reduction of vegetative growth and promote fruits and seed sinks. ABA has been observed to be required for primary root elongation of maize at low water potentials

(Sharp et al., 1994). All these changes facilitate the survival and adaptation of plants to stress. The relative activity of the K-shuttle can therefore support rapid growth when the supply of minerals, water and solar energy are abundant (modern agricultural management) or contribute to the restraint of growth under limited supplies of water or minerals (arid and desert ecosystems).

LITERATURE CITED

Ben-Zioni, A., Lips, S.H. and Vaadia, Y. 1970. Correlation between nitrate reductase, protein synthesis and malate accumulation. *Physiol. Plant.* 23: 1039–1047.

Ben-Zioni, A., Vaadia, Y. and Lips, S.H. 1971. Nitrate uptake by roots as regulated by nitrate reduction products of the shoot. *Physiol. Plant.* 24: 288–290.

Bloom, A. J., Caldwell, R.M., Finazzo, J., Warner, R.L. and Weissbart, J. 1989. Oxygen and carbon dioxide fluxes from barley shoots depend on nitrate assimilation. *Plant Physiol.* 91: 352–356.

Botella, A., Cerda-Cerda, A. and Lips, S.H. 1993. Dry matter production, yield and allocation of Carbon-14 assimilates by wheat as affected by nitrogen source and salinity. *Agron. J.* 85: 1044–1049.

Botella, M.A., Cerda, A., Martinez, V. and Lips, S.H. 1994. Nitrate and ammonium uptake by wheat seedlings as affected by salinity and light. *J. Plant Nutr.* 17: 839–850.

Bourgeais-Chaillou, P., Perez-Alfocea, F. and Guerrier, G. 1992. Comparative effect of N-sources on growth and physiological responses of soybean exposed to NaCl-stress. *J. Exp. Bot.* 254: 1225–1233.

Burström, H. 1945. Studies on the buffer systems of cells. *Ark. Bot.* 32A: 1–18.

Cao, T.M., Bismuth, E., Boutin, J.P., Provot, M. and Champigny, M.L. 1993. Metabolite effects for short-term nitrogen-dependent enhancement of phosphoenolpyruvate carboxylase activity and decrease of net sucrose synthesis in wheat leaves. *Physiol. Plant.* 89: 460–466.

Cramer, M.D., Lewis, O.A.M. and Lips, S.H. 1993. CO_2 fixation and metabolism in maize roots as affected by nitrate and ammonium nutrition. *Physiol. Plant.* 89: 632–639.

Cramer, M.D. and Lips, S.H. 1995. Enriched rhizosphere CO_2 can ameliorate the influence of salinity on hydroponically grown tomato plants. *Physiol. Plant.* 94: 425–432.

Cramer, M.D., Schierhold, A., Wang, Y.Z. and Lips, S.H. 1995. The influence of salinity on utilization of root anaplerotic carbon and nitrogen metabolism in tomato seedlings. *J. Exp. Bot.* 46: 1569–1577.

Cruz, C., Soares, M.I.M., Martins-Louçao, M.A. and Lips, S.H. 1991. Nitrate reduction in carob (*Ceratonia siliqua* L.) seedlings. *New Phytol.* 119: 413–420.

Cruz, C., Lips, S.H. and Martins-Louçao, M.A. 1993a. Effect of root temperature on carob growth: nitrate versus ammonium nutrition. *J. Plant Nutr.* 16: 1517–1530.

Cruz, C., Lips, S.H. and Martins-Louçao, M.A. 1993b. Nitrogen assimilation and transport in carob plants. *Physiol. Plant.* 89: 524–531.

Cruz, C., Lips, S.H. and Martins-Louçao, M.A. 1993c. Uptake of ammonium and nitrate by carob (*Ceratonia siliqua* L.) as affected by root temperature and inhibitors. *Physiol. Plant.* 89: 532–543.

Cruz, C., Lips, S.H. and Martins-Louçao, M.A. 1993d. The effect of nitrogen source on photosynthesis of carob at high CO_2 concentrations. *Physiol. Plant.* 89: 552–556.

Cruz, C., Lips, S.H. and Martins-Louçao, M.A. 1995. Uptake regions of inorganic nitrogen in roots of carob seedlings. *Physiol. Plant.* 95: 167–175.

Davis, W.J. and Zhang, J. 1991. Root signals and the regulation of growth and development of plants in drying soil. *Ann. Rev,. Plant. Physiol. Plant Mol. Biol.* 42: 55–76.

Dijkshoorn, W. 1958. Nitrate accumulation, nitrogen balance and cation-anion ratio during regrowth of perennial ryegrass. *Ned. J. Agric. Sci.* 6: 211–221.

Foyer, C.H., Noctor, G., Lelandais, M., Lescure, J.C., Valadier, M.H., Boutin, J.P. and Horton, P. 1994. Short-term effects of nitrate, nitrite and ammonium assimilation on photosynthesis, carbon partitioning and protein phosphorylation in maize. *Planta.* 192: 211–220.

Fromard, L., Babin, V., Fleurat-Lessard, P., Fromont, J.C., Serrano, R. and Bonnemain, J-L. 1995. Control of vascular sap pH by the vessel-associated cells in woody species. Physiological and immunological studies. *Plant Physiol.* 108: 913–918.

Gao, Z.F. and Lips, S.H. 1996 (in press). Assimilate allocation priority as affected by nitrogen compounds in the xylem sap. *Plant Physiol. and Biochem.*

Horgan, J.M. and Waring, P.F. 1980. Cytokinins and growth responses of seedlings of *Betula pendula* Roth. and *Acer pseudoplantanus* L. to nitrogen and phosphorus deficiency. *J. Exp. Bot.* 31: 525–532.

Hubick, K.T. 1990. Effects of nitrogen source and water limitation on growth transpiration efficiency and carbon-isotope discrimination in peanut cultivars. *Aust. J. Plant Physiol.* 17: 413–430.

Ishuzuka, J. 1977. Function of symbiotically fixed nitrogen for grain production in soybean. *Proc. Int. Seminar on Soil Environment and Fertility Management* in *Intensive Agriculture*, Tokyo, Soc. Sci. Soil Manure. Showado Press, Kyoto, pp. 618–624.

Kaminek, M., Trckova, M., Motyka, V. and Gaudinova, A. 1994. The role of cytokinins in the control of wheat grain development and the utilization of nutrients. *Biol. Plant.* 36(S): 315.

Kende, H. 1965. Kinetin-like factors in the root exudate of sunflowers. *Proc. Nat. Acad. Sci. USA* 53: 1302–1307.

Kevekordes, K.G., McCully, M.E. and Canny, M.J. 1988. Late maturation of large metaxylem vessels in soybean roots: significance for water and nutrient supply to the shoot. *Ann. Bot.* 62: 105–117.

Koshiba, T., Saito, E., Ono, N., Yamamoto, N. and Sato, M., 1996. Purification and properties of flavin- and molybdenum-containing aldehyde oxidase from coleptyles of maize. *Plant Physiol.* 110: 781–789.

Lea, P.J., 1985. Ammonia assimilation and amino acid biosynthesis. In: *Techniques in Bioproductivity and Photosynthesis*. (J. Coombs, D.O. Hall, S.P. Long and J.M. Scurlock, eds.). Pergamon Press, UNEP, pp. 173–179.

Leidi, E.O., Soares, M.I.M., Silberbush, M. and Lips, S.H. 1991. Salinity and nitrogen nutrition studies on peanut and cotton plants. *J. Plant Nutr.* 15: 591–604.

Leinhos, V. and Savidge, A. 1993. Isolation of protoplasts from developing xylem of *Pinus banksiana* and *Pinus strobus*. *Can. J. Forest Res.* 23: 343.

Lips, S.H. and Beevers, H. 1966a. Compartmentation of organic acids in corn roots. I. Differential labeling of two malate pools. *Plant Physiol.* 41: 709–712.

Lips, S.H. and Beevers, H. 1966b. Compartmentation of organic acids in corn roots. II. The cytoplasmic pool of malic acid. *Plant Physiol.* 41: 713–717.

Lips, S.H., Ben-Zioni, A. and Vaadia, Y. 1970. K$^+$ recirculation in plants and its importance for adequate nitrate nutrition. In: *Recent Advances in Plant Nutrition* (R.M. Samish ed.). Gordon and Beach Science Publishers, New York-London-Paris. Vol. 1: 207–215.

Lips, S.H., Leidi, E.O., Silberbush, M., Soares, M.I.M. and Lewis, O.A.M. 1990. Physiological aspects of ammonium and nitrate fertilization. *J. Plant Nutr.* 13: 1271–1289.

Martins-Loução, M.A., Pereira, M.F. and Correia, P.M. 1996. The role of plant/soil microorganisms interactions in carob development. *3rd International Carob Symp.* Cabanas-Tavira, Portugal, 14 pp.

McCully, M. 1995. How do real roots work? Some new views of root structure. *Plant Physiol.* 109: 1–6.

Omarov, R., Sagi, M. and Lips, S.H. 1998. Regulation of aldehyde oxidase and nitrate reductase in roots of barley (*Hordeum vulgare* L.) by nitrogen sources and salinity. *J. Exp. Bot.* 49: 897–902.

Pearen, J.R. and Hume, D.J. 1981. ^{14}C-labelled assimilate utilization by soybeans grown with three nitrogen sources. *Crop. Sci.* 21: 938–942.

Rabie, R.K., Arima, Y. and Kumazawa, K. 1980. Uptake and distribution of combined nitrogen and its incorporation into seeds on nodulated soybean plants as revealed by ^{15}N studies. *Soil Sci. Plant Nutr.* 26: 427–436.

Rajagopalan, K.V. and Johnson, J.L. 1992. The pterin molybdenum cofactors. *J. Biol. Chem.* 267: 10199–10202.

Robinson, J.M. 1988. Spinach leaf chloroplast CO_2 and NO_2-photoassimilation do not compete for photogenerated reductant. *Plant Physiol.* 80: 676–684.

Rodriguez, D.N., Santamaria, C., Temprano, F. and Leidi, E.O. 1995. Effect of *Rhizobium* spp. inoculation on growth and xylem sap composition of bean cultivars. IV Portuguese-Spanish Meeting of Plant Physiology, Estoril.

Sagi, M., Savidov, N.A., Lvo'v, N.P. and Lips, S.H. 1996. Nitrate reductase and molybdenum cofactor in annual ryegrass as affected by salinity and nitrogen source. *Physiol. Plant.* 99: 546–553.

Silberbush, M. and Lips, S.H. 1991a. Potassium, nitrogen, ammonium/nitrate ratio, and sodium chloride effects on wheat growth. I. Shoot and root growth and mineral composition. *J. Plant Nutr.* 14: 751–764.

Silberbush, M. and Lips, S.H. 1991b. Potassium, nitrogen, ammonium/nitrate ratio, and sodium chloride effects on wheat growth. II. Tillering and grain yield. *J. Plant. Nutr.* 14: 765–773.

Sharp, R.E., Wus, Y., Voetberg, G.S., Saab, I.N. and Lenoble, M.E. 1994. Confirmation that abscisic acid accumulation is required for maize primary root elongation at low water potentials. *J. Exp. Bot.* 45: 1743–1751.

Stallmeyer, B., Nerlich, A., Schiemann, J., Brinkmann, H. and Mendel, R.R. 1995. Molybdenum-cofactor biosynthesis: the *Arabidopsis thaliana* cDNA *cnx1* encodes a multifunctional two-domain protein homologous to a mamalian neroprotein, the insect protein Cinnamon and three *Escherichia coli* Proteins. *The Plant J.* 8: 751–762.

Ting, I.P. and Dugger, W.M. 1967. CO_2 metabolism in corn roots. I. Kinetics of carboxylation and decarboxylation. *Plant Physiol.* 42: 712–718.

Torre, A. de la, Begon, A., Delgado, B. and Lara, C. 1991. Nitrate dependent O_2 evolution in intact leaves. *Plant Physiol.* 96: 898–901.

Trekova, M., Kaminek, M. and Zmrhal, Z. 1992. Grain formation and distribution of nutrients in wheat plants after the application of the synthetic cytokinin N^6-(Metahydroxybenzyl) adenosine. In: *Physiology and Biochemistry of Cytokinins in Plants* (M. Kaminek, D.W.S. Mok and E. Zazimalova eds.). SPB Acad. Publ. The Hague, pp. 241–244.

Ullrich, A. 1941. Metabolism of non-volatile organic acids in excised barley roots as related to cation-anion balance during salt accumulation. *Amer. J. Bot.* 28: 526–537.

Veen, B.W. and Kleinendorst, A. 1985. Nitrate accumulation and osmotic regulation in Italian ryegrass (*Lolium multiflorum* Lam.). *J. Exp. Bot.* 36: 211–218.

Wegner, L.H. and Raschke, K. 1994. Ion channels in the xylem parenchyma of barley roots. A procedure to isolate protoplasts from this tissue and a patch-clamp exploration of salt passageways into xylem vessels. *Plant Physiol.* 105: 799.

Chapter 9

SLOW-RELEASE NITROGEN FERTILIZERS AND PLANT NUTRITION

Masahiko Saigusa

I. Introduction
II. Concept and Merits of Slow-release Nitrogen Fertilizer
III. Types and Properties of Slow-release Nitrogen Fertilizer
    A. Coated Fertilizer
    B. Uncoated Fertilizer
    C. Supergranules and Others
    D. Stabilized Fertilizers
IV. Merits of Slow/Controlled-release Nitrogen Fertilizer on Nitrogen Metabolism
    A. Smooth Assimilation of Absorbed Nitrogen
    B. Utilization of Aimed Form of Nitrogen in the Soil
V. Application Methods of Slow/Controlled-release Nitrogen Fertilizer
    A. Time, Rate and Placement of Application
    B. *Co-situs* Application of Seeds and Fertilizers
    C. Single Basal Application
    D. Multicropping System
VI. Case Studies on Application of Slow-release Nitrogen Fertilizer
    A. Rice
    B. Soybean
    C. Corn and Sorghum
    D. Wheat and Barley
    E. Forage
    F. Turfgrasses
    G. Vegetable Crops
    H. Container Plants
VII. Environmental Aspects

VIII. Conclusions and Future Prospects

Literature Cited

I. INTRODUCTION

Nitrogen is one of the essential nutrients for plants and its practical management as the major element in intensive agriculture for plant production is an important aspect. Plants respond quickly to available nitrogen in soil and nitrogen absorbed adequately increases photosynthesis, vegetative growth and eventually high yield. Nitrogen deficiency causes drastic reduction in dry matter production in many crops, whereas its oversupply causes excessive vegetative growth, plant lodging and susceptibility to disease and insect pests. Therefore, an optimally adequate supply of nitrogen is required to be maintained throughout crop growth for better plant yield and crop production.

Breeders select high-yielding varieties, adaptable to a heavy dressing of nitrogen fertilizer for maximizing crop yield. Therefore, application of nitrogen fertilizer to supply enough nitrogen for crop production is a common practice for farmers nowadays. On the other hand, intensive use of nitrogen fertilizer in developed countries has caused serious environmental problems, such as groundwater pollution, eutrophication of streams and lakes, destruction of the stratospheric ozone layer, greenhouse effects, soil acidification, etc. (Shoji and Gandeza, 1992). Consequently, considerable attention has been paid to the dispersion of fertilizer nitrogen applied to agricultural lands, i.e., leaching of nitrate nitrogen into groundwater, emission of nitrous oxide by denitrification into the tropospheric atmosphere and volatilization of ammonia from the surface application of urea.

Both maximizing crop yields and reducing environmental pollution caused by agricultural practices are urgent subjects, considering the continuing rapid increase in world population and the reduction of increasing rate of world arable lands. It may be possible to solve these problems by developing effective application methods of easily soluble fertilizer, or intensive use of slow-release fertilizer. However, recent experiences in agricultural practice in Japan clearly show that the use of slow-(controlled-) release fertilizer for crop production is much more feasible than developing the effective use of conventional fertilizers (Shoji, 1995). The ideal slow/controlled-release fertilizers should release their nutrients in adequate amounts at a certain time when they are needed by the crops.

Table 9.1. Comparative recovery of basal N by rice plants in north-eastern Japan (Shoji and Kanno, 1994)

Placement	Fertilizer	Recovery, %	Reference
Broadcast	Ammonium sulphate or urea	22–23	Shoji and Mae, 1984
Broadcast	POCU-100	48–62	Ueno et al., 1991
co-situs	POCU-S100	79–83	Kaneta et al., 1994

II. CONCEPT AND MERITS OF SLOW-RELEASE NITROGEN FERTILIZER

Rapidly available nitrogen fertilizers, or easily soluble nitrogen fertilizers, such as ammonium sulphate, urea, potassium nitrate, etc. refer to materials that are dissolved in water immediately after application. On the other hand, slow release, slow acting, metered release, controlled release, controlled availability and delayed release are terms used with materials that release their nitrogen into the soil at rates and amounts that match the need of the growing plant (Hauck, 1985). Although the nuances among these technical terms differ, the most popular terms 'slow release' or 'controlled-release nitrogen fertilizer' are used in this chapter.

According to the Association of American Plant Food Control Officials (1995), slow-release fertilizer is defined as that fertilizer containing a plant nutrient in a form which delays its availability for plant uptake and use after application, or which extends its availability to the plant significantly longer than a reference 'rapidly available nutrient fertilizer' such as ammonium nitrate or urea, ammonium phosphate or potassium chloride.

The release rate of nitrogen from fertilizer to soil can be controlled by manufacturing the following materials (Hauck, 1985; Trenkel, 1996): (1) materials with impermeable or semi-permeable coating through which dissolved nutrients diffuse, (2) materials with low solubility containing a form of nitrogen available to the plant, (3) materials with low solubility which, during their chemical and/or microbial decomposition, release materials available to the plant, (4) water-soluble or relatively water-soluble materials that gradually decompose and (5) materials releasing nitrogen in delayed form due to a small surface-to-volume ratio.

In comparison with rapidly available nitrogen fertilizer, slow-release nitrogen fertilizers have the following potential benefits (Allen, 1984; Trenkel, 1996) due to their unique characteristics (namely, delay of initial availability and/or extended time of continued availability).

1) Efficient Use of Nitrogen by the Crop

Generally, the release pattern of slow-release nitrogen fertilizer is synchronized with the growth rate of the crop and recovery of this fertilizer by the crop is much higher than that of a rapidly soluble one. As shown in Table 9.1, the recovery of ammonium sulphate nitrogen applied by broadcasting in paddy fields is 22–23%. On the other hand, the recovery of basal nitrogen was doubled by broadcasting slow-release fertilizer (POCU) and was nearly four times by *co-situs* application with rice seeds (Shoji and Kanno, 1994). Thus, it is highly possible to decrease the application rate of nitrogen fertilizer in rice cultivation about 20–30% by using a slow-release fertilizer.

2) Reduction of Nitrogen Losses and Environmental Pollution

Nitrogen uptake by crops is greatly increased by use of slow-release fertilizer. That is to say, environmental pollution caused by nitrogen fertilizer, such as nitrate contamination of groundwater, emission of nitrous oxide and volatilization of ammonia, is noticeably reduced.

3) Reduction of Salt and/or Ammonia Toxicity of Crops

The crops grown with slow-release nitrogen fertilizers are generally free from burning even if fertilizer is applied in relatively great quantities, while rapidly soluble nitrogen fertilizers often cause salt and/or ammonia toxicity in plants resulting from their quick dissolution (Saigusa et al., 1993).

4) Lasting Nitrogen Supply

The rate of release of nitrogen from slow-release fertilizers can be controlled from a period of several days to several years.

5) Developing Innovative Farming Systems

Slow-release nitrogen fertilizers permit a heavy application of fertilizer with no burning of the crop. Therefore, a single basal application of total nitrogen fertilizer (Sato and Shibuya, 1991; Saigusa et al., 1993), no-tillage farming with single *co-situs* fertilizer application (Kaneta et al., 1994; Kaneta, 1995; Saigusa et al., 1996), and multicropping vegetable farms, which significantly save labour, working hours and fossil fuels, can be realized in Japan (Shoji and Gandeza, 1992).

6) Reducing Application Cost

The innovative farming systems mentioned above reduce application cost. For example, no-tillage rice transplanting cultivation with single basal fertilization decreased farming costs by 65% compared to that of conventional rice cultivation (Kaneta et al., 1994).

7) Reducing Soil Acidification Originating from Heavy Application of Fertilizer

Rapidly soluble acidic nitrogen fertilizers such as ammonium sulphate, ammonium chloride increase the soil solution of SO_4^{2-} and Cl^-. Adding to these anions, NO_3^- formed by nitrification of NH_4^+ encourages soil acidification. However, some slow-release fertilizers such as resin-coated urea, condensation products of urea and aldehyde do not encourage soil acidification.

8) Smooth Assimilation of Absorbed Nitrogen

Rapid accumulation of ammonium ions and amino acids in plants by the application of soluble nitrogen may cause plant lodging and susceptibility to disease and insect pests. On the other hand, nitrogen absorbed from slow-release fertilizers is smoothly assimilated into the plant.

Slow-release nitrogen fertilizer has many potential benefits as summarized above. However, the most serious possible disadvantage is that the price of slow-release fertilizers is several times higher than that of conventional soluble fertilizer (Trenkel, 1996). Thus, internationally the application of this fertilizer is limited to several plants, such as horticultural plants, lawns and high-cash crops (Maynard and Lorenz, 1979). However, in Japan, slow-release fertilizers are extensively applied to many crops and are called 'Environment friendly fertilizers'. Even rice farmers are familiar with the merits of this fertilizer. Therefore, this chapter will deal mainly with the Japanese experience, especially experience with polyolefin-coated fertilizers (POCF).

III. TYPES AND PROPERTIES OF SLOW-RELEASE NITROGEN FERTILIZER

Release of nutrients from the fertilizer to the soil can be controlled in numerous ways, as mentioned above, and slow/controlled release and stabilized synthetic fertilizers are grouped as follows (Maynard and Lorenz, 1979; Hauck, 1985; Trenkel, 1996):

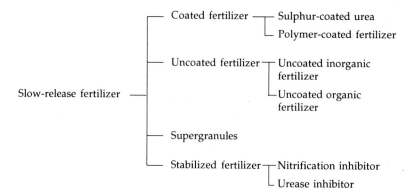

A. Coated Fertilizer

Various materials such as sulphur, polymers (e.g. PVDC-based copolymers, polyolefine, polyethylene, polyesters, alkyd resins, etc.), fatty acid salt, latex, rubber, wax, urea-formaldehyde resin, phosphogypsum, peat, etc. have been used as coatings for rapidly available nutrients, especially nitrogen. However, sulphur and polymers are the most commercially important. Therefore, mainly sulphur-coated and polymer-coated nitrogen fertilizers are discussed here.

1. *Sulphur Coated Urea (SCU)*: SCU was developed in 1961 by the Tennessee Valley Authority (now called the National Fertilizer Development Center). Both urea and sulphur are relatively inexpensive materials and sulphur itself is also one of essential elements for plants. Thus, sulphur-coated urea has gained the greatest importance within coated fertilizers (Trenkel, 1996).

Urea granules are coated with sulphur (impermeable membranes which slowly degrade through microbial, chemical and physical processes) and then with sealant wax and conditioner. A typical final product contains 36% N (as urea granules), 10 to 16% S and 5% sealant. However, the concentration of nitrogen and its release rate varies with the thickness of the sulphur coating. Currently manufactured products contain up to 42% N and 16% S.

SCU-10, SCU-30 and SCU-40 release 10%, 30% and 40% of their nitrogen within seven days respectively. Factors affecting the dissolution reaction of SCU were summarized by Allen (1984) as follows:

(1) the nitrogen release rate is accelerated in warm soil temperatures suggesting that biodegradation is involved in coating failure; (2) moisture tension accelerates the release of nitrogen; (3) dissolution is more rapid with surface placement than when mixed or layered in the soil; (4) root action accelerates dissolution of SCU; (5) dissolution is not affected by soil pH in the range of 5 to 8; (6) biodegradation of coating materials proceeds slowly, forming H_2SO_4. Thus, SCU is an effective source of both slow-release nitrogen and sulphur, and may also contribute to soil acidity.

SCU costs about 35% to 40% more than urea and is thought to be too expensive for general agricultural use. The effectiveness of SCU for rice production under rainfed conditions and for the production of ornamental crops, turfgrass, cranberries, etc., has been widely recognized (Maynard and Lorenz, 1979; Hauck, 1985).

2. *Polymer- or Resin-coated Controlled-release Nitrogen Fertilizer*: In this type of fertilizer, impermeable or semi-permeable membranes with tiny pores are used as polymer coatings. Coating resins are divided into two

groups: thermoplastic resin (polystyrene, polyethylene) and thermohardening resin (alkyd, phenol).

The choice of coating materials and technical coating process varies greatly among the producers. However, polymer-coating technology is evaluated to be much more advanced than that of sulphur coating (Trenkel, 1996), mainly because a much more accurate control of nitrogen release rate is obtained in polymer-coated fertilizers.

For instance, the release of urea from polyolefin-coated urea (POCU) particles is well controlled by changing the ratios of ethylene vinyl acetate, EVA (water permeable) to polyethylene, PE (water impermeable) as shown in Table 9.2 and Figure 9.1 (Fujita et al., 1977, 1983). The number of days required to release 80% of the urea from fertilizer particles in water at 25°C is 1300 days, if it is coated with 100% PE, and 98 days for coatings of PE:EVA=50:50. Furthermore, the release rate of nitrogen from POCU should be matched to the growth rate of crop. Thus, Fujita et al. (1983) adjusted the temperature dependency, Q_{10} of POCU by about 2 by adding talc to the resin in the range of 50–60%, because biochemical reaction in plants is about 2.

The release patterns of nitrogen from fertilizers are divided into two or three groups as shown in Figure 9.1: ordinary release groups (linear type and exponential type) and delayed release groups (sigmoid type) (Shoji and Gandeza, 1992; Shiozaki, 1996). The ordinary release POCU dissolves urea at a specific rate starting at the time of application while the delayed release POCU dissolves urea 30 or 40 days after application in water at a constant temperature of 25°C. In the ordinary release group, the POCU with a long release duration tends to be a linear release type, while the POCU with a short release duration tends to be an exponential type.

Table 9.2. Relation of the amount of polyethylene (PE) and ethylene vinyl acetate (EVA) in the fertilizer coating to the duration of N release (Fujita et al., 1983)

Composition of coating material (%)		Duration of Release* (days)
PE	EVA	
50	50	98
60	40	135
65	35	187
70	30	260
80	20	330
90	10	410
100	0	1300

* No of days to release of 80% of urea-N from the coated particles in water at 25°C.

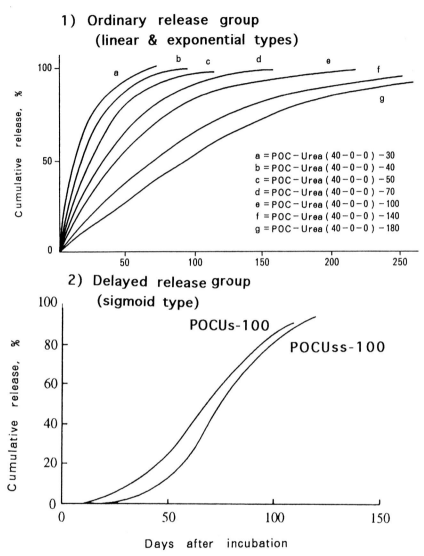

Fig. 9.1. Release of Nitrogen from different types of polyolefin coated urea(POC) in water at 25°C (Fujita, et al., 1983).

In both ordinary and delayed groups, POCFs contained either mono-, di-, tri-, or multinutrient elements (POCFs with release durations in the range of 30 to 360 days are available).

In Japan, there are more than ten companies which are interested in slow/controlled-release coating fertilizers and are marketing their own products, as shown in Table 9.3 (Shiozaki, 1996). The consumption of

Table 9.3. Selected formulation of commercially available slow/controlled-release coated nitrogen fertilizers in Japan (Shiozaki, 1996)

Nutrient element	Commercial name	Releasing type	No. of days for 80% dissolution at 25°C	Nutrient Content (%)	Company name
N	LPcoat	exponential	30,40,50,70,100,140,180	40-0-0	Chisso
N	LPcoat-S	sigmoid	60,100	40-0-0	
N	LPcoat-SS	sigmoid	100,140	40-0-0	
NPK	Long424	exponential	40,70,100,140,180,270,360	14-12-14	Asahikasei
NPK	Long331	exponential	100,140,180,270,360	13-3-11	
NPK	Long250	exponential	40,70,100,140,180	20-5-10	
NPK	Long426	exponential	70,100,140,180	24-2-6	
NK	SuperNKLong203	sigmoid	100,140,180	20-0-13	
NK	NKLong203	exponential	70,100,140,180	20-0-13	
NPK-Mg-ME	LongTotal	exponential	40,70,100,140,180,360	13-11-13-2	
NPK	SuperLong424	sigmoid	100,140,180	14-12-14	
NPK-Mg-ME	MicroLongTotal	exponential	40,70,100	12-10-11-2	
NCa	Longshokaru	exponential	40,70,100,140	12-0-0-23	
N	Seracoat-U	sigmoid	S(40),M(70),L(120),LL(140)	N:36-40	Central Glass
NK	Seracoat-CX	exponential	S(40),M(70),L(120)	15-0-15	
NPK	Sigmacoat-S200	sigmoid	2.5M*,4M,6M	12-10-10	Katakura Chikkarin
N	Sigmacoat-U	sigmoid	No2,3,4,5,6	N:34-40	
N	Coopcoat-U	sigmoid	2.5M,4M	40-0-0	Coop-Chemical
NPK	Coopcoat-F202	sigmoid	2.5M,4M	12-10-12	
NPK	Coopcoat-Fs202	sigmoid	2.5M,4M	12-10-10	
NPK	Coopcoat-Fsn331	sigmoid	2.5M,4M	13-3-11	
N	Mcoat-L	linear	60,140	41-0-0	Mitsubishi Kasei
N	Mcoat-S100	sigmoid	100	41-0-0	
N	Mcoat-S140	sigmoid	140	41-0-0	
NK	Coating NK	exponential	100	8-10-8	Kyowahakkou
N	UCcoat		T,H type	40-0-0	Ubekosan
NPKS	SCcoat	exponential	H,L	12-12-12	Mitsui Toatsu
NPK	Nippirincoat	exponential	50	8-12-8	Nihonhiryo

* M: month

polymer-coated fertilizers increased by 470% during the period 1985 to 1994 in Japan. Increasing rate of consumption of polymer-coated fertilizers in Japan (about 47% per year) is extraordinarily higher than that in the USA (about 10%) or in Europe (6.5%) (Trenkel, 1996).

B. Uncoated Fertilizer

The release rate of nitrogen from low-solubility nitrogen fertilizers is closely related to their exposed surface area, namely their size and density (Maynard and Lorenz, 1979). Thus, the larger, harder particles need a much longer time to dissolve than the smaller, softer ones.

1. *Uncoated Inorganic Fertilizer*: The general formula of uncoated inorganic nitrogen fertilizers is expressed as $MeNH_4PO_4 \cdot xH_2O$, where Me is a divalent metal ion (Me^{2+}). Among these materials, magnesium ammonium phosphate, MagAmp, was first suggested to be useful as a slow-release fertilizer in the middle of the 19th century and is the most common. Magnesium ammonium/potassium phosphate [7-40-72-12(Mg), 7-17-6-12(Mg)] are marketed as a trade name MagAmp (Hauck, 1985). The soil conditions affecting the release of nitrogen from fertilizer particles are nitrification rate, soil acidity and soil moisture, but not soil temperature in the range of 10°C–48°C.

2. *Uncoated Organic Fertilizer*: Urea is a kind of organic nitrogen fertilizer but belongs to rapidly available nitrogen fertilizers. However, many condensation products of urea and aldehydes, such as urea-formamide (UF), isobutylidine diurea (IBDU), crotonylidene diurea (CDU), etc. are sparingly soluble in water and useful as slow-release nitrogen fertilizers. Other uncoated slow-release organic nitrogen fertilizers are triazines (urea pyrolyzates), oxamide (diamide of oxalic acid), guanylurea, dicyandiamide, thiourea, processed waste products, etc. Among these uncoated organic fertilizers, the discussion here focusses on UF, IBDU and CDU, because of their importance in commercial use as slow-release nitrogen fertilizers.

a) Urea-formaldehyde (UF)

Urea-formaldehyde, ureaform, is formed by reacting urea with formaldehyde to produce a mixture of unreacted urea and methyleneureas with different long-chain polymers. A typical product contains 38% total nitrogen. UF polymer-mixture is fractioned as follow: Fraction I = cold-water soluble, CWS (25°C), Fraction II = hot-water soluble, HWS (100°C) and Fraction III = hot-water insoluble, HWI. The nitrogen efficiency is determined by the activity index: $AI = II/(Ii+III) \times 100 = (CWI - HWI)/CWI \times 100$, where, CWI = cold-water insoluble.

The American Plant Food and Control Officials (APFCO) have set an AI of 40 as a minimum with at least 60% of its CWI-N and total nitrogen content of at least 35%. Unreacted urea-N is usually less than 15% of total-N (Trenkel, 1996). The release of nitrogen from UF seems to be much more related to biological activity (temperature sensitive) than particle size and hardness of fertilizer. The UF reaction products have the largest share of the slow-release nitrogen fertilizer market in the world.

b) Isobutylidene Diurea (IBDU)

Isobutylidene diurea is a condensation product of urea and isobutylaldehyde. Its total nitrogen content is 32.18% theoretically; the APFCO definition requires a minimum of 30%, of which 90% is CWS-N (Trenkel, 1996). In upland conditions, water-insoluble IBDU is gradually hydrolyzed into urea, and urea is transformed to ammonium ions and then to nitrate by soil bacteria. Thus, the release rate of nitrogen from IBDU can be controlled by its particle size. The rate of hydrolysis is also influenced by soil temperature, moisture and pH, but not microbial activity.

c) Crotonylidene Diurea (CDU)

CDU was synthesized for the first time by the condensation of urea and crotonaldehyde, but nowadays is formed by the acid catalysis of urea and acetic aldehyde. It contains about 32% total N. The decomposition of CDU in the soil is performed by both hydrolysis and microbial process. Therefore, the release rate of nitrogen is controlled by its particle size as IBDU, and is also influenced by soil temperature and soil moisture.

Recently, it was demonstrated that CDU application brings a marked increase in bacterial population and the bacteria/fungi (B/F) ratio by decreasing the fungi population, and can control soil-borne pathogens. In the case of Chinese cabbage, the average proportion of diseased plants (clubroot: *Plasmodiophora brassicae*) in CDU treatment was markedly lower than that in ammonium sulphate (AS) treatment (CDU plot: 17.3% and 7%, AS plot: 71% and 68%, respectively in 1980 and 1981). The percentage of diseased plants is closely correlated to soil pH, bacterial population and B/F ratio. The disease-suppressing effects of CDU were also observed in Fusarium wilt in cucumber, clubroot in turnip, Pythium root rot in ginger, Fusarium wilt in strawberry, etc. (Chinen, 1995).

C. Supergranules and Others

The release rate of conventional soluble fertilizers is controlled by manufacturing compact and/or large particle fertilizers which means a relatively small surface-to-volume ratio.

IBDU and UF themselves are slow-release nitrogen fertilizers, but by making bigger particles (e.g. 16–17g/tablet), the releasing effect of the nitrogen can last for 2–4 years (Mitsubishi Chem. Indust., 1986). Materials such as supergranules, briquettes, tablets and sticks are preferably used in fertilizing trees and shrubs, turfs, some vegetables, ornamental plants, etc. (Trenkel, 1996).

D. Stabilized Fertilizers

Fertilizers containing nitrification and urease inhibitors are not classified as slow-release nitrogen fertilizers in the strict sense of the word, because many plants can absorb ammonium ion and urea themselves. However, these inhibitors can reduce the loss of nitrogen by leaching in nitrate form (NO_3^-), nitrous oxide (N_2O) emission or ammonia volatilization by inhibiting the following processes:

$$NH_4^+ \xrightarrow{\text{Nitrification inhibitor}} NO_2^- \xrightarrow{} NO_3^-$$
$$\textit{Nitrosomonas} \qquad \textit{Nitrobacter}$$
$$(NH_2)_2CO \xrightarrow[\text{Urease}]{\text{Urease inhibitor}} 2NH_3 + CO_2 + H_2O$$

Many chemicals have been reported as useful in inhibiting the nitrification process in soil; however, up to now only two types have gained practical and commercial importance: Nitrapyrin (2-chloro-6-(trichloromethyl)-pyridene) and DCD (dicyandiamide). On the other hand, only NBTPT or NBPT (N-(n-butyl) thiophosphoric triamide) are thought to be promising products as urease inhibitors (Trenkel, 1996).

IV. MERITS OF SLOW/CONTROLLED-RELEASE NITROGEN FERTILIZER ON NITROGEN METABOLISM

Because of their slow/controlled-release characteristics, slow-release nitrogen fertilizers have the following effects on nitrogen metabolism in field crops.

A. Smooth Assimilation of Absorbed Nitrogen

Heavy application of rapidly available nitrogen ferilizers often causes ammonium and/or nitrate accumulation in plant tissue. Ammonium may be toxic to plant tissue but absorbed ammonium is rapidly detoxified by conversion to amino acids and amides. However, even though excess nitrogen causes elongation of the vegetative phase and the entire plant life cycle, it also causes abnormal internode elongation and increased succulence, resulting in grain loss through delayed maturity, lodging, and susceptibility to disease and pest attack. Most plants tolerate nitrate accumulation within reasonable amounts, but excess nitrate can be toxic to some plants and to animals fed with grass containing it.

On the other hand, orchard grass grown by controlled-release nitrogen fertilizer assimilate absorbed nitrogen smoothly and did not accumulate high amounts of ammonium and/or nitrate (Saigusa et al., 1996). However, grass grown with urea accumulated more than 0.5% of nitrate after 15 days, which is much higher than the critical level (0.2%) of nitrate toxicity to cattle.

B. Utilization of Aimed Form of Nitrogen in Soil

Generally, in upland conditions, nitrogen applied in ammonium form is easily converted to nitrate by the activities of *Nitrosomonas* and *Nitrobacter* bacteria, whereas in paddy soil conditions, nitrogen in ammonium form is much more stable because of reductive conditions. Thus, it is very difficult to supply nitrogen in ammonium form for upland crops and in nitrate form for rice using conventional rapidly available nitrogen fertilizer. However, it is highly possible to supply such forms of nitrogen to the crops if they are applied as slow-release nitrogen fertilizers, especially polymer coated ones, because nitrogen released from fertilizer is directly absorbed by the plant root without injury.

In spinach cultivation in the field, accumulations of both oxalic acid and nitrate in plants are serious problems to human health. On the other hand, in solution cultures the total oxalic acid content of spinach can be decreased by increasing the ratios of ammonium to nitrate in the solution. Therefore, reduction of both oxalic acid and nitrate contents of spinach were studied in the field by growing with slow-release fertilizers: coated urea (UC) and coated ammonium phosphate (CAP) (Takebe et al., 1996). Compared to the ammonium sulphate plot, the contents of both oxalic acid and nitrate in spinach in the UC and CAP plots were significantly decreased, while the contents of sugar and ascorbic acid in these plots tended to increase. Thus, application of slow-release nitrogen fertilizers induced desirable effects on the spinach quality.

On the other hand, the rates of seed germination and seedling growth of rice in direct sowing cultivation with coated ammonium nitrate increased much more than those grown with coated urea and conventional ammonium sulphate fertilizer (Tashiro et al., 1996).

V. APPLICATION METHODS OF SLOW/CONTROLLED-RELEASE NITROGEN FERTILIZER

A. Time, Rate and Placement of Application

The efficiencies of applied conventional nitrogen fertilizers are remarkably influenced by time, rate and placement of application. Many application methods of nitrogen fertilizers have been devised: basal or top-dressing, a single basal or split, broadcasting, spot or band or broadcasting, surface or deep, etc. However, the ideal fertilizer should supply nutrients in adequate amounts at certain times when they are needed by plants.

Generally, slow-release or controlled-release nitrogen fertilizers are much more synchronized with the growth rate of plants and show much higher recoveries in plants in comparison with conventional fertilizers. For instance, the recoveries of nitrogen from polyolefin coated urea (POCU) in rice transplanting cultivation (61–83%) were remarkably higher than those from ammonium sulphate (9–33) regardless of fertilizer placement, as shown in Table 9.1 (Kaneta, 1995 a,b). Among the POCU plots, the highest recovery of fertilizer nitrogen (83%) was shown in *co-situs* (contact) application of seeds and fertilizer in a nursery box in which it was impossible to use rapidly available nitrogen fertilizer because of burning.

B. *Co-situs* (contact) Application of Seeds and Fertilizers

Rapidly available nitrogen fertilizers applied in the same site as seeds or in the intensive rooting zone, often cause salt and/or ammonium injuries to the plants because of their high solubility. Therefore, only small amounts of rapidly available fertilizer are used as seed-placement or contact-placement to give the crop a rapid start. However, controlled-release nitrogen fertilizers can be applied with seeds or near roots in large amounts with no burning (Fig. 9.2). To distinguish the seed-placement or contact-placement of small amounts of rapidly available fertilizer, Shoji and Gandeza (1992) proposed to name the application method '*co-situs*', placements' or '*co-situs* application', which means the existence of both fertilizers and seeds in the same sites (*situs*).

The central concept of *co-situs* application is to apply the nutrients in the intensive rooting zone with controlled-release fertilizers which have nutrient release patterns synchronized to the demand of the plant over the whole growing season (Shoji and Kanno, 1994). Therefore, *co-situs* application of controlled-release fertilizer increases the absorption of nutrients by the plants to a great extent through minimizing the reaction and interaction of the nutrients with the soil. A nitrogen supply

Fig. 9.2. *Co-situs* application of rice seed and POCUs-100 fertilizer in nursery box.

which lasts over the whole growing season from the controlled release fertilizer eliminates the need to top-dress and leads to the development of a single basal application of the total nitrogen fertilizer required of the plants. The *co-situs* application method demands only a narrow (about 5 cm width and 5 cm depth) spacing for seed and fertilizer and, consequently, leads to the establishment of no-tillage direct sowing with a single basal fertilization, which saves both labour and fossil energy (Saigusa et al., 1996).

C. Single Basal Application

Japanese agriculture has been facing serious problems, such as the rapid rise of labour costs, liberalization of trade, overproduction of rice, advanced age of farmers, decrease in successors, etc. Farmers are forced to reduce labour and production costs and increase productivity. Therefore, farmers are greatly interested in the single basal application method of total nitrogen required by crops using controlled-release nitrogen fertilizer. This application method is being extensively applied to many kinds of crops, such as rice, corn, wheat, rush, potato, strawberry, eggplant, spinach, onion, grass, etc. (Shoji, 1995; Shoji and Gandeza, 1992).

Achievement of the single basal application method with satisfactory high yield levels requires the following detailed information of each crop: desirable pattern of nitrogen uptake, assessment of natural nitrogen supply and recovery of soil and fertilizer nitrogen.

1. *Desirable Pattern of Nitrogen Uptake*: Both the amount and pattern of nitrogen uptake of crops influence their yield and quality to a great extent. Thus, a desirable pattern of nitrogen uptake of crops has first priority in the single basal application of total nitrogen fertilizer requirements. For example, Ueno et al. (1991) summarized the nitrogen uptake process of high-yield and high-quality rice (*Oryza sativa* cv. Sasanishiki) by transplanting, from the data of 280 samples in the past 15 years (1973–1988) and concluded that 6 t/ha of high-quality brown rice requires the uptake of 120 kg N/ha, and the processes of nitrogen uptake are: 40 kg N/ha from transplanting to 30 June, 60 kg N/ha from 30 June to heading, 20 kg N/ha from heading to maturation.

The necessary amount of 120 kg N/ha for 6 t/ha rice production should be supplied by soil nitrogen and fertilizer nitrogen. Single basal application of total nitrogen fertilizer is achieved by slow/controlled-release fertilizer or a bulk blend fertilizer of both conventional and controlled-release fertilizers, depending on the desirable pattern of nitrogen uptake of rice.

2. *Assessment of Natural Nitrogen Supply*: The amount of natural nitrogen supply is estimated by subtracting methods, ^{15}N-tracer method, the incubation method, etc. Among these methods, the incubation method is being extensively examined in estimating natural nitrogen supply both in upland and paddy soils (Sugihara et al., 1986). The following three nitrogen mineralization models were proposed to simulate nitrogen mineralization process in soils:

①simple type model
$$N=N_0 \{1-\exp(-k.t)\}$$
②two simple type models
$$N=N_0^1 \{1-\exp(-k_1.t)\}+N_0^2\{1-\exp(-k_2.t)\}$$
③simple type model combined with immobilization
$$N=N_{im}\{1-\exp((-K_{im}.t)\} + N_0\{1-\exp(-k.t)\}+C$$

where N(mg/100 g soil or %); amounts of mineralized nitrogen
N_0 (mg/100 g soil or %); N-mineralization potential (N_0^1, N_0^2: 1 and 2 are the indexes of the different components. N_0^1 and N_0^2 show a rapidly decomposable component and a moderately decomposable one respectively.
N_{im} (mg/100g soil or %); amounts of immobilized nitrogen

k, k_1, k_2 (day^{-1}): mineralization rate constant
k_{im}: immobilization rate constant.

The nitrogen mineralization pattern of paddy soil including air-drying effects is expressed by the two simple type models.

3. *Recovery of Soil and Fertilizer Nitrogen*: In order to determine the rate of fertilizer application in the single basal application of rice, recoveries of both soil and fertilizer nitrogen by rice were estimated by a ^{15}N-tracer study as follows (Ueno et al., 1991): 30–40% for rapidly available fertilizer (basal), 60–70% for controlled-release fertilizer (basal), 30–40% for initial ammonium nitrogen released by soil drying, 60–70% for ammonium nitrogen released at the middle and later growth stages and 30–40% for ammonium nitrogen mineralized in subsoil at the middle to late growth stage.

On the other hand, the recoveries of both soil and fertilizer nitrogen by upland crops are very diverse, because recoveries vary depending on soil condition, amount of precipitation, root system of crop, etc. The recoveries of controlled-release fertilizer nitrogen (46–58%) were relatively stable compared to those of rapidly available fertilizer nitrogen (7-48%) (Saigusa et al., 1990).

D. Multicropping System

Double cropping of lettuce and Chinese cabbage by single application of fertilizer was conducted in Nagano prefecture Japan (Sato, 1995). POC-NPK (14-12-14)-140 was applied on the ridges in the previous autumn to disperse labour in spring. Firstly, lettuce was transplanted in early spring. After lettuce harvesting in summer, Chinese cabbage was immediately transplanted on the same ridges without fertilization. Satisfactory yields of both lettuce and Chinese cabbage were gained by single application of POC-fertilizers. Multi-cropping of spinach was also conducted using both cell-grown-seedlings of spinach and POC-NPK fertilizer and saved labour costs to a great extent.

VI. CASE STUDIES ON APPLICATION OF SLOW-RELEASE NITROGEN FERTILIZER

A. Rice

Rice is one of the most promising crops for the 21st century because it can be planted in the most diverse environmental conditions, from sea level to 3000 m in Nepal, from tropic to cold temperate zones, from strongly acidic to neutral soil, from upland to wetland. Therefore, rice may show relatively stable or favourable production in climatic changes, such as higher temperature and higher CO_2 concentrations.

Rice is mainly cultivated in flooded soil and the soil-fertilizer regime is completely different from other crops, particularly with respect to the behaviour of applied fertilizer nitrogen (Allen, 1984). Namely, the nitrate contained in the fertilizer or converted from ammonia prior to flooding is entirely lost under flooded soil conditions through denitrification. Therefore, ammonium-N or amide-N-containing fertilizers, such as ammonium sulphate, urea, etc., have been used preferentially in fertilization of wetland rice. However, the recoveries of these fertilizers applied as basal treatment in paddy soil are not high (20–40%) (Shoji and Mae, 1984) and losses will be remarkably increased where alternate flooding and drying occur (in the areas dependent on natural rainfall, inadequate supply of irrigation water, treatment of midsummer drainage). Under alternate flooding and drying conditions, controlled-release fertilizers may be more effective than conventional fertilizers. Scientists were first interested in both sulphur-coated urea (SCU) and isobutylidene diurea (IBDU), but not in urea form (UF), because release of nitrogen from UF results from aerobic biodegradation (Allen, 1984). A large number of rice experiments using SCU were conducted in Asia and South America under conditions of intermittent flooding. SCU applied as a basal treatment was superior to urea in 46 to 56 experiments (Allen, 1984).

In recent years, there has been much interest in resin-coated urea for rice cultivation in Japan. Most of the experiments carried out with polyolefin-coated urea (POCU) proved that nitrogen recovery from POCU by rice is much greater than that from conventional urea or ammonium sulphate as already shown in Table 9.1 (Kamekawa, 1990; Ueno et al., 1991; Wada et al., 1991; Sato et al., 1993; Kaneta, 1995). Therefore, it was proved that reduction of application rate of POCU-N is feasible without decrease in brown rice yields (Kaneta, 1995): 5760 kg brown rice /ha for 100 kg N/ha of conventional ammonium sulphate, 6040 kg brown rice/ha for 41 kg N/ha of POCU and 6540 kg brown rice/ha for 62 kg N/ha of POCU.

Rice-transplanting cultures with a single basal application of total fertilizers using POCU or bulk blending fertilizer of POCU and conventional compound fertilizer are being extensively conducted throughout the country and are thought to be promising methods of fertilization in rice farming in Japan. In south-west Japan where the soil is thermic and udic, POCU-140 (about 80% of total urea is released within 140 days in water at 25°C) is recommended by the Japan Agricultural Association, while in north-east Japan where the soil is mostly mesic and udic, POCU-100 is recommended (Shoji and Gandeza, 1992).

Even in Japan, POCU fertilizer is much more expensive than urea or ammonium sulphate fertilizers (2 or 3 times respectively), but it can contribute to the innovation of fertilizer applications and the farming system, whereby the total farming costs can be notably reduced (Shoji and Gandeza, 1992; Shoji and Kanno, 1994; Trenkel, 1996). For example, no-tillage transplanting of rice cultures using seedlings with single basal fertilization of POCU to nursery boxes can preclude the labour of plowing, puddling, fertilizer application in the paddy field, and midsummer drainage, consequently reducing more than 20% of the total rice-farming cost (Kaneta, 1995). In this system, the rice seedlings are raised using delayed type of POCU (POCUs) (Sato and Shibuya, 1991) to avoid burning the rice seedlings in nursery boxes with heavy fertilization.

The most desirable rice-farming system is direct seeding by no-tillage and single basal fertilization using POCU, which can also reduce the growing process of rice seedlings. This rice-farming system is being conducted in well-drained paddy soil in Japan and requires less than 50 hours of labour/ha (Kimoto et al., 1995). Recovery of nitrogen from POCU-100 for basal application (63.2%) in this system was much higher than that from conventional ammonium sulphate (AS) for basal application (8.5%) (Sato et al., 1993). A drastic reduction of mineral nitrogen in the AS-plot was due to denitrification just after submerging because most of the AS-N applied was converted to nitrate-N during dry-soil conditions. Recoveries of top-dressed AS-N applied at the panicle formation stage and full heading stage were 52.8% and 41.5% respectively. Consequently the brown rice yield in the POCU plot (5.71t/ha) was higher than that in the AS plot (3.91 t/ha) reflecting the difference of total uptake of fertilizer nitrogen. Therefore, controlled-release nitrogen fertilizer can resolve most serious problems in non-tilled rice culture such as low efficiency of conventional N fertilizer, frequent application of nitrogen fertilizer throughout the growing season and environmental loading of N_2O.

From these Japanese experiences, Trenkel (1996) has suggested that controlled-release fertilizer, such as sophisticated polyolefin-coated fertilizer, may be widely used even for low cash-value crops if their farming system can be innovated by employing controlled-release fertilizers. Serious agro-environmental problems will also stimulate such innovation.

B. Soybean

It has been considered that most of the nitrogen in soybean is fixed by symbiotic bacteria in nodules on the roots if the proper strain of

Rhizobium bacteria is present in the fields, or if the seed is properly inoculated and thus responds little to nitrogen.

Activity of symbiotic bacteria appears around 4 weeks after germination and increases with advance in plant growth; activity decreases significantly in the late growth stage (Hardy et al., 1968). The decrease of N_2 fixation activity (nodule senescence) in this stage may be attributed to the competition between the vegetative part (N_2 fixation) and reproductive part (seed growth) for photosynthates. Therefore, appropriate top-dressing of nitrogen fertilizer may be able to supplement the nitrogen lacking in the maturation stage. Recently, alleviations of the depressive effect of nitrogen application on nodulation and N_2 fixation have been examined using controlled-release fertilizers (Aragaki, 1989; Takahashi et al., 1991a,b; 1992).

Effects of top-dressing of controlled-release nitrogen fertilizer (75 kg N/ha) on seed yield and total nitrogen accumulation in soybean plants were studied in comparison with those of ammonium sulphate (Aragaki, 1989). Seed yield in the polyolefin-coated urea-70(POCU-70) plot (3.88 t/ha) was superior to that in the ammonium sulphate (AS) plot (3.57 t/ha) and that in the control plot (3.21 t/ha). Uptake of nitrogen from AS increased notably in stage II, whereas the amount of natural nitrogen fixed by nodule bacteria considerably decreased. This effect by AS decreased in stage III and virtually disappeared in stage IV. On the contrary, the absorption of N by plants in the POCU plot continued by the end of the maturing stage. POCU slightly depressed N_2 fixation in stage II and III but increased in stage IV (pod-filling and ripening stage). The difference in depression effect in N_2 fixation between the AS plot and the POCU plot was supported by the amount of allantoin-N in the petiole, which is a transport form of N fixed by nodule bacteria. The recoveries of top-dressed POCU-N and AS-N were 67% and 54% respectively and thus POCU may also contribute to reducing the groundwater pollution caused by nitrogen fertilizer.

To improve the depressive effect of nitrogen application on N_2 fixation, the deep placement of controlled-release nitrogen fertilizer in soybean cultivation was examined in detail by Takahashi et al. (1991 a,b; 1992). Polyolefin-coated urea-100 (POCU-100, 100 kg N/ha) was applied as basal fertilizer immediately under the seed-placement lines along the ridge at a depth of about 20 cm. The results show that accumulation of dry matter and total nitrogen are much improved in plants with deep-placement treatment because of the promotion of leaf growth and retardation of senescence in the maturation stage. Consequently, the seed yield in this treatment (4.24 t/ha) was about 14% higher than that in the control (3.73 t/ha), reflecting the difference in total accumulation of nitrogen in the R7 stage: 393.4 kg/ha for the deep-placement plot and

330.4 kg/ha for control. The recoveries of nitrogen from POCU by plant were 49–62%, which were much higher than that reported for readily available nitrogen fertilizer (about 30%).

Acceleration rather than depression of N_2 fixation by deep placement of POCU-100 was found throughout the growth stages until maturity. The percentage contribution of fixed N_2 to the total N accumulation in the shoots, estimated by determination of the relative ureide-N abundance in the root sap, was lower in the plants with deep placement (64%) than that in control plants (74%). However, the amount of fixed N_2 in the deep placement at the R7 stage was 7 kg/ha higher than that in the control. Therefore, N_2 fixation by nodule bacteria tended to be accelerated in the deep-placement plot and it may be concluded that deep placement of slow/controlled-release nitrogen fertilizer for soybean cultivation is a feasible method to obtain a high seed yield compatible with N_2 fixation.

C. Corn and Sorghum (Tall crop)

Allen (1984) noted that corn and small grain crops absorb large amounts of nitrogen early in their growth cycle (luxury consumption), store it in leaves, and transfer it to the developing seed; this results in very efficient use of rapidly available nitrogen fertilizer. Consequently, slow-release nitrogen fertilizers are thought to have little benefit for these crops, except when there are large leaching losses soon after application. However, the amounts of nitrogen absorbed by dent corn until the critical time of machinery top-dressing (about 1 m of plant height) was only 1/10 of that at harvesting time (Saigusa et al., 1993). Furthermore, if top-dressing time coincides with the rainy season, which is normal in Japan and part of the tropical zone, it is often hindered by rain and wet-soil conditions, forcing the farmer to forego machinery top-dressing and resort to manual application. Even if the farmer can apply the rapidly available nitrogen fertilizer at the appropriate time, the recovery of applied nitrogen is low in Japan because of the usual occurrence of heavy continuous rain in June and July. Therefore, it is too difficult to maintain the nitrogen nutrition during the late growth stage of tall crops such as corn, sorghum, sugar-cane, etc. using rapidly available nitrogen fertilizer. The use of manures for corn cultivation is the traditional way to supply nitrogen gradually, but their limited availability and the great labour required to make and spread them make them less practical than inorganic nitrogen fertilizer. A single basal application of the total nitrogen using slow/controlled-release fertilizer seems to be the most promising application method for tall crops in maintaining the nitrogen nutrition of the late growth stage, saving labour for top-dressing, nullifying the uncertainties of climate and improving the recovery of fertilizer.

To improve the growth of corn both in early and late growth stages, a single basal application of total nitrogen in combination with rapidly available nitrogen fertilizer (ammonium sulphate: AS) and controlled-release fertilizer (polyolefin-coated urea-70: POCU-70) at the rate of 5 kg N/ha and 10 kg N/ ha, respectively was examined (Shoji et al., 1991; Saigusa et al., 1993).

Dent corn grown on the plot of single basal application of POCU and AS, showed equal or greater nitrogen uptake and yield to that grown on a conventional fertilizer plot; it also showed higher lodging resistance (Saigusa et al., 1993). According to Shoji et al. (1991), recoveries of POCU-70 applied as basal fertilizer were 49% in 1988 and 66% in 1989, while those of AS were 30% in 1988 and 44% in 1989. The recoveries of POCU increased notably (56% in 1988 and 78% in 1989) when they were calculated in a base of dissolved POCU. Therefore, a single basal application of POCU in tall crop cultivation is a practically useful method not only for labour saving and maintaining the growth of corn in the late growth stage, but also for lowering the impact of nitrogen fertilizer on the global environment. Furthermore, single basal application of slow/controlled-release fertilizer on non-tilled direct seeding culture of corn seems to be the most desirable farming system in the near future, in which the farmer will be required to practice both high-yielding crop production and environment-friendly farming.

D. Wheat and Barley

Nitrogen fertilizers are sometimes applied in the previous fall rather than in the spring for fall- or spring-sown small cereal grains such as wheat and barley in North America, for the sake of more uniform distribution of labour and equipment demand, time limitation in the spring and lower fertilizer prices in the fall (Malhi and Nyborg, 1992; Yadvinder-Singh et al., 1994). However, fall application of nitrogen fertilizer is generally inferior to spring application because of overwinter loss of nitrate through denitrification. In these areas, large urea granule application with a chemical nitrification inhibitor such as dicyandiamide lowers the rate of nitrification to a great extent, reduces over-winter nitrogen loss, and consequently causes further improvement in grain yield and nitrogen uptake of cereal crops (Yadvinder-Singh et al., 1994). Slow-release nitrogen fertilizers can reduce the transformations, immobilization, denitrification and leaching (Frye, 1977), but sulphur-coated urea did not function well as a fall- or spring-applied nitrogen fertilizer in spring-sown barley (Malhi and Nyborg, 1992).

In north-east Japan, wheat and barley are sown in middle fall and harvested early the following summer. Substantial nitrification occurs in the fall and low evapotranspiration in fall and melting snow in winter cause nitrogen loss by both leaching and denitrification. Therefore, several top-dressings during the late growth stage are necessary to increase the nitrogen supply to these crops when rapidly available fertilizers are used (Saigusa et al., 1983). Slow/controlled-release fertilizer can be basally applied as a nitrogen source for winter crops, since it can supply the nitrogen requirements of these crops until the following summer with no significant losses. Single basal applications of total nitrogen with a combination of polyolefin-coated urea (70 kg N/ha as POCU-70) and ammonium sulphate (30 kg/ha) could eliminate the repetition of top-dressing, improve plant growth in the early growth stage, and increase grain yield by 20% without lodging; it is recommended as a practical fertilization method of fall-sown wheat by the Iwate Agricultural Experiment Station (Gandeza and Shoji, 1992).

E. Forage

The reviews by Hauck (1985) and Allen (1984) discuss numerous reports on the use of slow-release nitrogen fertilizers on forage grasses. Grasses require continuous nitrogen supply over the growing season because their growing seasons are relatively long compared to other upland crops. Grassland farming is typical extensive agriculture and needs the maximum yield of acceptable quality forage with the least cost. For these reasons, the effects of SCU : sulphur-coated urea (the least expensive slow-release nitrogen fertilizer) on both yield and nitrogen uptake of forage grasses were intensively examined (Allen, 1984). SCU application generally achieved a greater efficiency of nitrogen use by controlling both the luxury uptake of the plant and the losses through leaching or volatilization compared to urea application. Consequently, grasses supplied with SCU showed greater increases in both nitrogen uptake and dry matter production.

A single application of total nitrogen fertilizer in spring on an orchard grass (*Dactylis glomerata* L.) meadow using controlled-release polyolefin-coated urea (POCU) showed relatively more uniform growth throughout the growing season and improved both nitrogen uptake and dry matter yield of grass compared with a 4-time split application of conventional fertilizers (Saigusa et al., 1994). However, the release rate of nitrogen from POCU applied surfacially to grassland was about 30% faster than that applied in a plow layer in upland conditions.

Slow-release nitrogen fertilizer may greatly contribute to high dry matter production, due to smooth regrowth of the aerial part after clipping, preservation of botanical composition, labour saving for fertilization, prevention of nitrate toxicity to cattle and reduction of nitrogen fertilizer impact through leaching of nitrate, emission of nitrous oxide and volatilization of ammonia to the global environment both in meadow and pasture.

F. Turfgrasses

Turfgrasses are mainly grown for both aesthetic purposes such as home lawns, parks and gardens, and aesthetic-utilitarian purposes such as golf courses, horse-race courses and many kind of athletic fields (Maynard and Lorenz, 1979). Continuous uniform growth, maintaining high-quality conditions such as good colour and vigour throughout the growing season, is essential for turfgrasses to offer pleasing and durable qualities for dwellers, visitors and players. Turfgrasses require more frequent clipping and nitrogen fertilization to maintain that pleasing quality compared with grasses for meadow and pasture, and thus may be more susceptible to disease and insect pests. On the other hand, turfgrasses are a relatively high-value crop and the fertilization cost may be less important than the other agronomic crops (Allen, 1984). Therefore, many scientists have taken great interest in slow-release nitrogen fertilizers which are less dangerous in burning and reduce the number of top-dressings.

Golf greens generally have good drainage and are frequently clipped to maintain a smooth and pleasing surface condition. Thus slow-release nitrogen fertilizers seem to be much more effective than rapidly available ones, because of lesser amounts of leaching loss and burning. However, only condensation products of urea and aldehyde such as ureaforms, isobuthylidene diurea, and crotonylidene diurea are used, while coated nitrogen fertilizers such as sulphur-coated urea (SCU) and resin coated ones are not recommended for greens because close mowing may fracture granules and destroy the slow release effect (Allen, 1984).

On the other hand, SCU has been rated favourably and is now widely used for fairways in golf courses and other sports grounds in combination with liquid fertilization. Furthermore, resin-coated ureas were recently shown to be much better controlled-release nitrogen sources for turfgrass fertilization compared to SCU. The release rate of nitrogen from resin-coated urea is mainly temperature dependent and is well predicted by meteorological data; thus a single-spring application of resin-coated urea-100 is feasible for turfgrass fertilization. It provides superior colour to split-spring and fall application throughout the growing season except early spring. The uniform delayed release of nitrogen from resin-coated ureas make them well suited for fertilizer programmes that are based on one application per year (Hummel, 1989).

G. Vegetable Crops

Most vegetable crops are economic crops and are required to be not only high in yield but also of high quality (colour, size, shape, succulent sugar concentration, etc.). For these purposes, farmers are forced to maintain a high and continuous level of fertility throughout their growing process by frequent application of rapidly available fertilizers. On the other hand, the relative cost of fertilizer nitrogen in vegetable production seems to be less important compared with that for cereal or forage production because of relatively high returns. Therefore, the use of slow-release nitrogen fertilizers in vegetable crop production is highly possible and numerous experiments have been conducted to evaluate the effects of slow-release nitrogen fertilizers, as reviewed by Hauck (1985). According to these reviews, widespread use of slow-release nitrogen fertilizers in vegetable production is limited to areas with severe leaching conditions. However, the use of slow-release fertilizers in vegetable production has many advantages, such as reducing nitrogen loss through leaching, denitrification or volatilization, providing the ammonium form of nitrogen in upland conditions, eliminating crop damage by burning, lowering high labour and energy requirements, etc. In fact, slow/controlled-release nitrogen fertilizers are commonly used in vegetable production in Japan where extraordinary high quality and thoughtful consideration to environmental problems are required in vegetable crop production.

Leaf vegetables, such as spinach, lettuce, celery, etc. must be liberally fertilized throughout their growth period to ensure green colour, tenderness, succulence and high yields. Adding to these common characteristics, oxalic acid content of spinach is also very important for the consumer, because it may be one of the causes of calculus. Takebe et al., 1996 clarified that there is a close relation between the production of oxalic acid and the reduction of nitrate in spinach; the content of oxalic acid in spinach in the solution culture decreased along with increasing ratios of NH_4-N to NO_3-N in the solution. However, in upland conditions it is impossible or virtually impossible to maintain the ammonium form of nitrogen in soil solutions by applying rapidly available ammonium compounds, because ammonium is easily converted to nitrite ions by *Nitrosomonas* bacteria and is eventually converted to nitrate by *Nitrobacter* bacteria in the soil. It may be possible to provide the ammonium form of nitrogen to the plants grown on upland conditions by applying slow/controlled-release fertilizers containing ammonium or urea, which can supply ammonium nitrogen or urea directly to the plant root with no burning.

A field experiment was conducted for the purpose of decreasing the content of oxalic acid in spinach by growing it with controlled-release fertilizer in Ando soil. The fertilizers were applied 6 cm below the seeds at 10 cm width. The results showed that the content of oxalic acid in spinach was remarkably decreased in the coated-urea and coated-ammonium phosphate plots (47–72% of the value found in the ammonium sulphate plot), while both sugar and ascorbic acid tended to increase in these plots (Takebe et al., 1996). Therefore, application of controlled-release fertilizers can provide at least a part of the ammonium or urea directly to the roots of the plant before nitrification in the soil, and consequently may induce a desirable effect on the quality of spinach and other similar vegetable that accumulate oxalic acid.

Most fruit vegetables maintain both vegetative growth and reproductive growth, and require a continuous supply of nutrients. However an oversupply of nitrogen often cause excess vegetative growth and thus retards or ceases reproductive growth. A single basal application of controlled-release nitrogen fertilizers can contribute not only to the supply of a continuous level of nitrogen to the plant throughout the growing period, but also to labour saving.

Tomato is one of the typical fruit vegetables showing simultaneously both vegetative growth and reproductive growth. Numerous experiments have been conducted to evaluate rapidly available fertilizer and slow/controlled-release fertilizer. However, the results of both field and greenhouse experiments were somewhat varied with no consistent horticultural advantage to the use of slow-release fertilizers such as sulphur-coated urea, resin-coated urea, ureaform methylene urea, isobuthylidene diurea, etc. (Maynard and Lorenz, 1979; Csizinszky, 1994). Controlled-release fertilizers with a more accurate release rate, polyolefin-coated fertilizers, were recently developed in Japan (Fujita et al., 1977; 1983) and a single basal application of total fertilizer on tomato cultivations was conducted using this polyolefin-coated fertilizer (POC-NPK-140) and compound fertilizer (Sato, 1995). In this innovative cultivation plot, the marketable yield of tomatoes increased 5%, regardless of the 20% decrease in fertilizer nitrogen, compared to a conventional cultivation plot. Furthermore, this method saved labour cost for fertilization to a great extent by eliminating 7 repetitions of top-dressing.

Similar results were obtained in greenhouse cultures of strawberries by a single basal application of controlled-release fertilizers, such as Nutricote together with organic fertilizers: 8% increase in marketable yield and elimination of 8 repetitions of top-dressing (Shibuya, 1992). From an economic point of view, a single basal application of controlled-release fertilizer is superior to conventional fertilization methods in terms of both saving labour cost for fertilization and reducing the dangers of salt injury often occurring in greenhouses.

H. Container Plants

In container nurseries, the soil volume as reservoir of water and nutrient is quite small and requires frequent irrigation, resulting in a loss of plant nutrients and groundwater pollution. Furthermore, heavy application of rapidly available fertilizers may cause salt injury to plants. Therefore, most container nurseries in Ohio and the north-eastern United States have developed a fertilizer programme based on slow-release fertilizer; rapidly available fertilizers are used only as supplements (Smith and Treaster, 1992). Nowadays, slow-release fertilizers are widely used in container nurseries of ornamental landscape plants such as azalea, juniper, cotoneaster, spindle tree, Alberta spruce, sweet fern, holly *Thuja*, etc. (Smith and Treaster 1992; Struger and Lumis, 1992; Lumis and Khanna, 1992). Among the various slow-release nitrogen fertilizers, in general resin-coated ones showed more consistent effects on the growth of azalea, cotoneaster and spindle tree (Smith and Treaster, 1992). Two-season slow-release fertilization for *Juniperus* and *Thuja* were also examined by Lumis and Khanna (1992). According to their results, long release pre-incorporated Nutricote 16-10-10, Type 270 provided acceptable growth of *Thuja* but not of *Juniperus*, which is more sensitive to the rate of fertilizer.

VII. ENVIRONMENTAL ASPECTS

Rapidly available nitrogen fertilizers are known to be one of the main causes of serious agro-environmental problems, such as groundwater and stream water pollution, excess accumulation of nitrate or nitrite in plant, ozone layer destruction, 'ozone holes' and soil degradation by salt accumulation or acidification (Gandeza and Shoji, 1992; Shaviv and Mikkelsen, 1993).

In upland conditions, any mineral nitrogen source is eventually converted to nitrate via nitrite by microbial transformation. However, most soils have little amount AEC (anion exchange capacity) and thus nitrate is leached or removed from the root zone into groundwater and stream water. It may be poisonous in humans and animals, particularly ruminants. Excess accumulation of nitrate in plant often occurs when heavy application of mineral nitrogen is done and results in the same detrimental effects for them.

VIII. CONCLUSIONS AND FUTURE PROSPECTS

Generally, the nitrogen release patterns of controlled-release fertilizers are synchronized with the growth rate of the crops and the recoveries of these types of fertilizer by the crops are remarkably higher than those of a rapidly available one. Therefore, environmental pollution caused by nitrogen fertilizer is noticeably reduced and controlled-release fertilizer is evaluated to be an environment-friendly fertilizer. Further investigation on the new function of these fertilizers supplying aimed form of nitrogen in soil conditions will be needed.

LITERATURE CITED

Allen, S.E. 1984. Slow-release nitrogen fertilizers. In: *Nitrogen in Crop Production* (R.D. Hauck et al., eds.). ASA, CSSA, SSSA Madison, Wisconsin, pp. 195–206.

Aragaki, K. 1989. Studies on the analysis of high-yielding factor and the method of nitrogen top-dressing in soybean plants grown under drained paddy field of alluvial soil in Yamagata prefecture. *Spec. Bull. Yamagata Agric. Exp. Stn.* 16: 1–42 (in Japanese with English summary).

Association of American Plant Food Control Officials (AAPFCO). 1995. Official Publication No. 48. AAPFCO Inc., West Lafayette, Indiana.

Chinen, H. 1995. CDU fertilizer. In: *Sinnoho eno chosen* (S. Shoji, ed.). Hakuyusha, Tokyo, pp. 81-92 (in Japanese).

Chisso-Asahi Fert. Co. CDU, slow-release nitrogen fertilizer. Chisso-Asahi Fert. Co. Ltd., Tokyo, pp. 1–12.

Csizinsky, A.A. 1994. Yield response of bell pepper and tomato to controlled-release fertilizers on sand. *J. Plant Nutri.* 17: 1535–1549.

Frye, W.W. 1977. Fall-versus spring-applied sulphur coated urea, uncoated urea and sodium nitrate for corn. *Agron, J.* 69: 278–282.

Fujita, T., Takahashi, C., Ohshima, M., Ushioda, T. and Shimizu, H. 1977. Method of producing coated fertilizer. United States Patent 4,019,890.

Fujita, T., Takahashi, C., Yoshida, S. and Shimizu, H. 1983. Coated granular fertilizer capable of controlling the effects of temperature upon dissolution-outrate. United States Patent 4,369,055.

Gandeza, A.T. and Shoji, S. 1992. Application of polyolefin-coated fertilizers to crops in northeast Japan. In: *Controlled-Release Fertilizers with Polyolefin Resin Coating.* Kanno Printing Co. Ltd., Sendai, Japan, pp. 43–62.

Hardy, R.W.F., Jackson, E.K. and Burns, R.C. 1968. The acetylene-ethylene assay for N_2 fixation: laboratory and field evaluation. *Plant Physiol.* 43: 1185–1207.

Hauck, R.D. 1985. Slow-release and bioinhibitor-amended nitrogen fertilizer. In: *Fertilizer Technology and Use.* 3rd ed. (O.P. Engelstad, ed.). pp. 293–322.

Hummel, N.W. Jr. 1989. Resin-coated urea evaluation for turfgrass fertilization, *Agron, J.* 81: 290–294.

Kamekawa, K., Nagai, T., Sekiya, S. and Yoneyama, T. 1990. Nitrogen uptake by paddy rice (*Oryza sativa* L.) from ^{15}N labelled coated urea and ammonium sulphate. *Soil Sci. Plant Nutr.* 36: 333–336.

Kaneta, Y. 1995a. Single application of controlled availability fertilizer to nursery boxes in non-tillage rice culture. *JARQ* 29: 111–116.

Kaneta, Y. 1995b. Single application of controlled availability fertilizer to nursery boxes in non-tillage transplanting rice culture. In: *Shinnoho eno chosen*. (S. Shoji, ed.). Hakuyusha, Tokyo, pp. 203-220 (in Japanese).

Kaneta, Y., Awasaki, H. and Murai, Y. 1994. The non-tillage rice culture by single application of fertilizer in nursery box with controlled-release fertilizer. *Jap. J. Soil Sci. Plant Nutr.* 65: 385-391 (in Japanese with English summary).

Kimoto, H., Okatake, S. and Tomiyasu, Y. 1995. Direct seeding of rice in no-tilled well drained paddy soil. Nobunkyo, Tokyo, pp. 1-161 (in Japanese).

Lumis, G. and Khanna. 1992. Two-season slow release fertilization for *Juniperus* and *Thuja*. In: *Controlled Release Fertilizers* (S. Shoji and A.T. Gandeza, eds.) Konno Printing Co. Ltd., Sendai, Japan, pp. 78-81.

Malhi, S.S. and Nyborg, M. 1992. Recovery of nitrogen by spring barley from ammonium nitrate, urea and sulphur-coated urea as affected by time and method of application. *Fert. Res.* 32: 19-25.

Maynard, D.N. and Lorenz, O.A. 1979. Controlled-release fertilizer for horticultural crops. *Hortic. Rev.* 1: 79-140.

Mitsubishi Chem. Indust. 1986. Long-lasting nitrogen fertilizer IBDU. Mitsubishi Chem. Indust. Ltd., Tokyo, pp. 1-21.

Saigusa, M. 1995. Single basal application of total nitrogen fertilizer in no-tillage cultivation of upland crops. In: *Shinnoho eno chosen* (S. Shoji, ed.). Hakuyusha, Tokyo, pp. 255-274.

Saigusa, M., Shoji, S. and Sakai, H. 1983. The effects of subsoil acidity of Andosols on the growth and nitrogen uptake of barley and wheat. *Jap. J.Soil Sci. Plant Nutr.* 54: 460-466 (in Japanese with English summary).

Saigusa, M., Shibuya, K. and Abe, T. 1994. Single application of fertilizer in spring on orchard grass (*Dactylis glomerata* L.) meadow using controlled release coated urea. *J. Jap. Grass. Sci.* 40: 95-100 (in Japanese with English summary).

Saigusa, M., Taki, N. and Shibuya, K. 1996. Influence of applied nitrogen form on nitrate accumulation of grasses in grazing pasture. *Tohoku J. Crop. Sci.* 39: 87-88 (in Japanese).

Saigusa, M., Shoji, S., Goto, J., Kodama, H. and Abe, T. 1990. Nitrogen fertility of acid Andisols and nitrogen application to upland crops. *Trans. 14th Inter. Cong. Soil Sci.* IV: 638-639.

Saigusa, M., Kodama, H., Shibuya, K. and Abe, T. 1993. Single basal application of nitrogen fertilizer on dent corn cultivation using controlled release urea. *J. Jap. Soc. Grass. Sci.* 39: 44-50 (in Japanese with English summary).

Saigusa, M. Md, Z.,Hossain Tashiro, T. and Shibuya, K. 1996. Maximizing rice yield with controlled availability fertilizer and no-tillage culture in controlling environmental degradation. *Proc. Intn. Symp. Maximizing Sustainable Rice Yields through Improved Soil and Environmental Management*. Khonkaen Thailand, 1: 75-85.

Sato, K. 1995. New Fertilizaton methods of Vegetable crops — single basal application of total fertilizer using controlled availability fertilizer. In: '*Shinnoho eno chosen*' (S. Shoji, ed.). Hakuyusha, Tokyo, (1995), pp. 275-297 (in Japanese).

Sato, T. and Shibuya, K. 1991. One time application of total nitrogen fertilizer at nursery stage in rice culture. *Rep. Tohoku Br. Crop Sci. Soc. Japan* 34: 15-16 (in Japanese).

Sato, T., Shibuya, K., Saigusa, M. and Abe, T. 1993. Single basal application of total nitrogen fertilizer with controlled-release coated urea on non-tilled rice culture. *Jap. J. Crop Sci.* 62: 408-413 (in Japanese with English summary).

Shaviv, A. A. and Mikkelsen, R.L. 1993. Controlled-release fertilizer to increase efficiency of nutrient use and minimize environmental degradation. A review, *Fert. Res.* 35: 1-12.

Shibuya, T. 1992. Case studies on application of polyolefin-coated fertilizer. In: *Controlled Release Fertilizers* (S. Shoji and A.T. Gandeza, eds.). Konno Printing Co. Ltd., Sendai, Japan, pp. 63-67.

Shiozaki, H. 1996. Application technique of coated fertilizer. *Hiryojiho* 380: 16–31 (in Japanese).
Shoji, S. 1995. *Shinnoho eno chosen.* Hakuyusha, Tokyo, 381 pp.
Shoji, S. and Mae, T.1984. Dynamics of inorganic nutrients and water. In: *Ecophysiology of Crops.* Buneido, Tokyo, pp. 97–172 (in Japanese).
Shoji, S. and Gandeza, A.T. 1992. *Controlled Release Fertilizers with Polyolefin Resin Coating.* Kanno Printing Co. Ltd., Sendai, Japan, pp. 1–92.
Shoji, S. and Kanno, H. 1994. Use of polyolefin-coated fertilizers for increasing fertilizer efficiency and reducing nitrate leaching and nitrous oxide emissions. *Fertilizer Research* 39: 147–152.
Shoji, S., Gandeza, A.T. and Kimura, K. 1991. Simulation of crop response to polyolefin-coated urea: II. Nitrogen uptake by corn. *Soil Sci. Soc. Amer. J.* 55: 1468–1473.
Smith, E.M. and Treaster, S.A. 1992. Nutricoete — a comparison of slow release fertilizers. In: *Controlled Release Fertilizers* (S. Shoji and A.T. Gandeza, eds.). Konno Printing Co. Ltd., Sendai, Japan, pp. 72–75.
Struger, S. and Lumis, G. 1992. A study of Nutricoete 16-10-10, rate and ratio. In: *Controlled Release Fertilizers* (S. Shoji and A.T. Gandeza, eds.). Konno Printing Co. Ltd., Sendai, Japan, pp. 76–77.
Sugihara, S., Konno, T. and Ishi, K. 1986. Kinetics of mineralization of organic nitrogen in soil. *Bull. Nat. Inst. Agro-Environmental Sci.* 1: 127–166 (in Japanese with English summary).
Takahashi, Y., Chinushi, T., Naguno, Y., Nakano, T. and Ohyama, T. 1991a. Effects of deep placement of controlled release nitrogen fertilizer (coated urea) on growth, yield and nitrogen fixation of soybean plants. *Soil Sci. Plant Nutr.* 37: 223–231.
Takahashi, Y., Chinushi, T., Nakano, T., Hagino, K. and Ohyama, T. 1991b. Effect of placement of coated urea fertilizer on root growth and rubidium uptake activity in soybean plant. *Soil Sci. Plant Nutr.* 37: 735–739.
Takahashi, Y., Chinushi, T., Nakano, T. and Ohyama, T. 1992. Evaluation of N_2 fixation and N absorption activity by relative ureide methods in field-grown soybean plants with deep placement of coated urea. *Soil Sci. Plant Nutr.* 38: 699–708.
Takebe, M., Ishihara, T., Ishi, K. and Yoneyama, T. 1995. Effect of nitrogen forms and Ca/' K ratio in the culture solution on the contents of nitrate, ascorbic acid and oxalic acid in spinach (*Spinacia oleracea* L.) and Komatsuna (*Brassica campestris* L.). *Jap. J. Soil Sci. Plant. Nutr.*, 66: 535–543 (in Japanese with English summary).
Takebe, M., Sato, N., Ishi, K. and Yoneyama, T. 1996. Effects of slow-releasing nitrogen fertilizers on the contents of oxalic acid, ascorbic acid, sugars and nitrate in spinach (*Spinacia oleracea* L.). *Jap. J. Soil Sci. Plant Nutr.* 67: 147–154 (in Japanese with English summary).
Tashiro, T., Saigusa, M. and Shibuya, K. 1996. Effects of the form of nitrogen fertilizer on the establishment of seedlings in no-tilled direct seeding culture of rice. *Tohoku J. Crop Sci.* 39: 3–4 (in Japanese).
Trenkel, M.E. 1996. Controlled-release and Stabilized Fertilizers. Present Situation and Outlook. Report to FAO (1996), pp. 1–190.
Ueno, M., Kumagai, K., Togashi, M. and Tanaka, N. 1991. Basal application technique of whole nitrogen using slow-release coated fertilizer based on forecasting of soil nitrogen mineralization amount. *Jap. J. Soil Sci. Plant Nutr.* 62: 647–653 (in Japanese).
Wada, G., Aragones, R.C. and Ando, H. 1991. Effect of slow release fertilizer (Meister) on the nitrogen uptake and yield of the rice plant in the tropics. *Jap. J. Crop Sci.* 60: 101–106.
Yadvinder Singh, S., Malhi, S., Nyborg, M. and Beauchamp. E.G. 1994. Large granules, nest or bands: Methods of increasing efficiency of fall applied urea for small cereal grains in North America. *Fert. Res.* 38: 61–87.

INDEX

1, 10-phenanthroline 8
1-Naphthoxy acetic acid 66
13 S globulin 8
2 S Globulins 5
7 S Globulins 5
α-difluoromethyl arginine 219
α-difluoromethyl ornithine 220
α-ketoglutarate 255
α-oxoglutarate 30, 31
Allocasuarina torulosa 175, 186
Amyloplast 65
α, β, γ and δ zeins 7
Abiotic stress 269
Alium graveloux 219
Abscisic acid 15, 295, 298, 301
Acacia saligna 290
Acer rubrum 117
Acetosyringone 217
Acetylene 79, 80, 184, 187, 188 190, 192
Actinomycete 174, 180, 183
Actinorhiza 180, 183, 191
Actinorhizal nodules 79
Activators, transcriptional 54
Acyl chain 69
Acylated oligo-N-acetylglucosamine 55
Adhesion 50
Adhesives 50
Aerobic 46, 86, 175
Aerodynamic 111, 112, 113, 116
Aerosol 109
Agar 49
Agrobacterium tumefaciens 47, 217, 219
Agroforestry 196
Agrostis alba 216
Alanin 4, 37, 70, 207, 208, 211, 215, 217, 218, 220, 223
Alberta spruce 332

Albizia stipulata 62
Albumin 4
Albumin polypeptide 4
Albumins 3
Alcaligenes latus 193
Aldehyde oxidase 295, 299
Aleurone layers 3
Alginate 183
Alphatic amines 211
Alkyd resins 311, 312
Allantoin 222
Allelopathic 233
Allocasuarina 175, 176, 177, 181, 184, 186
Allocasuarina lehmaniana 184
Alnus 185, 190
Alnus glutinosa 81
Alnus rubra 190
Alnus glutinosa 65, 80, 83
Alnus incana 83, 190
Alnus-Frankia 192
Alpha (α) zein 4
Alpha-amylase 16
Amides 222, 293, 294, 296, 297, 299, 317
Amination 134
Amino acids 2, 31, 32, 35, 37, 62, 72, 75, 78 121, 125, 133, 134, 137, 140, 156, 190, 191, 206, 208, 215, 222, 294, 295, 297, 310, 317
Amino nitrogen 14
Amino-terminal 73
Aminobutyric acid 215
Aminoethoxy vinylglycine 65, 81
Aminopeptidase 7, 9, 180, 181
Aminotransferase 133, 215
Ammonia 8, 28, 30, 31, 46, 51, 59, 76, 78, 84, 121, 133, 174, 206, 211, 329
Ammonia Toxicity 309

Ammonium 24, 25, 27, 28, 29, 30, 31, 32, 34, 36, 37, 63, 72, 81, 82, 118, 125, 130, 141, 143, 207, 208, 209, 210, 211, 220, 221, 223, 232, 233, 234, 235, 243, 244, 245, 247, 248, 249, 250, 251, 252, 255, 256, 258, 259, 260, 262, 264, 265, 266, 267, 269, 271, 286, 288, 289, 293, 294, 295, 296, 297, 298, 299, 317, 318, 322
Ammonium chloride 209, 309
Ammonium injuries 319
Ammonium ions 310, 317
Ammonium nitrogen 35, 331
Ammonium nitrate 79, 106, 308
Ammonium phosphate 308
Ammonium sulphate 210, 308, 309, 323, 324, 325, 327, 328, 331
Amylases 3
Amyloplast 62, 65
Anaerobic 288
Anapleurotic 253, 293
Anatomy 62
Angiosperms 140
Antibiotic 51, 67, 178
Antifungal 178
Antirhizobial bacteriocin 84
Apical 30
Apical meristem 54
Apoplasmic 270
Apoplast 120, 157, 263
Apoplastic 24, 109, 119, 120
Apopletic 24
Apoproteins 299
Arabidopsis thaliana 26
Arabinogalactan 70
Arachis hypogacea 66
Arginine 3, 5, 25, 37, 140, 175, 211
Arginine decarboxylase 218, 219, 220
Aromatic 233
Ascorbate 119
Ascorbate peroxidase 158
Ascorbic acid 64, 318, 331
Asparagine 8, 31, 37, 82, 121, 137, 140, 208, 211, 218, 293, 297
Asparaginyl endopeptidase 8, 9
Asparagus 207
Asparagus officinalis 207
Aspartate 134, 140, 297
Aspartate Protease 7
Aspartate residues 7
Aspartic acid 208
Aspergillus nidulans 262
Atmosphere 24, 134, 136, 156, 174, 284, 298

Atmospheric 46, 133, 151, 157, 159, 187, 188, 192, 288
nitrogen 191
Atmospheric environment 13
Atrichoblasts 57
Autoregulations 60, 61, 65, 78
Autotrophic 267
Auxin 57, 64, 66, 81, 83, 86, 209, 217, 219, 223, 284
Avena fatua 14
Avena sativa 25, 146
Azalea 332
Azide 84
Azorhizobium 47, 51
Azorhizobium caulinodans 47
Azorhizobium sesbaniae 85
Azorhizobium 47, 68
Azospirillum 86
Azotobacter chrococcum 84
Azotobacter vinelandi 85

Bean 13
Bacillus radicicola 46
Bacteria 78
Bacteriocin 84
Bacteroid 59, 61, 67, 71, 72, 76, 77, 85
Bacteroidal 63
Barley 10, 12, 15, 16, 137, 327, 328
Barley aleurone layer 7, 16
Barley seeds 7
Bauxite 50
Bean 13
Bentonite 50
Beta(β) zein 4
Beta vulgaris 29
Beta-Conglycinin 5
Betaine phosphate 216
Betaines 52
Betula pendula 130
Bicarbonate 293
Biodegradation 311
Biological nitrogen fixation 46
Biomass 155, 193, 234, 258, 270, 271, 284, 286
Biophysical factors 17
Biotic stress 269
Blastococcus sp. 181
Boronia megastigma 243
Boundary layer 111, 113
Bradyrhizobium 47, 49, 51, 67, 68, 74, 77, 78
Bradyrhizobium japonicum 54, 55, 68, 69, 74, 75, 84, 185

Brassica compestris 221
Brassica napus 5, 27, 37
Bryophytes 127
Buckwheat 10

Calcicola 156
Calcifuge 156
Calcium 54, 69
Calcium silicate 50
Calvin cycle 251, 252
Capsicum annuum 125, 132
Carbamoyltransferase 218
Carbohydrate 35, 65, 245, 246, 247, 252
Carbon 28, 35, 36, 51, 73, 82, 133, 178, 179, 192, 193, 206, 244, 247, 250, 252, 253, 255, 256, 258, 267, 269, 270, 271, 293, 294, 295, 301
Carbon dioxide 80
Carbon skeletons 80, 81, 157
Carbon-nitrogen 232, 233
Carbonic anhydrase 252
Carbonyl 25
Carboxy terminus 73
Carboxylation 252
Carboxypeptidases 9
Carrier protein 25, 36, 38
Carrot 56
Castor endosperm 8
Casuarina 174, 175, 177, 178, 179, 180, 181, 182, 184, 185, 186, 188, 189, 190, 191, 192, 193, 194, 195, 196
Casuarina equisetifolia 175, 177, 179, 181, 182, 183, 185, 188, 191, 193, 195, 196
Casuarina cunninghamiana 179, 180, 183, 186, 189, 190, 194
  symbiosis 194
Casurina-Frankia 174, 177, 194, 195, 196
Casuarina glauca 175, 180, 190, 191, 193,
Casuarina littoralis 195
Casuarina obesa 193
Casuarina ologodon 196
Catabolic 135
Catabolism 134
Cathepsin D 7
Cathepsin H 7
Ceanothus americanus 190
Celery 330
Cell division 15
Cellulase 180
Cellulose 86
  fibrils 54
Ceratonia siliqua 243, 288, 298
Cereals 2, 15
Ceuthostoma 175, 177

Chalcone 54
Chemolithoautotrophic 193
Chinese cabbage 322
Chitin oligosaccharide 56
Chitin synthetase 68
Chito-oligosaccharide 68
Chlorate 84
Chlorella 260
Chlorides 63
Chlorogenic acid 66
Chloroplast 30, 130, 131, 133, 142, 143, 146, 148, 157
Chlorophyll 146
Choline 178
Cholinekinase 72
Chromatin 209
Chromatograph 187, 188
Chromosome 67
Chrysanthemum morifolium 220
Cinnamomum 115, 136
Cis-vaccenyl 69
Citrulline 191
Clostridium 76
Clubroot 316
Coated Fertilizer 311
Coated-ammonium nitrate 318
Coated-ammonium phosphate 331
Coated-urea 318, 331
Colletia cruciata 189
Colloids 24
Comptonia peregrina 177, 189, 190
Concentration gradient 113, 114, 116, 117, 119, 132
Conductance 113, 125, 149
Coniferous 118
Container nurseries 332
Contamination of groundwater 309
Controlled-release fertilizers 322, 324, 325, 327, 331, 333
  nitrogen fertilizer 308, 318
Conventional ammonium sulphate 318
  fertilizers 307
Corn 320, 326
Cotoneaster 332
Cotton seeds 6
Cross-inoculation 47
Crotonylidene diurea 315, 316, 329
Cucumber 316
Cucumis sativus 115, 136
Cucurbita maxima 126
Cucurbita pepo 8
Cycloheximide 10, 178
Cyclohexylamine 219

Cysteine 2, 78
    endopeptidase 8
    protease 3, 7, 9, 16
Cytokinin 15, 57, 66, 83, 184, 209, 210, 219, 250, 284, 285, 295, 297, 298, 299, 301
Cytoplasmic nitrate 27, 28, 35
Cytosolic 30, 252
    nitrate reductase 260

D-glucosamine synthetase 68
Daidzein 52
Datisca 174
Datisca glomerata 190
Daucus carota 206
Deamidase 8, 9
Deamidation 8
Decarboxylase 219
Decarboxylic acids 85
Deciduous 113
Denitrification 104, 323, 328, 329, 330
Diamine oxidase 220
Diadzein 54
Diazotrophs 85
Dicyandiamide 327
Diethylstilbestrol 25
Differentiation 59
Diffusion coefficient 110, 116
Difluoromethylornithine 219
Dinitrogen 78, 79, 187
    pentoxide 104
Dinitrogenase reductase 76, 85
Dithiothritol 8
Diurnal rhythm 37, 114
DNA sequencing 47
Dolomite 50
Dormant seeds 2, 9, 10, 17
Droxyproline-rich 70
Durum wheat 9

Eally signals 52
Edaphic 62
Eddy correlation 107
Egg 320
Elaegnus 184, 186
    angustifolia 190
    communata 190
    triflora 186
Electorphoretic 181
Electrical resistance 111
Electrogenic uniport 265
Embryogenesis 208, 209, 210, 211, 215, 217, 218, 219, 220, 223
Embryonic axis 6, 11, 12, 13, 14, 15, 16, 17

Endocytosis 59
Endogenous nitrate 14
Endopeptidases 6, 7, 9
ENDO gene 56
Environment friendly fertilizers 310
Epigeal 50
Epiphytic lichens 119
Equisetifolia 179, 180, 181
Eriophorum vaginatum 233
Essential amino acids 2
Esterases 181
Ethylene 65, 66, 83, 158, 187, 221
Euonymus 136
Eutrophication 46
    of streams 307
Evapotranspiration 328
Exopeptidases 6

Fagopyrum esculentum 8
Fagus sylvatica 130
Fatty Acids 147, 148, 311
Fe-protein 76
Ferredoxin 30, 31, 133
Ferricyanide 262
Ferredixin gene 75
Fertilizer controlled release 331
Fertilizer nitrogen 321
Fimbriae 54
Fix genes 54, 77, 85
Flavonoids 52, 54, 85
Forage 328
    grasses 328
Formamide 315
Fossil fuel 104
Frankia 174, 175, 177, 178, 179, 180, 181, 182, 183, 184, 185, 186, 189, 192, 193, 194
Fumigation 113, 114, 116, 118, 121, 125, 126, 127, 129, 130, 132, 133, 135, 157, 158
Fusarium wilt 316

Galactolipids 148
Galnisoga parviflora 149
Gama (γ)-zein 4
Gelatin 49
Geodermatophilus 181
Geotrophic 184
Gibberellic acids 83
Gibberellins 15
Ginger 316
ω-Gliadin 6
Global environment 327, 329

Globulins 3, 4, 5
Glomus mossae 181, 182
Glucosamine 55, 68
Glucose oxidation 221
Glutamate 30, 133, 134, 137, 217, 297
Glutamate dehydrogenase 133, 206, 210
 pyruvate transaminase 134
 synthase 31, 206, 215, 256
Glutamic acid 3, 206, 207, 208, 210, 211, 215, 218, 223
Glutamine 4, 30, 31, 37, 72, 121, 133, 134, 140, 206, 207, 210, 211, 215, 218, 220, 223, 255, 260, 293, 297
 synthetase 30, 72, 206, 211, 275
 Synthase 72
Glutamyl residue 8
Glutenin 3, 4
Glycine 15, 70, 121, 217
Glucine betaine 217
Glycin max 5, 25, 61, 136, 149, 151
 wightii 63
Glycinin 5
Glycolipids 147
Glycolytic 221
Glycoprotein 5, 70, 82, 217
Gossypium 217, 218
Granulated peat 50
Grass 320
Green-house effects 307
Groundwater pollution 307
Guanidinium 25
Guanylurea 315
Gum arabic 50
Gymnostoma 174, 175, 177, 186, 188
Gymnostoma papuanum 186, 188, 196
Gynostoma australiensis 186

Haemorprotein 78
Helianthus annuus 5, 115, 116, 121, 131, 133, 136, 137, 150
Hemisphere 174, 178, 185
Hemoprotein 77
Hempseed 10
Hexaphosphate 3
Hippophae rhamnoides 180, 190
Histidine 208
Histodifferentiation 220
Homeostasis 266
Homodimer 76
Hordeum vulgare 26, 51, 121, 125, 132, 137, 140, 142, 146
Hydrogenase 71, 188, 189, 192, 193
Hydrolysis of proteins 6

Hydrolytic 60
 enzymes 15
Hydrophobic 25, 117
Hydroponic 243
Hydroponically 248
Hydrosphere 24
Hydroxyl-radicals 105
Hyperpolarization 25, 27
Hyphae 182, 183
Hyphal 178
Hypopleurophyte 259

IAA oxidase 64
Idoacetate 7
Immobilization 327
Immunoglobulin G 34
Immunological 72
In vitro 67, 126, 178, 194, 206, 207, 208, 210, 211, 215, 217, 220, 221, 222, 223, 253
In vivo 126, 206, 210, 222, 223, 251, 253, 298
Infectible zone 56
Innovative Farming Systems 309
Inositol 3
Insect pests 329
Isobutylidene diurea 315, 316, 323, 329, 331
Isoleucine 2
Isotopic nitrogen tracers 108
Ipomoea 208

Jack-bean 8
Jasmonic acid 16
Juniper 332

Kidney bean 10
Klebsiella pneumonia 74, 76, 77, 78, 81, 84

Lablab purpureus 61
Lactuca sativa 26, 149, 155
Latex 311
Lathyrux 52
Lectin 3, 52, 73
 related proteins 3
Leghaemoglobin 59, 63, 65, 67, 71, 77, 80, 83, 190
Legume 2, 4, 8, 9, 52, 54, 59, 60, 64, 65, 67, 68, 72, 77, 81, 83, 84, 85, 86, 87, 156
 nodules 82
Leguminosae 150
Leguminous 51, 52, 71, 79
 seeds 3

Legumins 5
Lemna 263
Lettuce 322, 330
Leucine 4, 208, 218
Lignin 185
Limestone 50
Lipid bilayers 69
Lipids 119, 147, 175
Lipo-oligosaccharide (Glycolipid) 54
Lipoligosaccharides 70
Lithosphere 24
Lodging 327
Lolium perenne 25, 28, 134
Lotus corniculatus 191
Lupines 6, 10, 15
Lupinus albus 10
Luteolin 52, 54
Lycopersicon 147
Lycopersicon esculetum 115, 121, 125, 131, 132, 136, 155
Lysine 2, 3

Macronutrients 3, 244
Maize 13, 14, 16
    kernels 13
    seeds 6
Malate 81, 263, 286, 288, 294, 295
    dehydrogenase 72
Malondialdehyde 147
Mannitol 49
Medicago 73
    sativa 63, 66, 142
    truncata 80
Metallic ions 8
Metalloprotease 8
Methionine 2, 4
Methyl Jasmonate 16
Methylgyoxal-bis (guanylhydrasone) 219
Methylglyoxal-bis 219
Microaerophilic 175
Microbial decomposition 308
    transformation 332
Microbiota 266
Micrometerological 106, 107, 108, 109, 112, 114, 115, 160
Micropropagation 194
Microsymbiont 63, 83, 85
Mo-pterin cofactor 299
MoFe protein 76
Multicropping System 322
Mutagenesis 67
Mutagenic 86
Mutant 58, 60, 80

Mycelium 175
Mycorrhizae 233, 289, 298
Mycorrhizal 181, 270
Myoglobin 71
Myrica gale 174, 184, 185, 190

N-acetylglucosamine 55, 68
N-acyl-D-glucosamine 55
N-ethylmaleimide (EMI) 7, 8
N_2 fixation 325, 326
Naphthylic acetic acid 66
Naphthyl phthalamic acid 57
Naringenin 52, 54
Nerium 136
Neutral endopeptidase 10
Niche 233
Nikel 192
Nicotiana glutinosa 131
    tabacum 36, 115, 116, 136, 206
Nif genes 74, 75, 77, 85, 87
Nitragin 49
Nitrate 14, 24, 25, 26, 27, 28, 29, 30, 31, 32, 34, 35, 36, 37, 38, 46, 64, 80, 81, 84, 104, 106, 107, 111, 116, 117, 118, 119, 120, 121, 125, 126, 127, 129, 130, 132, 133, 134, 135, 137, 140, 142, 146, 149, 150, 151, 156, 157, 158, 206, 207, 208, 209, 210, 211, 221, 223, 232, 233, 234, 235, 243, 245, 246, 247, 248, 250, 251, 252, 255, 256, 259, 260, 262, 264, 265, 266, 267, 269, 270, 271, 284, 286, 288, 289, 290, 291, 292, 293, 294, 295, 296, 297, 297, 298, 299, 301, 302, 317, 318
    accumulation 318
    ion 14
    nitrogen 307
    reductase (NR) 14, 29, 34, 80, 121, 206, 243, 267, 299
    reductase activity 13
    toxicity 317, 318, 329
    uptake 14
Nitric acid 104
Nitrification 266, 315, 328, 328, 331,
Nitrite 24, 46, 77, 80, 119, 120, 121, 127, 130
    reductase 30, 130
Nitrobacter 318, 330
Nitrogen 24, 25, 26, 27, 31, 33, 34, 36, 37, 38, 47, 51, 64, 68, 69, 72, 74, 76, 77, 78, 79, 80, 81, 82, 84, 85, 107, 111, 117, 118, 119, 120, 121, 125, 126, 127, 129, 131, 132, 133, 134, 135, 136, 137, 140, 142, 146, 148, 149, 150, 155, 156, 157, 158,

159, 160, 174, 181, 185, 186, 188, 189, 191, 192, 194, 195, 196, 206, 207, 208, 210, 211, 215, 217, 218, 220, 221, 222, 223, 234, 235, 244, 246, 247, 249, 250, 252, 253, 255, 256, 259, 260, 262, 264, 265, 266, 267, 269, 270, 271, 284, 285, 286, 290, 293, 294, 296, 307, 308, 317, 319, 326, 327
 content 11
 deficiency 307
 fertilizer 307, 332
 fixation 51, 59, 63, 75, 83, 86, 87
 fixing 73
 loss 330
 losses 309
 mineralization 321, 322
 mobilization 11, 15, 17
 transfer 9, 10, 11, 12, 13, 14, 15, 16, 17, 32
Nitrogen metabolism 317
 uptake 309, 327, 328
 fixing 62
Nitrogenase 59, 71, 75, 76, 77, 78, 79, 80, 81, 82, 83, 85, 86, 149, 179, 184, 185, 187, 188, 189, 190, 192, 193
Nitrogenous 209, 296
 salts 14
Nitrosomonas 318, 330
Nitrous oxide 104, 307, 309, 329
Nocturnal 114
Nod 54
Nodular 83
Nodulation 51, 61, 62, 64, 84
Nodulation (nod) genes 52
Nodule 58, 59, 61, 63, 80, 81, 324
 Initiation 56
Nodulin 26, 57, 60, 69, 70, 71, 73
 gene 59
Non-dormant seeds 14
Non-legumes 86
Non-leguminous 81, 87
Non-reducing 55
Non-structural carbohydrate 36

Oligosaccharides 52
Onion 320
Ontogeny 61
Opaque-2 mutant 3, 12
 protein 3
Open reading frame 75
Operons 68, 78
Orchard grass (*Dactylis glomerata*) 318, 328
Organic 82, 118, 119, 121, 133, 175

Organic nitrates 105
Organic nitrogen 13, 142
Organogenesis 52, 55, 59, 60
Ornithine 211, 218
 decarboxylase 219, 220
Oryza sativa 28, 250, 321
Osmium tetroxide 178
Osmotic 82, 298
Osmoticum 63
Ougeinia dalbergioides 62
Oxalate 81
Oxalic acid 318, 330, 331
Oxamide 315
Oxidative pentose phosphate 30
Oxidized atomspheric 149
 nitrogen 109, 114, 129, 143
Oxoglutarate 133
Oxygen 51, 59, 70, 71, 78, 82, 86, 185, 188, 192
Oxygen-sensing repressor 78
Ozone 64, 105, 148, 149, 158

P-chlorophenoxy-isobutyric acid 66
p-coumaric acid 221
Palmitic acid 178
Parachloro-mercuric benzoate 8
Paradoxically 47
Parametrization 108, 113, 114
Parasponia 190
Pea 10
Peat 50, 311
Pedospheric 159
Pelleting 50
Penicillin 67
Pentaglucosamine 55
Pentose phosphate 221
PEP carboxylase 31, 32
Pepstatin 7
Peptapeptides 70
Peptidases 6
Peptide 84
 bond 7
Peptidyl glutamyl peptidase 11
Peribacteroid 59, 77
 membrane (PBM) 72
Pericycle 55
Permeases 259
Peroxidase 70, 148
Peroxides 77
Petunia hybrida 146
Phagocytotic 59
Pharbitsnil 147
Phaseoli 84

Phaseolus vulgaris 5, 8, 32, 65, 73, 115, 117, 125, 126, 128, 131, 132, 133, 134, 136, 143, 146, 147, 148, 149, 150, 151
Phenol 66, 312
Phenolic 52
Phloroglucinol 185
Phosphatase 253, 262
Phosphate 51
   synthetase 253
   dehydrogenase 148
Phosphatidic acid 147
Phosphatidyl 178
Phosphoenol pyruvate carboxylase 72, 253
Phosphogypsum 311
Phospholipase D 147
Phospholipids 147, 148
Phosphorus 81, 193, 194, 285
Phosphorylates 78
Photochemical smog 105
Photophosphorylation 249
Photorespiration 30
Photochemical 114
Phytin 3
Phytohormones 51, 56, 57, 58, 64, 65, 66, 83, 209, 250
Phytotoxic 64
   potential 108
Picea abies 114, 115, 128, 132, 133, 134, 135, 137, 140, 142, 148
   glauca 250
   rubens 128, 129, 150, 158
   sitchensis 130
Pinus banksiana 148
   pinaster 5
   Ponderosa 117, 118
   radiata 206
   sylvestris 126, 140, 142
Pisum 52
   sativum 28, 128, 134 148
Plant Growth Regulators 15
Plant lodging 307, 310
Plasmalemma 24, 25, 27, 28, 37, 77, 109, 119, 120
Plasmid 67
Plasmodesmata 63
Plasmodiophora brassicae 316
Plastids 30, 130
Plumule 2, 6
Polyolefin-coated urea 312
Poa pratensis 151
Polar bonds 4
Polyamine 218, 219, 220, 221, 223
Polyester 311

Polyethylene 311, 312
   glycol 86
Polymer coatings 311
   technology 312
Polymers 311
Polyolefin-coated fertilizer 310, 324, 331
   urea 70, 100, 323, 325, 327, 328
Polyolefine 311
Polypeptide 3, 27, 75
Polyphenol 246
Polysaccharide 52, 54, 159
Polystyrene 312
Polyunsaturated 148
Populus 115, 136
   deltoides 129
Potato 320
Predators 3
Primordia 57, 61, 70, 73, 184
Primodium 57, 58
Prolamine 3, 4
Proline 4, 16, 70, 207, 208, 214, 215, 216, 217, 220, 222, 223
   rich 73
   rich protein 70
Prosopis 63
Protease 3
   activity 13
   C_1 8
Proteases 6, 7, 9, 10, 15, 17
Protein bodies 3, 10
   deamidases 9
   deamidation 8
   engineering 2
   hydrolysis 9, 13, 15, 17
   nitrogen 221
   phosphorylation 253
   storage vacuoles 3
   synthesis 15
   tannin 233, 246
Proteinase 9, 180
Proteolysis 8, 12, 15, 16
Proteolytic 8
   activity 15, 16
   enzymes 3
Proteosome 10
Proton motive force 27
Protoxylem 70
Pseudonodules 66
Pseudotsuga menziesii 143
Pumpkin cotyledons 10
Purafil 158
Purine nucleosidase 72
Putative signal peptide 73

Putrescine 219, 220
Pyridine nucleotide 30
Pyruvate 179
Pythium root rot 316

Quercus 136
   alba 117
   palustris 118
   robur 130

Radiotracer 260
Raphanus sativus 115, 136
Rapidly available nutrient fertilizer 308
Reporter gene 71
Repression 37
Reserve mobilization 15
Resin-coated Controlled-release Nitrogen Fertilizer 311
Resin-coated urea 310, 323, 331
Resistance 112, 113, 115, 116, 119, 327
Resistant 132
Restriction fragment length polymorphism 180
Rhicadesin 54
Rhizobacteria 51
Rhizobia 46, 47, 49, 50, 51, 52, 54, 56, 58, 63, 65, 67, 74, 75, 78, 84, 86
Rhizobial 55, 78
Rhizobial Nitrogenase 76
Rhizobium 46, 47, 48, 49, 50, 51, 52, 54, 55, 56, 58, 59, 61, 62, 63, 67, 68, 69, 73, 74, 77, 78, 85, 86, 87, 325
   japonicum 76, 80
   leguminosarum 47, 52, 54, 55, 67, 68, 69, 84
   lupini 76
   tropica 84
   meliloti 54, 55, 63, 67, 68, 69, 74, 75, 84, 85
   phaseoli 75, 85
   sesbania 86
   legume symbiosis 73
Rhizosphere 52, 54, 62, 67, 235, 259, 263, 271
Rhododendron 221
Rhodopseudomonas palustris 47
Rice 3, 4, 15
   cultivation 309
   grain 13
   seedlings 324
   farming 324
   transplanting 323
Ricinus communis 4, 29

RNA polymerase 78
Robinia pseudoacacia 80
Root Environment 12
Root protein 36
   shoot ratio 38
Roots 14
Rubber 311
Rubisco 251, 258
Rubra 190
Rush 320

S-adenosyl methionine decarboxylase 219
Salicylic acid 16, 66, 83
Salinity 81, 291, 298, 299
Saprophytes 175
Scenedesmus minutum 255
Screening 49
Scutella 14
Secondary signals 56
Seeds 2
   germination 2
   proteins 2, 3
Seedlings formation 12, 13, 16, 17
   growth 8, 9, 12, 15, 318
Seedlings 32
Semi-permeable coating 308
Senescence 67
Senescence zone 62
Serine 8, 70, 140, 217, 218, 220, 223
   Protease 8
Sesbania cannabina 86
   rostrata 47
Shepherdia canadensis 190
Shikimate pathway 221
Sigma factor 78
Signal exchange 46, 51
Signal transduction 58, 86
Signalling 54, 59
Signalling respector 57
Signals 54, 65
Single basal application 320, 321, 322
Sisymbrium officinale 14
Slow release nitrogen fertilizer 309, 311, 317, 319, 321, 326, 327, 328, 329
   fertilizer 307, 308, 309, 310, 331, 332
Slow/controlled-release nitrogen fertilizer 309, 310, 316, 317, 318, 319, 320, 325
Sodium hypochlorite 178
Soil acidification 307, 310
   acidity 311, 315
   bacteria 316
   moisture 315
   nitrogen 321
   temperature 315

Solanum tuberosum 221
Solanum nigrum 149
Somatic embryos 56
Sorghum 326
  vulgare 115, 136
Soybean 5, 8, 114, 137, 324
  embryos 5
Spermine 219, 221
Spermidine 219, 221
Spinacea oleracea 131
Spinach 318, 320, 330, 331
Spinaca oleracea 131, 133
Spindle tree 332
Sporangia 175, 183
Spores 177
Squash seeds 13
Stachydrine 52
Starchy endosperm 6
Stomata 109, 113, 114, 115, 117, 127, 284, 298.
Stomatal 19, 120, 125, 128, 132, 149, 155
  Conductance 116, 143, 158
Storage 34
  organs 6, 13
  proteins 3, 4, 6, 8, 9, 12, 13
  tissue 11, 12, 15, 17
Stratosphere 104
Stratospheric ozone 307
Strawberries 331
Strawberry 316, 320
Stress 62, 63, 81, 83, 298, 302
  metabolism 10
  induced
Stresses 299
Strigonellina 52
Subambient 253
Subepidermal 60
Succinate 72, 81, 179
Succinic semi-aldehyde 218
Sucrose 35, 36, 50
  Synthetase 72
Sugar 52, 55, 221, 318, 331
Sugar-cane 326
Sulphydryl 148
Sulphide 252
Sulpholipids 147
Sulphur 64, 76, 311
  coated urea 323, 327, 329, 331
Sunflower seeds 4
Supernodulation locus 84
Superoxide radical 77
Supergranules 316
Surface 111
Sweet fern 332

Symbionts 51
Symbiosis 52, 58, 72, 85, 87, 174, 181, 188, 189, 190, 192, 196
Symbiotic 24, 46, 51, 52, 59, 60, 61, 65, 67, 75, 76, 82, 84, 85, 86, 175, 185, 191, 243, 271
  bacteria 324
  megaplasmid 74
  nitrogen 149
  protein 71

TDZ 215
Terminal oxidases 77
Thioproline 215, 216
Thiourea 315
Thuja 332
Tissue hydration 14
Tomato 331
Torulosa 175, 186
Toxic proteins 3
Transamination 31, 133
Transcuticular 113
Transducers 259
Transepoxysuccinyl-L-leucylamido (4-guanidino) butane (E64) 7
Transformations 327
Transgenic 73
Transglutaminase 9
Transpiration 114, 115, 151
Transporters 259
Transglutaminase 8
Transthylakloid 143
Tricarboxylic acid 255
  cycle 31
Trichoblasts 57
Trichomes 56
Trifolitoxin 84
Trifolium 52
Triodobenzoic acid 57
Triose Phosphates 253
Triticum aestivum 115, 125, 132, 134, 137, 158
Troposhpere 104
Troposheric 105, 116
Tryptophan 3
Tungstate 34, 36
Turbulent 106
Turfgrasses 329
Turnip 316
Tyrosine ammonia 221

Ulmus americana L. 118
Uncoated Fertilizer 315

Uncaoted Inorganic Fertilizer 315
Uncoated Organic Fertilizer 315
Upland crops 322
Uptake 112
Urea 46, 307, 308, 315, 317, 318, 323, 324, 331
　formamide 315
　pyrolyzates 315
　formaldehyde-resin 311
　granule 327
Ureaform Methylene urea 331
Ureaforms 329
Urease inhibitors 317
Ureids 82
　biosynthesis 72
Uricase 72
Uridine 58

Vaccinium macrocarpon 29
Vacuole 28
Vacuoles 35, 127, 156
Valine 208, 218
Vegetable Crops 330
Vesicular-arbuscular mycorrhizae 67
Viburnum 115, 136

Viciae 52
　faba 5, 32, 64, 159
Viciae 69
Vigna radiata 7, 10, 64, 67, 83
Volatile 118, 157
Volatilization 46, 330
　of ammonia 309

Wax 311
Wheat 6, 8, 9, 114, 320, 327, 328
　bran 7

Xanthine dehydrogenase 72, 299
Xanthobacter 47
Xenopus 56
Xylem 28, 31, 32, 35, 37, 38, 82, 140, 191, 265, 284, 288, 290, 291, 293, 294, 295, 296, 297, 299, 301
　parenchyma 298

Zea mays 25, 115, 131, 136
Zein 4
Zelkova 136
Zinc-finger sequences 73
Zobium leguminosarum 52